Dedication

This book is dedicated to my wife, Lynette, without whose dedication and commitment this book could not have been written. Not only did she support us financially during the research and writing, but she inspired, advised and assisted in the writing, especially concerning the spiritual matters.

Foreword by Erich von Däniken

Toward a New Paradigm!

Humanity's Extraterrestrial Origins

ET Influences on Humankind's Biological and Cultural Evolution

by
Dr. Arthur David Horn
with
Lynette Mallory Horn

Illustrated by Jack Hoffman

||||||||||||||||||||||| SILBERSCHNUR |||||||||||||||||||||||

Text and illustrations copyright © 1994
by
Dr. Arthur David Horn
and
Lynette Mallory Horn

All rights reserved. No part of this publication may be reproduced or transmitted in any form or by any means, electronic or mechanical, including photocopy, recording, or any information storage and retrieval system, without permission in writing from the publisher.

Fur further information or book orders, please write or call:
A. & L. Horn
P.O. Box 5572
Lake Montezuma, Arizona 86342
(415) 789-8529

ISBN: 0-9649845-0-4

First Printing: February, 1996
Second Printing: August, 1997

Illustrations: Jack Hoffman and Stefan Huber
Graphic Artwork: Lynette Mallory Horn
Layout: Gary Tannatt and Ulla Schmid

Acknowledgments

Once or so in a lifetime, individuals stand beside you to connect with a deeper level of Truth, without hesitation, realizing there would be neither glitz nor glory in return. We have received this amazing level of support from several people over the erudite course of researching and publishing this book.

The most notable to be accounted, literally gave this book life. J. Leonard Nitsch, with his gentle and humble nature, sacredly connected with us, demonstrating his generous capacity to reach for that deeper level of Truth. He took the responsibility to engage, act and create a difference. The difference between the rest of humanity and Leonard, is simply - choice. He stood up for a higher purpose. Our whole intent, here and now, is to honorably recognize him with an accountable richness, gratitude and grace from the Great Spirit.

We are especially grateful for all of our Mothers' and Fathers' support throughout our lives.

The other individuals who participated in "keeping the ship afloat", during those critical years of 1993-95, must also be accounted. Our richest and most loving support has been forthcoming and going for some time now, from our daughter, Michele, from Norman Cohan, Janet Barrows Connaughton, Zenon Michalak, and most lovingly, Linda and Tom Medvitz. Where would we have been and what would we have done, if... Great Spirit's many blessings on their magnificent Souls! They stood up and got counted upon our most precipitous alphas and omegas. We heartily acknowledge the greatness in each one of you.

Special thanks must also go to Tom Hockemeyer, Gary Tannatt and Ulla Schmid for all the work they have put into producing this book.

Lastly, we wish to thank the thousands of people dedicated to the Truth of human existence, some of whose works we have used and referenced here, especially the many people who have recommended and/or loaned us books over the past few years.

Any mistakes and shortcomings in our own striving for the Truth of human existence expressed here are, of course, our own.

Humbly, we now bow down in gratitude to acknowledge our Oneness in Spirit with all of those generous Souls who have committed to live and breathe the interconnectedness with and for the Truth.

A Note About the Authors

Arthur D. Horn, Ph.D. grew up in small towns of eastern Kansas. Raised as a Christian, he became disillusioned with religious dogma as a young man and became a Darwinist. After he received his B.A. degree from the University of Colorado, he earned his Ph.D. in physical anthropology from Yale University in 1976. He was a professor of biological anthropology for fourteen years, mostly at Colorado State University. Dr. Horn specialized in the study of non-human primate ecology and human evolution.

After becoming aware of an existence beyond the three-dimensional, materialistic science of today, he resigned from his job and science in 1990. Soon after his resignation, he and his wife, Lynette, moved to Mount Shasta, in northern California, where he resumed his research into human origins, a life-long interest. Dr. Horn's research into human origins soon revealed the extraterrestrial (ET) phenomenon, and he eventually realized that ETs were intricately involved in the origin and development of humans. Through his consistent exploration, reading and experiencing, along with Lynette's loving emotional and physical support, they gathered the faith between them that moves mountains to preserve and richly sustain a simple lifestyle.

When Professor Horn's family gently inquired as to what he would do after leaving academia, he responded with this, "I want to write a book on human origins and grow a garden!" He and Lynette both felt that to write, produce and promote publications that might serve as introductory manuals, or educational guides upholding the scientific and sacred teachings, as well as to grow a garden of fresh organic food, while living in the magnificent Mount Shasta setting, would be their ultimate chosen lifestyle. Dr. Horn's documented research has produced a profound book, *Humanity's Extraterrestrial Origins, ET Influences on Biological and Cultural Evolution*, which was released in early summer of 1994 as a prepublication edition and was initially printed to supply copies of the book to major investigators, abductees, book stores, and for those participants who continue to deeply explore human origins and evolution. The Horns released the premier first edition in February, 1996.

Lynette Mallory Horn is a strong-willed, L. A. born woman, who came to the San Francisco-Marin County area to champion the provisions of private and public alternative health care for children and for those whom she calls the "Ageless Ones". She has been researching and developing alternative approaches for mental and physical primary health care most of her life. Lynette's publishing endeavors served as one process to provide the public with comprehensive, and oftentimes controversial publications, and are paramount to her mission and higher purpose in Life.

She was raised spiritually as a Christian, with an early penetrating education from her Father that honored the spiritual teachings from every human walk. Lynette was taught as a child to listen to her heart. Her Father was very instrumental in awakening the teachings within her Spirit. Whereafter, she proceeded to personally and professionally train as a spiritual healthcare provider, using natural healing as her primary process. Lynette has breathed life into the intent of universalizing her wisdom and knowledge through herbs and herbal combinations, body therapies, plus varied healing techniques of mind over matter. Her writing and publishing endeavors developed to the point she realized they were mutually supportive in promoting more public awareness for natural healing and basic alternative health care.

During Lynette's wide-spread academic years, she experienced an expansive and liberal education, which included attending and graduating from the University of California at Santa Cruz, to study Humanistic Psychology. A tremendous roadmap of alternative experimenting with students was happening during the seventies while she attended the University. Santa Cruz was her birthcanal and the friends, professional associates and those who became family members acted as fairly masterful midwifes. What a celebration of Life they lived!

After meeting Arthur Horn in 1988, when Lynette had just returned from a European tour with one of her spiritual teachers, she immediately discerned that this partnership/union with Dr. Horn would not come easy! After all, biological anthropologists consider themselves scientists first, with no awareness of a Spirit/Soul! Lynette and Arthur went together to trek the California Trinity Alps on an anthropological expedition ...and soon were married, after a whirlwind Summer courtship of traveling back and forth between Colorado and California, remaining truly committed to their love (AHO! but, dear sweet Mother, it wasn't easy!) for each other ever since. Arthur and Lynette live today wherever the Spirit guides them, trusting that the Earth's revolution is proceeding to align all frequencies to focus them on the Truth!

Contents

	Foreword	xii
	Introduction	xv
1.	The Darwinian Myth of the Origin and Evolution of Life	1
2.	The Darwinian-anthropological Myth of Human Evolution	35
3.	The Ancient Mesopotamian Account of Human Creation	50
4.	The Creation of Heaven and Earth and All of Life	65
5.	Ancient Humans, Regressive and Progressive ETs	80
6.	The Anunnaki, Humankind and the Great Deluge	95
7.	The Advent of Human Food Production	107
8.	The Rise of Ancient Human Civilizations	122
9.	More "Heartless" Behavior of the "Gods"	142
10.	Esoteric Correlations with Historical Data	159
11.	Hoodwinked and Controlled	166
12.	The Late Twentieth Century UFO-ET Phenomena	183
13.	Some Biocultural Aspects of Humans and the ET Question	217
14.	The Reptoids	244
15.	The Greys	256
16.	Jehovah and the Ancient Hebrews	265
17.	Christianity and the Teachings of Jesus	286
18.	Possible ET Influences on Christianity	296
19.	Some Spiritual Distortions in Other World Religions	316
20.	Man's Inhumanity to Man	325
21.	Humanity's Golden Moment	338
	Bibliography	356
	Index	367

Illustrations

1.1	Amniotic egg of reptiles.	20
2.1	*Proconsul africanus.*	38
2.2	One of the more robust species of several species of *Australopithecus* of Africa.	40
2.3	*Homo erectus* cranium from Zoukoutien Cave, China.	43
2.4	Steinheim skull.	45
2.5	Neanderthal skull from La Chapelle-aux-Saints, France.	46
3.1	Map of ancient Sumer.	51
3.2	Evolution of a few words of Sumerian cuneiform writing.	53
3.3	An Assyrian cylinder seal illustration of the birth of the first human.	58
4.1	Evolution of the orangutan, gorilla, chimpanzee and the human according to current neo-Darwinian dogma.	77
4.2	Possible evolution of the orangutan, the gorilla, the chimpanzee and the human.	77
7.1	Ninurta granting the plow to humans.	118
8.1	An archeologist's view of the enormous ziggurat of Ur.	126
11.1	Illustration based upon eyewitness testimony of a "comet" observed in 1479 in Arabia.	173
12.1	Two pages of a twenty-one page "above top secret" affidavit released by the National Security Agency (NSA) after censoring it.	193
12.2	Depiction by an artist of an underground laboratory of a grey species, based on several eyewitness accounts of abductees.	211
13.1	A view of the brains of a fish, a reptile and two mammals demonstrating the size of the cerebral cortex relative to older parts of the brain.	227
13.2	A neuron.	232
14.1	A "lizzie", or reptoid species drawn from abductee eyewitness accounts.	249

14.2	Sagittal views of the brains of four mammals showing the extent of the limbic system relative to the cerebral cortex.	251
15.1	A drawing of a tall grey from a particular subspecies of grey (the Orion grey, type 1).	259
20.1	The approximate area occupied by Plains Indians in 1850.	330

Foreword

Did our ancestors receive visitors from space?

As an author of twenty-one books related to the topic of this book, as well as an elder gentleman with fifty-nine years of experience, I know that it is possible to argue about (almost) anything. About the archaeological findings and their interpretations, about the possibilities of interstellar space travel, about the evolution of the mind and spirit, and, of course, about the ancient legends and myths of our ancestors.

However, there is something that cannot be argued about: the existence of the holy scriptures. They undeniably exist. They have been passed on and discarded innumerable times for millennia. Even though it sounds like it, the *Avesta* is no investment fund, but contains the religious texts of the Parsees. The Vedas, too, are not a brand of fruitcakes. Instead, they constitute the oldest, religious transmissions of India. The Rigveda alone contains 1028 hymns. Even the *Mahabharata* is neither a train station in Pakistan, nor a rice dish, but a collection of holy scriptures. And the *Popul Vuh* is not a coffee bean, but the Bible and sacred teachings of the Mayans. I am even acquainted with the tongue-twisting texts of the Samaranganasutradhara and know that religious scriptures hide behind this monster of a word.

The scriptures exist alongside the book of the Prophet Enoch, the book of the Prophet Ezra and the holy or unholy texts of the Bible. The only really astonishing fact is, that many of these holy scriptures were written in the first person singular. And this creates a dilemma.

One may assume the point of view of the authors that these holy scriptures were teachers or writers, perhaps even dreamers of their time. But, now they talk about some teachers from the clouds speaking to them, instructing them, or they describe, in the first person singular, experiences they suffered because of these heavenly teachers. All just phantasies? Just foul lies then, because they have experienced nothing and no teacher has talked to them. Why - so one must ask - did we build our great religions from these falsely reported documents?

Stop! The believer screams. The texts are not lies, the authors have experienced all this. Agreed! But, then, who talked to them? And, the fact that there was talk cannot be seriously denied in view of the abundance of the holy scriptures. Who gave instructions to the people? Who manifested with smoke, fire and noise? With wheels and wings?

The dear Lord - who else! The naive individual replies indignantly. To state it clearly: personally I believe very much in God. I know that this grandiose Creation, and consequently the Spirit of God, does exist. I belong even to those who pray every day. And, the dear God I am referring to has no need to ride in a noisy vehicle that smokes, shakes and stinks, that displays rimmed wheels and rattling wings.

Therefore, if it was not the dear Lord who talked to the ancestors and who presented himself with a technical vehicle - who remains, then? Who - please?!

The texts exist, written down in the first person singular as an experiential account. They are not allowed to be considered fabricated, else the religions will collapse. And, in view of innumerable accounts, it could not be the dear Lord who is referred to. What solutions remain?

Amazingly, many holy texts report that the gods had originally created man in their image. What shall be gathered from that? Did we not learn from the teachings of Darwin that mankind became what it is today by means of a long, evolutionary process? Why should God or gods have created man? What gods? We all know that there are no gods. Do the ancient holy scriptures lie?

Thousands of years ago our ancestors were technologically underdeveloped. They could not understand space travel or genetic engineering. Then, alien space travelers descended from the clouds with smoke, fire and noise. They were the heavenly teachers of the legends. Our ancestors did not grasp this. And, so the space travelers became gods in the mythologies and transmissions of a budding mankind. A misunderstanding in the choice of words. Those space travelers changed something on good old planet Earth, but the authors, Arthur and Lynette Horn, will explain this to you themselves in their excellent book. I hope you will have an interesting time reading it.

Erich von Däniken
Feldbrunnen, Switzerland
March 12, 1994

"The Life which is unexamined is not worth living."
Socrates

Introduction

Search for the Truth

This book is about the truth, as I understand it. More specifically it is about the truth of our origins and the meaning of our lives here on this planet we call Earth. To a large extent, this document is about my own personal quest for truth, with an emphasis upon revealing the myths currently being perpetuated in every belief system of humankind, both secular and religious. I will demonstrate that virtually every current belief system that delves into the origins and development of humanity is full of untruths and distortions that serve to keep the reality of human origins concealed.

I have discovered, in addition, evidence of an enormous machination of long standing to keep humankind ignorant of its true origins in order that the human species on Earth can be controlled and "used". Who is responsible for this conspiracy on humankind? As I began to search in earnest for the truth of human origins a few years ago, I met again and again with the extraterrestrial, or ET, phenomenon. This was certainly not what I was looking for; I wasn't sure I even believed in ETs as recently as five years ago. But it soon became obvious to me that, in fact, ETs had much, if not everything, to do with human origins and development. Furthermore, I found strong evidence that groups of regressive ETs, always with the help of a few humans, have conspired and are conspiring to keep the knowledge of the ET influences in human evolution hidden from humankind. I discovered that the principal information that these regressive ETs have distorted and concealed, is the spirituality of humans. They do this to keep humankind from spiritually evolving to the point where their own dark influences and imprints would no longer have any effect upon humankind.

I will demonstrate that spiritual truths have been severely distorted in all world religions. I will concentrate, however, on exposing the enormous holes in the scientific belief system that asserts that

Introduction

humans have no spiritual existence at all, since that was my "expertise" and my belief throughout most of my adult life. While exposing the fraud of the current belief systems I will piece together, at least at an introductory level, the true story of the origins and existence of humankind with a very brief review of the spiritual truths of this universe.

Until 1990 I was a proponent of the scientific - what I will call the Darwinian-anthropological - belief system of human existence. For fourteen years I was a professor of biological anthropology, mostly at Colorado State University in Fort Collins. My specialties were 1) non-human primate behavior and ecology and 2) human evolution, both of which I taught almost every semester of my academic career in introductory and/or more advanced level courses.

The world's scientific view is part of what is known as the materialistic-humanistic point of view, which I would simply define as the belief that all of existence consists of matter. That is, the universe, life, humans and all of existence evolved from matter by chance circumstances following natural "laws", and that all life, including concepts of what we call God, and concepts of the spirit or whatever, arise solely out of the physical processes of matter. This materialistic-humanistic movement which we see represented in the Western world today by academia, the media, and other powers, had its beginnings in medieval Europe, when the story of the world's beginning and humankind's origin as contained in the book of Genesis in the Bible was being questioned.

At the age of 46 years, in 1988, I married Lynette, a woman who had developed an integrated spiritual practice throughout most of her life. In 1989, with Lynette and a woman friend of ours, I had a rich experience that convinced me beyond all doubt that there was a spiritual existence outside of matter. After this experience, I began to think long and hard about the existence and origins of humankind. If humans have a spiritual existence outside of matter, as my wife and all religions everywhere claimed, how does this spiritual existence tie in with our biological evolution? Was there a supreme divine being and, if so, what is Its nature? What is the meaning of life, anyway, in light of a spiritual existence? Could it be that our understanding of our origins and existence - at least my understanding - was partially or totally incorrect? These and many other questions were tumbling through my mind in the weeks after this "spiritual gnosis". While I had many questions in my mind, one thing was certain: I no longer believed what I was teaching. In order to maintain some sense of integrity, I resigned my position at Colorado State University in the Spring of 1990.

Since then, I have been investigating human origins and existence, and have discovered an extraordinary, fantastic story. The story of our existence involves inter-dimensional travel, time travel, reptile-like humanoids that helped "create" us only to then enslave and control us for thousands of years, secret society conspiracies, government conspiracies and duplicity, as well as the promise of an immortal life of evolving toward the blissful Divine. I am aware that some readers might find this statement unbelievable. But I urge you to read on and judge for yourself when you learn more of the facts. I do indeed present such a breathtaking, amazing story relative to the materialistic "humanistic" story permeating our society, that I'm anticipating that many of readers will not be able to see and believe the truth at first, just as was the case with me. However, I have uncovered evidence that we humans were first "created" through genetic engineering, then duped and controlled by alien races of other life forms, that are not originally of this Earth. The primary method employed by these ETs is denial and distortion of the truth of our spiritual existence, thereby hindering and ultimately distracting our collective and individual spiritual evolution. Nonetheless, we primitive humans have slowly been evolving mentally and spiritually, despite the best efforts of our creator "gods" to keep us divided, ignorant, and controlled. We have expanded our awareness even to the point that we can now understand the truth of our origins and existence, and are capable of evolving very rapidly to higher states of existence. In addition to our own mental and spiritual evolution, the aid of more evolved, positive ETs who do have humankind's best interest in mind have been helping the primitive human species since our human origins. The effect has served to prepare us humans to make a tremendous evolutionary leap in the next ten to twenty years.

Yes, I believe that my research has drawn near to an understanding of the astonishing truth surrounding the sensitive question of our biological and spiritual origins, and I would like to share it with you. Changes are beginning to happen very fast. There is not much time. The next few years are a time for rapid consciousness-spiritual evolution. Be aware of the radical theory of our origins and existence that I and many others have presented or will present. Pay attention to what is going on. Most if not all of us have evolved the mental capacity needed to observe the data and make logical and reasonable decisions regarding the facts of our origins and existence. Most, if not all of us have evolved mental capacities necessary to evolve to a higher state of being. In any case, it is time for our souls to learn more about the Divine and what to do to move closer to It and away from the domination-victim plots that are currently being played out on Earth.

Introduction

The message of this book is ultimately a very happy one, as I believe that in the next decade or two, humankind will have a magnificent opportunity for extremely rapid spiritual evolution toward the Light. On the other hand, apparently much of humankind is not quite ready for the great spiritual leap, and may suffer much hardship as we approach this period of rapid and great change. The only recourse of individual humans and humankind in general, I feel, is a rapid concerted effort on our parts to turn toward the Light. But how do we move toward the Light? How do we discern which or what are the forces of the Light, the Truth, and which are the forces of darkness, of non-truth. It is hoped that this review of our origins will assist humans around the world to develop their discernment, as we explore these murky waters that surround our origins and existence on Earth.

Let me tell you of some of the twists and turns I have gone through in my lifetime with regard to my knowledge and beliefs about human origins and the meaning of life. I feel the subject matter of this book will be better introduced to the reader through a recounting of my own personal consciousness evolution. The review will cover several viewpoints that I have adopted during my personal and professional journey. I hope readers can relate some of their own experiences and beliefs concerning our origins and existence to one or other of my experiences in this area.

The True Believer

I grew up in a small eastern Kansas town. My parents were both very religious, and my three sisters, my parents and I attended the small Episcopal Church in town at least once every Sunday. My parents were very devoted to our church and served as Sunday school teachers, choir members and participated in other adult activities. I was an acolyte - one who assists the priest in performing the worship service on Sundays. I came to love the ritual of the Episcopal Church services and I was convinced that our church was better than the Lutheran, Catholic, Methodist, or Baptist churches and all the other Christian varieties known to me. I was only vaguely aware that there were people in the world who were not Christians. I was your everyday, average, all-American boy from Kansas!

Naturally, I believed the story of creation contained in the Book of Genesis in the Old Testament of the Bible. I learned that God created, in six days, the heavens and the Earth and all life, including humans. This story was good enough for me at the time, but I must confess that I didn't have a burning concern to learn about our origins as a teenager. I was more interested in the batting average of Stan Musial, and the success of his baseball team, the St. Louis Cardinals. In fact, I became obsessed with the standard American sports of baseball, basketball and football. I noticed that as my participation in sports increased through high school, my interest in our church and religious matters generally decreased.

Religious Crisis

After graduating from high school, I attended Kansas State University to study civil engineering. I generally enjoyed my engineering studies and did well, but I began to find the curriculum of the engineering school rather confining and limited. As I remember, out of 144 hours required to graduate, only twelve of them were electives to be taken outside the engineering course requirements. Twelve hours were all I had in four or five years to become a more broadly educated individual - that's only three or four courses! During my junior year, I took my first elective, a literature appreciation course in which we read several classics of literature, including Plato, Dostoyevsky and Hemingway. I loved the course and was enthralled by the books we read. I had enjoyed reading to some extent before, but after this course I began to read voraciously. Certainly my small-town Kansas reality system began to expand dramatically through this joy of reading.

The summer following my junior year, I worked on a U.S. Forest Service surveying crew in Alaska, with about six young men like myself. Through most of the summer, we lived in a trailer located in a very remote area nestled in a beautiful mountain range. After work, the whole crew loved to read. We would read until our portable generator ran out of gasoline, then we would talk in the ensuing darkness about the books we were reading and other topics that interested us. It was during these discussions that I began to feel the full agony of my religious crisis which had been simmering for a few years.

For a number of years I had been bothered by all the different sects and churches in Christianity and the manner in which they all claimed that their way was the best way, if not the only way, to be living a good Christian life. As a youngster, I had no trouble with this because I "knew" the Episcopal Church was the superior Christian Church. But as I talked religious matters over with friends of different faiths, I began to have doubts. Furthermore, I gradually became aware that there were millions in the world that didn't even believe Jesus Christ was Lord! How could God the Father allow this to happen, if indeed, Jesus was Lord? What was worse, in my eyes, was how could an all-powerful, loving God, as God was depicted as being in the New Testament, allow so much misery to exist in the world. I had read enough of human history to know that wars, diseases, poverty, racial hatred and many other stupid and cruel behaviors were rampant on Earth. I began to view the plight of humankind with despair and disgust. I had kept my faith through my first couple of years of college and had attended church fairly regularly. But, by the end of my summer spent in Alaska, I declared myself an atheist and quit going to church altogether. I missed the comfort and support my faith had given me, but I did not see how a thinking person could believe in an omnipotent, loving God, as was depicted in the New Testament. Further, I felt a bit duped at having been brought up with such a narrow point of view.

The Formation of a Darwinist

In the fall, after my summer in Alaska, I decided to quit my engineering studies. I didn't know what I wanted to do, so my plans were to save some money and then travel in Europe. However, these plans were abruptly put aside when I was drafted into the United States Army in December, 1963.

I didn't care for the Army, but I had been raised during the McCarthy era with the cold war raging, and I was quite willing to defend America against those rotten Commies. The nicest thing that happened to me in the Army was that I was stationed in Thailand for a year, where I was a surveyor in a road construction company. We worked hard, but I occasionally had time to visit and get to know part of Thailand, a Buddhist country where orange-robed monks are a common sight. I took it upon myself to learn a bit about Buddhism and the general history of Thailand and that part of the world.

While I was serving my tour of duty in Thailand, my girlfriend in Colorado sent a book that was popular at the time, *African Genesis*, by the late Robert Audry. The book concerns human evolution, dealing primarily with the behavior of very early members of the human family, the australopithecines - small-brained, bipedal primates, who lived between four and one million years ago. This was my first exposure to the theory of evolution and Darwinism of any kind, as it was not taught in my high school. I remember well the one day that my high school biology teacher mentioned the theory of evolution. I was only vaguely aware that such a theory existed. It sounded interesting, but the teacher, after talking about the theory for five or ten minutes, never returned to the subject.

Trying to explain the existence and behavior of modern humans in terms of our biological evolution appealed to me very much. I read other books on the subject and soon became convinced that human origins and existence could best be explained by the theory of biological evolution as presented by Darwin and his followers. The fossil evidence for natural evolution seemed overwhelming to me.

After reading about biological evolution for several weeks, I could hardly believe that the creation myth of Genesis was still believed by millions in the twentieth century. I personally began to look upon the Genesis myth of creation with scorn; how could anybody believe the Earth and everything on it, including humans, were created in six days in the face of overwhelming evidence that the Earth was billions of years old? To me, all religious teachings began to seem absurd, including the idea that each human had an immortal soul.

After being honorably discharged from the Army, I worked for a year in Colorado in order to save money and be eligible for in-state tuition. I enrolled in the University of Colorado at Boulder in 1967, as an anthropology major with the intention of specializing in the study of human biological evolution.

I loved my two-plus years at Boulder, particularly the anthropology courses. I wanted to get into a good graduate school program, so I worked hard during the week in order to earn good grades. I even studied hard on many weekends!

Upon graduation from C.U. in 1969, I began my graduate studies the same year at Yale University in New Haven, Connecticut, entering the Biological Anthropology program.

I was quite enthralled with Yale, its traditions, scholarship and cultural events. I enjoyed association with my fellow students as well as my professors, with whom I became friends. I loved my classes and the more I learned about evolution, in particular primate evolution, the more I longed to learn. However, at the risk of offending some old Yalies, I must confess that my favorite part of Yale was the gothic-style gymnasium and the ubiquitous pick-up basketball games.

The highlight of my studies at Yale did not occur in New Haven, but in Zaire, a central African country. I conducted my Ph.D. research in Zaire studying wild chimpanzees and monkeys for more than two years. Those of us who were interested in primate evolution wanted to know all we could of living monkeys and apes, so that we could better interpret the fossil record of primate evolution. My stay in Zaire was difficult because I was alone, and the chimpanzees I was trying to study, known as the Bonobo chimpanzee, were rare and hard to see in the rain forest habitat area where I tried to study them. After two years in Zaire, I returned to New Haven and wrote my dissertation, thereby completing the requirements for a Ph.D.

Introduction

I spent the next fourteen years as a professor of biological anthropology, mostly at Colorado State University (CSU) at Fort Collins. I enjoyed Colorado, the state I had come to love more than any other. And, high on my list of personal priorities, CSU had a noon-hour basketball program for the faculty and graduate students.

While I had many interests other than evolution and science, even some in metaphysics, I became convinced that we humans were strictly biological beings, special animals with language capabilities and a complex culture. Like most scientists, I was somewhat disturbed and frustrated that more students and more of the general public didn't fully accept our animal heritage and try to understand it more. I was convinced that the better we understood our biological roots, the better we could organize our society to eliminate wars, racial prejudice, tyranny and a host of other problems that have plagued human civilizations since ancient times.

However, I must say that I did have some doubts about the scientific explanations of human existence, centered around the extremely rapid evolution and development of humans in the past few thousand years. Fossil and cultural remains show that human evolution progressed very slowly for hundreds of thousands of years, then began to advance quite rapidly about 35,000 years ago. This rapid advancement through what anthropologists call the Upper Paleolithic took a giant leap, beginning at about 10,000 BC into what is called the New Stone Age or Neolithic period, when humans began to develop an entirely new life style. For hundreds and thousands of years primitive forms of humans (*Homo erectus* and their like) lived in caves or simple shelters eating only wild plants and what animals they could hunt or scavenge with their primitive stone and wooden tools. Then they suddenly began to develop in the Near East as farmers and herders of domesticated animals. A few thousand years later at around 4,000 BC, a few human populations established civilizations in Sumer (in what is now southern Iraq and Kuwait) and Egypt. These civilizations were remarkable - they had towns and cities with monumental architecture, metallurgy, a vast foreign trade, the first writing, extensive scientific knowledge of the "heavens", and a host of other sophisticated features. I explained to my students that with the appearance of modern humans, complete with language, more and better memory, and forethought capabilities, humans expanded their life-styles dramatically. I explained that my colleagues, the archeologists, argued that humans pulled themselves up through domestication of plants and animals and civilization by their own bootstraps with these dramatically expanded capabilities. I accepted and taught this account, as there seemed to be no other viable explanation. Still I was puzzled -

Introduction

there were many questions in my mind. I wondered, for example, why the ancient Egyptians and Sumerians seemed to have been more advanced, in many ways, than the much later Romans of the Roman Empire (~200 BC to ~AD 500) and the Europeans of the Dark Ages (~AD 500 to ~AD 1000) and Medieval times (~AD 1000 to ~AD 1500).

In the early 1980s, I read the book *Chariots of the Gods*, by Erich von Däniken. As most readers know, Mr. von Däniken wrote a series of best-selling books in the 1960s and 1970s, in which he claimed that the early humans received a jump start in the formation of these civilizations from ancient astronauts, or extraterrestrials who were from civilizations that had evolved on other planets. Von Däniken suggested that the ancient astronauts, in addition to giving rise to the first human civilizations, had somehow given human biological evolution a boost that would allow humans to be able to maintain a civilization. From a scientific point of view, the data (or observations) von Däniken used to support his case were not as solid or as well-documented as one would like. My archeologist associates scorned his work. Nevertheless I was somewhat intrigued by his ideas, but it was not fashionable in my profession to be reading von Däniken, so I pushed aside his pioneering works with a promise to myself to look into the matter of ET influences in the development of humankind at a later time, if more evidence became available. So, I left von Däniken and his supporters behind and concentrated on biological anthropology.

In the summer of 1988, while on vacation in California, I met and fell in love with my wife, Lynette. I didn't realize it at the time, but my belief system and life were about to change dramatically. As mentioned, Lynette was a firm believer in God and the spirituality of all humankind. As a professional Darwinist-anthropologist and atheist, I thought her belief system to be misguided and a bit strange, to say the least. In fact, like most scientists, I viewed the belief in any spiritual existence as out-dated and a waste of time. I was somewhat impressed with Lynette's belief system, however, as she was strictly nondenominational, and she borrowed freely from many of the great master "teachers", as she called them, from Jesus, to Buddha, to many whom I didn't know. She maintained that the spiritual teachings of all the great teachers contained many divine truths, although their teachings had been much distorted and misunderstood by humans. She believed that we humans are in the process of evolving spiritually, individually and collectively, and that some day all humans would arrive at a deeper understanding of our spirituality. I respected her beliefs primarily because they were neither divisive nor judgmental; her belief system considered all humans divine, no matter what

Introduction

religion or culture they lived in. As I mentioned earlier, one of my biggest objections to formal religions is that they tend to divide and alienate humans according to religion, and each religion or sect claims to be the best or the only avenue to the divine - to God.

In the summer of 1989, while we were visiting a friend of my wife's in California, I experienced my first spiritual transmission, or channeling session, to which I referred earlier. Most scientists, including myself at the time, consider the so-called "separate entities" that speak through channelers as a part of the channeler's own mind. Most scientists would argue that channeling ultimately derives from the subconscious personalities of the individual transmitting the channeling. But here was this supposedly subconscious personality who seemed to know all about me. "It" knew things about me that nobody else in the world knew, including Lynette. "It" reviewed some events of my life as a young man and showed how certain events and personality traits helped me to maintain a very positive outlook on life, despite the fact that I did not believe in God. I was astounded and overwhelmed. I was astonished and somewhat embarrassed that any entity, physical or non-physical, knew so much about me. I had revealed to no one some of the events and experiences this entity talked about. I finally had to trust my own experience and I began to comprehend that this was a separate, non-physical entity that had addressed me.

This experience had a profound influence on my life. The very existence of an entity beyond our three dimensions, I reasoned, would cause the Darwinist-anthropological belief system, on which I had based my life, to become hopelessly inadequate as an explanation of human existence. I began to take seriously Lynette's belief system, and I read and studied voraciously all sorts of metaphysical books. Also, Lynette and I had many discussions about the spiritual nature of existence. We began to meditate together. Gradually, I myself, began to feel "the Spirit within".

During the next school year, I experienced much difficulty in my teaching. How could I teach bright young college students the Darwinian-anthropological view of human origins and human evolution, when I didn't believe it myself? I began to feel that in order to maintain my integrity, I would have to quit what I had heretofore considered my life's profession. I enjoyed being a professor and I loved my specialty, but I now knew the approach of all biologists and anthropologists to be very limited and incorrect.

At the end of the spring semester of 1990, I resigned my tenured position at Colorado State University. My relatives, friends and colleagues were very surprised and anxious as to why I had resigned

Introduction

with only seven years before I could receive retirement benefits. I'm sure many of them wondered about my sanity. I told our friends many reasons why I was fed up with academia. But, I didn't tell them the main reason I quit - that it is presently extremely difficult in academia, especially in anthropology departments, to hold what are considered radical views that involve a belief in some sort of spiritual existence of humankind.

Soon after my resignation, my wife and I moved to the small town of Mount Shasta, California. Mount Shasta itself is a beautiful, dormant volcano in Northern California, only about forty miles south of the Oregon state border. The area around Mount Shasta is quite sparsely populated by California standards, and it is a beautiful, tranquil place to live!

Shortly after we moved, I discovered a book, *Genesis Revisited*, by Zecharia Sitchin, which claimed that humans had been genetically engineered by ETs, by combining the genes of an archaic form of human with the genes of the ETs. Furthermore, according to Sitchin, these ETs were responsible for the beginnings of human civilization. Mr. Sitchin primarily used the ancient writings of Mesopotamia, the land between the two rivers, the Tigris and Euphrates, where the first writings and first human civilization began. Archeologists and other scholars consider the historical writings of the Sumerians and the people of later Mesopotamian civilizations as fantasies and myths, certainly not reliable history. The Mesopotamian epic tales tell of "gods" living for thousands of years, flying all about, and generally behaving in a manner not befitting gods, at least by modern standards. However, on the surface, Zecharia Sitchin's hypothesis made a lot of sense to me, considering the extremely rapid advances of humankind in the past few thousand years. Sitchin seemed to be on much more solid, historical ground than others who had advocated a similar hypothesis.

After reading *Genesis Revisited*, I decided to research this subject further, since human origins had been a subject of intense interest throughout most of my adult life. It wasn't long before I was again completely immersed in the subject of human origins, and I decided to continue the research and write a book on the subject. The more research I did, the more amazing and breathtaking the story of the origins of humankind became. I soon was obsessed with the whole project. In the meantime, I was experiencing my own evolution of consciousness. While most of this evolution took place on the mental plane at first, the more I learned, the more the world changed in my eyes. So this book is a representation of my own conscious and spiritual evolution over the past few years.

Introduction

Accessing a New Paradigm

A new, more intelligent paradigm is offered through the presentation of this book, hopefully, one that is accessible to all. Primarily, the story focuses on what I have discovered of the ET influences on the origins and existence of humankind. Expanding this much further, the book is also the story of the underhanded conspiracy to keep humankind ignorant of its origins and existence, especially concerning the spirituality of humans. Ultimately, this is a book about human spirituality, because one cannot understand the conspiracy and the truth of our existence without some knowledge of the spiritual and physical origins and existence of humankind.

Basically, I wish to inform readers that this book is an introduction to the story of human origins, the meaning of life, and the centuries-old conspiracy to keep humans ignorant. I'm sure everyone would agree that this presentation is an ambitious undertaking. Nonetheless, this is what I am attempting to accomplish, at least at an introductory level. This book is an introduction into these vast, convoluted subjects surrounding human existence. If there really is a conspiracy to keep humankind ignorant of our origins and nature, one would expect that written sources that divulge some of the truth would be very scarce, making the truth difficult to approach. I have found this to be the case. Many of the written sources used here are obscure and completely discounted by those in powerful positions in America and throughout the rest of the world, who advocate any one of the current secular or religious world views of human existence.

Nevertheless, all statements are documented and referenced to the maximum extent possible. Often, books that cover some of the subjects presented here have not been well-referenced, which allows the skeptic to simply dismiss them. Considering the references provided here, at least skeptics can become true skeptics, by becoming familiar with some of the sources and the materials used in this book.

Most readers will have no trouble with the historical and scientific references used. Also, the much more obscure and suppressed human historical references, whether one believes them or not, should not cause the reader too much dismay. However, some of the "esoteric" sources might cause some readers uneasiness. Most of what I am calling esoteric sources are human communications from entities that are not of this world. Most of the esoteric sources used here are "channeled" from a non-Earthly entity through a human by means of what we would call mental telepathy. As mentioned, virtually all scientists consider channeling a manifestation of the subconscious mind of the channeler. Dr. Arthur Hastings, however, has done a detailed

Introduction

study of channeling and demonstrated quite conclusively that most channeling is definitely not the result of overactive subconscious human minds. While Dr. Hastings observes that some channeling is literally nonsense, much of channeling is concerned with therapeutic advice which proved to be most helpful to people suffering from ill health, and spiritual teachings of a high order.[1] Hastings points out that channeling has been exhibited in humans since the beginnings of recorded history. I will present evidence that much channeling is from entities who do not have the best interests of humans in mind, and wish only to confuse and control humankind, as well as from entities who do have the best interest of humans in mind.

The question is then raised concerning my use of channeled sources: how do I know these sources have the best intended interests for humankind at heart? How do I know the channeled sources I use represent and convey the truth? First of all, I do not believe that any of the sources used here, channeled or not, represent the absolute truth, including this book. The materials covered here, ranging from ancient history, to reincarnation, to multidimensional existence and even more amazing phenomena are so complex to the primitive human mind, that it would be extremely difficult at this time to write and understand a detailed account of the truth of our existence, even if there were not a conspiracy to keep us ignorant of such matters. Higher entities, who may wish to convey the truth, must transmit through a human channeler or interpreter, which automatically leads to distortions and misinterpretation. The life experiences, imprints, conditioning and fears of the channeler might lead to errors in any transmission. Furthermore, the higher entity who transmits through a channeler may not wish to reveal every detail about a given subject. For example, one of the entities I use, known as The Awareness (see Chapter 10), has said that It purposely holds back some information that might get its followers into trouble, or that they would not understand. Also, It has stated a few times that information It provides sometimes becomes distorted simply by the process of going through a channel. The Awareness frequently tells its readers not to believe It on face value, but to question, explore, doubt, and discover for themselves the truth through their own investigations.

I have only used four esoteric sources in writing the book, one of which is not a result of channeling. I use esoteric sources primarily to gain insights and information into ancient human history. I use these esoteric sources because they generally correspond to and corroborate with ancient human historical sources of human origins and existence. We will see through the course of this book that every one of the few esoteric sources used here corroborates, to a large extent, the

Introduction

few ancient historical sources of human origins and existence that are available.

The first section of the book, is a somewhat detailed review of the Darwinian-anthropological myth of human origins. Materialistic scientists have generally been the loudest and most outspoken critics of those who have claimed that there was ET influence in the origin and development of humankind. A close look at the scientific view, that all life on Earth, including humans, occurred through random, chance circumstances, will reveal that there are gaping holes in this scenario. When one considers the available historical evidence of considerable ET involvement in human affairs, the scientific view becomes the most unbelievable, farfetched, implausible theory of human origins and development ever put forth by humankind, in my opinion. Indeed, I will argue that because of the false assumptions and massive denial of the data by scientists, that scientific references are the weakest and poorest sources one could use to try to learn of human origins and development.

There is a further purpose in reviewing the Darwinian-anthropological perspective of human existence than in merely giving scientists their due. Gallup polls measuring the belief systems of Americans consistently show that only about eleven percent of Americans believe that "pure Darwinian evolution" was responsible for human origins. These same polls reveal that an additional one-third of Americans subscribe to a form of Darwinism, but believe that God guided the evolutionary process. Almost half of Americans, on the other hand, believe in Creationism, or that human beings were created within the past ten thousand years by God with humans having no relationship to lower forms of life.[2] So, why spend time reviewing Darwinism if it receives such a weak vote of confidence from the people of my country? The answer to this question is found in the fact that, despite its poor standing among Americans, Darwinism is the official doctrine of the United States Government, even though this is not written as a law. Darwinism enjoys an unprecedented high status among American colleges and universities and, indeed, in virtually all of the major universities of the western world, Darwinism is the center piece of all sciences that deal with humans and other forms of life. All American scientists and medical researchers must adhere to a Darwinian point of view in order to receive funding for research projects doled out by multi-billion-dollar-funded government research agencies, such as the National Science Foundation and the National Institute of Health. Could it be that Darwinism is part of a massive cover-up of the ET influence on humankind? The answer is yes, this could be the case, and this is the main reason I wish to demonstrate

Introduction

the massive holes in a Darwinian-anthropological perspective of humankind. After demonstrating many of the extreme weaknesses in the Darwinian-anthropological perspective of human origins, I will turn to the few ancient human historical records that have survived to reveal the ET influence in the origin and development of humans. The best historical records we have of the ancient development of humankind come from the first post-deluge human civilization, that of ancient Sumer and later Mesopotamian dynasties. Zecharia Sitchin and others have demonstrated that the Mesopotamian historical records, written by even more primitive and naive humans than we primitive and naive humans of the twentieth century, can now be understood because modern humans have developed primitive capabilities of space travel, nuclear weapons, and genetic engineering. Mesopotamian historical records and other sources, including esoteric sources, indicate that ETs genetically engineered modern humans and instigated the rise of civilization. We will see that both positively-oriented and negatively-oriented ETs were involved with the development of humankind from the beginnings of this Earth, but that negatively-oriented ETs have largely been in control of the Earth for the past several thousand years.

The quite sudden rise of human food production, farming and domestic animal herding, and the beginnings of the first civilizations, are covered in Chapters 7 and 8. Again, those who have previously maintained that there was an ET influence in these crucial stages of the development of humankind, have presented very little of the materialist, scientific viewpoint. A close look at what archeologists and modern materialistic historians have to say of these critical advances of humankind, compared with what the ancient Mesopotamian historical accounts say, reveals the utter absurdity of the materialistic approach to the rise of human civilizations.

An examination of ancient human history reveals that the negative ETs that were directly involved with human affairs at that time had little or no compassion for their creations, the human species. The evidence indicates that many of the negative ETs evolved spiritually to a more positive orientation, and left Earth, or were driven from the Earth by negative ETs between 2,000 and 600 BC. Other negative ETs maintained their regressive orientation and literally went underground, at about the same time, where they have secretly manipulated and controlled humankind until the present day. We will review evidence that negatively-oriented ETs, from their underground strongholds, have sought to keep humankind divided throughout history along political, racial, religious and monetary lines in order to render humans more vulnerable, malleable and controllable. The

Introduction

regressively-oriented ETs have maintained this control always with the help of a small group of humans, usually of secret societies or secret governments. We will review other methods the negatively-oriented ETs use to control humankind, including negatively influencing individual and collective early human development (imprints) and subsequent conditioning.

The main method negatively-oriented ETs have employed to keep humankind divided and controlled is by distorting and hiding spiritual truths. The negatively-oriented ETs realize that humans can escape their manipulations by evolving to spiritual states that are higher than they have obtained. The world religions will be briefly reviewed to demonstrate that they all contain spiritual distortions and false spiritual teachings.

Finally, I will review the good news - that positively-oriented ETs and other positive entities are trying to help humans expand their consciousness and spiritual understanding to the point when we no longer need to play the victim in the controller-victim game that has been played on Earth for thousands of years. We will review evidence that indicates that, through non-judgmental love and compassion, humankind is now entering the phase of its development where very rapid spiritual development is possible.

An explanation of some of the terminology used is necessary. Most books dealing with the ET phenomena use the term extraterrestrial, or ET, to indicate a visitor from outer space. They use the term alien to refer to beings that have lived for a considerable period of time on Earth, usually in underground dwellings, but who are descended from ETs. However, in California and elsewhere, the media often refers to foreigners who come to the United States illegally to work, as aliens. Therefore, I have settled with the term ETs to indicate species of entities whose origin and most of their development was not on this Earth, whether they've had a lengthy stay on Earth, or not.

Also, I would like to comment on - and modify my usage of the polar terms, positive and negative. I'm well aware that no organization, entity, or group of entities is completely positive or completely negative. For example, although I will argue that all world religions contain many false spiritual teachings and spiritual distortions, I'm aware that these religions have given comfort to many humans over the years. Some humans, in fact, have actually obtained a high degree of spiritual enlightenment by following the rituals of their world religion. Similarly, the ET "gods" of the ancient Sumerians (the Anunnaki) may have behaved mostly in a negative, non-compassionate, unloving way toward most of their human subjects, but they did aid in the founding of human civilization, which has allowed humans to slowly progress to a point where rapid spiritual evolution is possible. There are many different levels of comprehension, some of which are beyond the understanding of most humans at the moment. Therefore, as I write about your particular religion or secular belief system in a negative way that may be offensive to you, remember that I am aware that at some levels of understanding, no belief system is completely negative or positive. Similarly, when I write about negative ETs or positive ETs, I am using these terms for readability, rather than such terms as less-than-positive ETs or mostly benign ETs. In most cases, I will use the terms progressive ETs and regressive ETs, rather than positive ETs and negative ETs. Progressive ETs are ETs that are progressing toward the Divine Creator through love and compassion, and regressive ETs are not. Furthermore, as I will elaborate in the last chapter, we all must expand our consciousness to the point where we can accept all the negative aspects of this universe, as well as the positive aspects, without judgment. While we are all limited, to some extent, in our communications by our languages, "progressive" and "regressive" are meant to awaken fewer judgmental attitudes than the adjectives "positive" and "negative."

Let us begin our journey toward the Truth of human beginnings by reviewing the Darwinian-anthropological myth of human origins. Many readers will already know that the Darwinian-anthropological story of the origins and evolution of life and humans is flawed. However, this review of the Darwinian-anthropological view of the origins of modern humans, in addition to revealing its numerous flaws, will serve as an introduction to the true origins of humankind.

---

[1] Hastings, A. 1991. *With the Tongues of Men and Angels: A Study of Channeling*. Holt, Rinehart and Winston, Inc., Fort Worth, Texas.

[2] Gallup Organization. 1993. "Half of U. S. believe creationism." *San Francisco Chronicle*, September 13, p. A5.

> *"From creation to the creator,*
> *we move and have moved*
> *and shall move ever, ever,*
> *ever returning and growing in love."*
> Ray Pitsker, *Love is Forever*

Chapter 1

The Darwinian Myth of the Origin and Evolution of Life

The Theory of Evolution

The theory of evolution was first publicly presented in 1858. Today, in the late twentieth century, the theory, with some modifications and much more biological knowledge, is the predominate origin "myth" of Western societies, if not the world. Neo-Darwinism, as the theory is known today, has unprecedented status among scientists and many other intellectuals. There is little debate over the premise of the neo-Darwinian theory among biological scientists today. While some scientists do observe a few biological peculiarities that seem to present some difficulties to the theory, a vast majority of these scientists view these difficulties as trivial anomalies that will eventually be explained within the framework of neo-Darwinism. Neo-Darwinism is the official belief system of the United States Government, as all American scientists who apply for government research grants from any government agency must not deviate very far from Darwinian dogma. As a result of its almost unquestioned status, the neo-Darwinian dogma influences public policy, such as modern medicine and the multi-billion dollar industry of American medical and scientific research, and countless other aspects of modern life.

But does the neo-Darwinian theory which propounds to explain the origin and purpose (or lack of purpose) of all life on Earth deserve such high and unquestioned status? In this chapter, we will review some of the deficiencies in the neo-Darwinian theory, after we briefly review the principles of the theory of evolution. To understand the impact that the theory of evolution has had on human consciousness, one must understand the origin myth that came before. Prior to the theory of evolution, the biological aspects of humans were not even considered in origin myths. Humans had no understanding of their relation to other forms of life. Despite its shortcomings, the Darwinian movement of the past one hundred and thirty odd years has demonstrated the importance of our biological entities, and the neo-Darwinian theory does explain some aspects of biological evolution rather well. Possibly the reason Darwinism has such high status today is because we are in the midst of awakening to our biological origins. Despite the beliefs of creationists, the evidence for some sort of biological descent with modification (biological evolution) is overwhelming.

The Darwinian movement of the past one hundred and thirty-five years represents a still primitive species attempting to understand and define biological evolution. With no previous comprehensive theory of evolution it would be very surprising if humans had been able to formulate a correct theory of evolution on the very first try. Even more surprising is that there have been few advances in the theory of evolution since the time of Darwin and Wallace, despite the deficiencies of the theory that Darwin and Wallace themselves recognized. At present there is no alternative to the neo-Darwinian theory in Western societies, except the creationist story in the Book of Genesis. The story of the six-day creation of the world and all life, including man and woman by an omnipotent god, is absurd in face of the evidence that the world and life is millions and, perhaps, billions of years old.

Most neo-Darwinists today react as if an attack on the current theory is an attack on evolution itself. Hopefully, in a few years there will be a viable alternative to neo-Darwinism that will explain our biological evolution, as well as taking into account our spiritual existence. In fact, some of the foundations for a new theory of evolution are already being laid.

The theory of evolution is best understood with a brief historical review. Before the theory of evolution, not only did the majority of the people of Western societies believe in a six-day creation, they believed that all species of plants and animals were fixed and could not change. The immutability of species ideas were very prevalent and deep-seated in the early eighteenth century, and these ideas dated back to ancient Greek philosophers that lived a few hundred years B.C.

In the early part of the seventeenth century, Archbishop Ussher, an Irish prelate and scholar, calculated from Genesis that the world had been created in exactly 4004 B.C. Even today many Bibles emphasize

The Darwinian Myth of the Origin and Evolution of Life

this date in the Book of Genesis.

Belief in a six-day creation, unchanging species, and a world less than 6,000 years old, characterized the world of the young Charles Darwin (1809-1882), the son of a well-to-do English physician. There were a few men who had anticipated the theory of evolution, or had actually believed in evolution before Darwin, including his own grandfather, Erasmus Darwin. But nobody had developed a believable, comprehensive theory of evolution.

Darwin's father, wanted his son Charles to be a physician. However, after two years of study in Edinburgh, Darwin dropped-out of medical school. Charles was far from a brilliant student, as he was usually more interested in hunting and fishing than in academic achievement. After leaving the field of medicine he eventually enrolled in Cambridge University as a theology student. He graduated from Cambridge in 1831 and would probably have become an obscure country parson had he not been afforded the opportunity to sail around the world.

As a naturalist and companion of the captain, Darwin was invited to sail aboard the *H.M.S. Beagle* to map the Southern shores of South America. Darwin was frequently allowed to go ashore where he made collections of- and made extensive notes of the flora and fauna he encountered. What had been a hobby - the observation of nature and natural history - became a serious endeavor for Darwin during the five years of the *Beagle's* voyage from 1831 to 1836. He observed many natural phenomena on this voyage that convinced him that creationism and the immutability of species must be in error. For example, on the Galapagos Islands, located about 600 miles off the West Coast of South America, Darwin observed similar species of finches (today known as Darwin's finches). Darwin reasoned that these thirteen similar species of finches might possibly have evolved from a common ancestor, instead of having been individually created by God.

By 1844, Darwin had completed a summary of his views on evolution which were very similar to the views he was to present to the world fifteen years later. But he felt he did not have sufficient data to support his views. He continued to read and accumulate facts until he received a sharp prod from Alfred Russel Wallace (1823 - 1913). Wallace, also an Englishman, read extensively and was a keen observer of nature like Darwin. Unlike Darwin he went to work at the age of fourteen and never received a formal higher education. As a young man, Wallace began collecting animal specimens in exotic, little known places. He was able to support himself by selling his exotic animal specimens, especially birds and insects. Like Darwin, he had wondered why some species of animals were often so similar to each other.

In 1858, while on an island in the Dutch East Indies, Wallace was suffering from one of his periodic attacks of fever. He "saw" a solution as to why there are similar varieties of the "original type" of plant or animal, and his ideas were remarkably similar to the theory of evolution Darwin had formulated years previously.

Darwin and Wallace, having a mutual interest in natural history, already periodically corresponded with each other. Therefore, after Wallace had written a paper concerning his insights, he sent it to Charles Darwin, the very man who had developed a similar theory. At first, after reading Wallace's paper, Darwin considered completely destroying his own work on the subject lest he be accused of dishonorable behavior. But friends of his, who were aware of his ideas and work on the subject, persuaded him to write a paper that would be presented to a group of naturalists along with Wallace's paper.

Hence, the theory of evolution was presented to a meeting of the Linnean Society in London in 1858. The first public presentation of the theory of evolution did not create a stir because most members of the society did not agree with Darwin and Wallace's concepts of evolution.

This incident prompted Darwin to come out of the closet, as it were, and publish the ideas he had begun to formulate more than twenty years previously, even if he didn't feel he had all the information he needed to support his views. Thus, a year later, in 1859, he published his famous book, *The Origin of Species by Means of Natural Selection, or the Preservation of Favoured Races in the Struggle for Life*, the title being usually shortened to *The Origin of Species*.

The theory of evolution as presented by Darwin in the book is really quite simple. The essential thesis presented in *The Origin of Species* is that species are mutable and that they do change over time. Darwin argued that all living organisms, including humans, are derived from other organisms through a gradual evolutionary process, the main process being that of natural selection. Darwin based his theory of natural selection on a few observations of the natural world and deductions he made from these observations.

Darwin observed that organisms are capable of producing many more offspring than can survive. For example, trees produce many seedlings, fish and amphibians often lay hundreds and thousands of eggs. Small mammals usually have large litters of offspring. Darwin further observed that despite the tendency of organisms to progressively increase, the population of these organisms remained more or less constant because the food supply of the organisms did not increase in a similar manner. Darwin deduced, therefore, that there must be a struggle among the offspring of organisms for survival. He further observed that organisms within a species vary in their physical (morphological) traits considerably. He deduced that those organisms of a species that possessed the most favorable variations in size, speed, or whatever, would be more likely to survive. Although the principles of inheritance (genetics) were completely unknown at that time, Darwin and most everyone knew that some variations of living organisms were somehow passed on to future generations. Darwin reasoned, therefore, that over long periods of time, successful morphological variations would produce large differences in organisms that would gradually result in new types - new species.

The anguish, turmoil and debate caused by the conflicting views of Darwin's theory of evolution and the accepted religious dogmas of the day that raged in the late 19th century and continues to the present day, is well-documented in numerous books and papers. Darwin and his supporters not only argued that species were mutable and had evolved, not been created, but they were also eliminating God, the human soul, and a spiritual existence from humankind. Evolution is a slow random process, Darwinists claimed, with no purpose and no direction. The theory supports a basically secular, materialistic view of humans and the universe. Naturally those of religio-spiritual inclinations objected strenuously.

But Darwinism was slow to be accepted by many naturalists and biologists as well. Many naturalists and biologists of the late 19th century didn't care for the randomness, the purposelessness, and the proposed lack of direction of the Darwinian theory. Some form of Lamarckism was still widely believed to be a primary selective force in evolution. Lamarckism, named after a French evolutionist who preceded Darwin, Jean Baptiste Lamarck (1744 - 1829), was a belief that what is known as "acquired characteristics" could be passed on to future generations. For example Lamarckism would hold that a man who, through much exercise, developed large muscles during his lifetime, would pass on this characteristic to his offspring. Darwin himself believed in some form of Lamarckism, as well, as he included Lamarckism in the first six editions of *The Origin of Species*.

What we today know as the science of biology did not actually have its beginnings until the late 19th century. Envious of the successes of the physical science's explanation of how the physical matter of the universe was organized, early biologists organized their discipline along the lines of physics and chemistry. Towards the end of the 19th century many esteemed American and European universities had laboratory space for their new biologists where scientific experiments could be carried out. Following the lead of the physical sciences, the new biologists would believe nothing unless it was demonstrated in the laboratories.[1] In these new laboratories many experiments designed to test the tenets of Lamarckism were carried out. An often-quoted experiment that was supposed to prove that Lamarckism did not exist was carried out by August Weismann, a late 19th century German biologist. Weismann amputated the tails of hundreds of mice over five generations, only to find that all the progeny grew normal, not even slightly shortened tails. Since no evidence was found in the laboratories of the new biologists that characteristics acquired during an organism's lifetime were passed on to future generations, belief in Lamarckism died.

Population Genetics

In the early part of the 20th century, the new discipline of genetics was studied in many laboratories of the new biologists. The principles of inheritance through genetics were completely unknown to Darwin and Wallace, even though the original discoverer of the principles of inheritance, Gregor Mendel (1822 - 1884) lived and carried out his own experiments at the same time Darwin was formulating and defending their ideas of biological evolution. Mendel became a monk and school teacher in the Austrian-Hungarian Empire in Brünn, which was later to become the country of Czechoslovakia. There, working mostly with pea plants, he formulated the basic principles of inheritance. His work was forgotten, until it was rediscovered by Dutch botanist, Hugo de Vries, around the turn of the century. In Europe and America, during the first decades of the 20th century, the science of genetics flourished and expanded. In America, the early geneticists carried out their experiments mostly with the common fruit fly, which has a very short generation length and is easily handled in the laboratory. They refined and expanded the knowledge and understanding of the principles of inheritance.

Still, Darwinism did not have a monopoly among biologists. There were still other claims besides natural selection as the primary agent of evolution. Finally, in the 1920's and 1930's, biologists from several countries who were trying to construct mathematical models of how evolutionary change could occur in populations of organisms, successfully merged the science of genetics with Darwinism. This merging, often called the "Modern Synthesis", the synthetic theory, population genetics or neo-Darwinism, finally gave Darwinism the status it enjoys today. The population geneticists theorized that mutations that rendered small genetic changes in a population of organisms, provided the variation from which natural selection selected those organisms most fit to survive. These small genetic changes in individuals, if favorable to the individual, were spread throughout the population through sexual reproduction. In animal populations individuals will often migrate as young adults or adults and become members of another population. For example, the species of forest monkeys I studied in Zaire, known as the black mangabey (*Cercocebus aterrimus*), lives in groups (populations) of fifteen to twenty individuals of both sexes at all stages of maturity (adult, adolescent, infants).[2] As young males of these small populations reach maturity they often migrate to another group. Young males of many primate species have often been observed migrating from their natal populations. In some species of primates (e.g., gorillas) the females often migrate from their natal population. Migrations are common in animal populations. Through migration and sexual reproduction, a favorable small genetic change spreads from population to population within a species until it has spread to all populations of a species. In this way neo-Darwinists see species as gradually changing and evolving.

Molecular Biology

I will briefly review a few principles of molecular biology here to illustrate the marvelous complexity of the biochemical basis of life. Later in this chapter we will consider the neo-Darwinists' arguments for the chance origins of life on this planet, and some knowledge of the biochemical basis of life is necessary to understand this controversy. More detailed accounts of what follows are contained in most college-level introductory biological texts for readers who wish to learn more about molecular biology.

In 1953, an American scientist and a British scientist co-discovered the molecular structure of the genetic material, deoxyribonucleic acid, or DNA. In 1953, James Watson and Francis Crick published a scientific paper reporting the double helical structure of DNA, and thus helped solve the centuries-old puzzle of heredity.

In the years since Watson and Crick revealed the chemical basis of heredity, there has been rapid growth in the knowledge of the biochemical basis of life. This knowledge has shown that all living organisms, from microscopic one-celled bacteria to considerably more complex organisms, such as ourselves, are very similar biochemically. The biochemical similarity, the neo-Darwinists argue, supports their theory that all life forms stem from one origin and that we are all interrelated.

The genetic material of all organisms, the DNA, can be thought of as a blueprint of the organism. The DNA are within the nuclei of every one of the trillions of cells in our bodies (with few exceptions). DNA is a nucleic acid made up of millions of nucleotides. Each nucleotide is made up of a sugar molecule, a phosphate molecule, and a nitrogen base. The sugar and phosphate molecules are identical in every nucleotide, but there are four different bases, known as adenine (A), thymine (T), guanine (G), and cytosine (C). So the DNA consists of two strands of millions of nucleotides strung together. The two strands are bound together through hydrogen bonds between the bases. The two strands are twisted around each other to form the so-called double helix.

The DNA holds the code for the assembly of many proteins. The proteins are assembled in the cytoplasm of the cell. The code (the information) for the assembly of proteins is transferred from a single strand of DNA (which has "unzipped") to the so-called RNA. A strand of DNA that codes for the assembly of a particular protein is known as a gene.

The RNA I just mentioned which is equally needed for the synthesis of proteins is a single strand of nucleic acid, but otherwise very similar to DNA. There are several types of RNA, but the type of RNA that picks up the code for the assembly of a protein in the nucleus of a cell from the DNA is known as messenger(m) RNA. After the code for the assembly of a protein is "transcribed" to such an mRNA, the mRNA moves from the nucleus of a cell to the cytoplasm where the protein is to be assembled. The actual assembly of the protein takes place at what is known as a ribosome. The ribosome is a complex organelle consisting of some fifty proteins and three chains of RNA, and acts as a decoder.

The proteins themselves are made up of a long string of amino acids. Amino acids are small organic compounds consisting of some ten to twenty atoms. Only twenty different types of amino acids are used by organisms in the assembly of proteins.

Three nucleotides in the DNA (which are known as triplets or codons) code for one amino acid. The code, however, is redundant. For example, AGA codes for the amino acid Serine, but so do AGG, AGT, AGC, TCA and TCG. These triplets (codons) are decoded at a ribosome, with the aid of another form of RNA known as transfer(t) RNA. Each tRNA molecule consists of a short strand of RNA about 100 nucleotides long which is folded into a compact hairpin looped structure. Each tRNA molecule can recognize a particular triplet, a codon, of the mRNA and the appropriate amino acid specified. The tRNA molecule attaches to the appropriate amino acid and transfers it to the area of the ribosome where the protein molecule is being assembled. After all the amino acids of a protein are strung together in the correct order, it is detached from the tRNA molecules, and, under the influence of various electro-chemical forces, automatically folds into a complex three dimensional (3-D) form. While each amino acid of a protein is bound to an adjacent amino acid in identical ways, each amino acid has a unique side chain. It is at these side chains of the amino acids of a protein where the work, or the atomic interactions of a protein takes place. The 3-D configurations of a protein occur in such a way as to bring about the maximum number of atomic interactions between the various amino acids.

The protein molecules are the ultimate stuff of life. Proteins make up most of what we are, from our fingernails, our skin, our internal organs, our hair, to our muscles, etc. These "basic" proteins are known as structural proteins. There are also proteins known as enzymes which speed up or carry out various chemical reactions. Proteins also carry out logistic and transport functions, such as the protein hemoglobin which transports oxygen from our lungs to other parts of our bodies.

Proteins are so small that they must be magnified a million times before they can be seen. But there are an enormous number of different protein functions, and a corresponding enormous number of different 3-D configurations. The 3-D configurations of each protein are dictated directly by the amino acid sequence of the protein. A protein usually has one hundred to five hundred amino acids. The substitution of just one amino acid may have little or no effect on the protein, or it may completely affect the function of the protein.

For example, in the beta chain of a normal hemoglobin molecule (there are alpha and beta chains), if mutation occurs in the sixth amino acid (out of 146 amino acids), the red blood cells tend to be misshapen. These misshapen blood cells do not carry oxygen as normal blood cells do and a person who carries this mutation in one gene (heterozygous) at the location (locus) of the gene of the beta chain of hemoglobin on the DNA, suffers some anemia, but has greater resistance to malaria. A person who carries this mutation in two genes (homozygous) at the locus of the beta chain gene of hemoglobin, usually dies before reaching maturity. This disease, called sickle cell anemia, does allow a person who has only one gene (a heterozygote), to have better resistance to a certain type of malaria, known as *falciparum* malaria, than those with normal blood. Not surprisingly, the presence of this mutation that causes sickle cell anemia is found most frequently among the blacks of Africa, where *falciparum* malaria is most common. But this disease is also found in low frequencies among southern Europeans and southern Asians where this type of malaria is also found. The DNA for normal beta hemoglobin at the sixth triplet, or codon, is CTC, which codes for the amino acid, glutamic acid. However, in sickle cell anemia, a small mutation, known as a point mutation, caused the base T to be changed to an A over the course of time. The CAC specifies the amino acid valine. This small change of one base on the DNA molecule affects the entire structure of a protein, and, consequently, the structure and efficiency of red blood cells, and in turn the well-being of the individual.

Today, molecular biology is completely incorporated in the neo-Darwinian theory of evolution. Small point mutations, as the one illustrated above in the so-called sickle cell trait, are theorized to gradually accumulate and to eventually yield new species, new orders and classes of organisms.

Reductionism and Neo-Darwinism

Neo-Darwinism and the science of biology have a reductionist, deterministic, and mechanistic view of life. This is due in large part to the attempts of the early biological scientists around the turn of the century to emulate the physicists of the 19th century who had a very reductionist, deterministic and mechanistic view of matter. This view advocates that everything, including life, is the result of the evolution of matter, from the big bang which physicists believe marked the origin of the universe some eight to twenty billion years ago, to the chance beginning of life on Earth some 3.5 billion years ago. Thus, the neo-Darwinian theory of evolution has become part of what is known as scientific materialism which was started by the early successes of the "hard sciences", physics and chemistry.

Reductionism in biology simply means that in order to understand an organism, one must understand the smaller parts of an organism. This is why as high school biology students, many of us had to dissect worms, frogs and other small animals. We had to understand the layout of the veins, arteries, organs, and other small interior components of the animal before we could understand how a living animal lived. Little or nothing, was said, at least in my high school, about how the animal actually lived in nature.

To carry biological reductionism beyond the high school level, as molecular biologists have done so successfully since the 1950s, one must understand the molecular level of an organization and operation of the organism. From this knowledge, biological organisms can be further reduced. For example, Richard Dawkins, a prominent ethnologist and neo-Darwinist, has claimed that it is ultimately the genetic material that evolves. The bodies of organisms are seen by Dawkins as merely "survival machines" for the genes.[3] And the chance evolution of the genetic material is viewed by neo-Darwinists as responsible for countless human achievements, from beautiful symphonies to marvelous scientific theories, such as the one discussed here. For example, zoologist Edward Wilson a prominent figure in evolutionary biology, writes that, after more research in the future, he believes that "the mind will be more precisely explained as an epiphenomenon of the neuronal machinery of the brain."[4]

Countless other examples could be given of how biologists have reduced large, complex phenomena, such as the human mind, to matter. The molecular and biochemical levels we have examined can further be reduced to the atomic and sub-atomic levels. Many scientists today believe that, as science progresses, all behavior, including human behavior, social movements, and history itself, will be reduced to sub-atomic events.

The mechanistic model of modern-day biology, including neo-Darwinism, is usually traced to the famous 17th century French philosopher and scientist, René Descartes. Descartes made the famous statement "Cogito, ergo sum", meaning, "I think, therefore, I am." In this way Descartes divided the universe into two categories, mind and matter. Descartes thought that only humans could think, as he believed that only humans had a mind. Everything else, including animals, was in the category of matter and therefore behaved in a machine-like manner. The machine model of animals (and plants) has endured since the days of Descartes and many modern biologists of many stripes believe animals behave as mechanical machines. We will review some examples of animal behavior in this and later chapters. The readers will be left to decide themselves how much animals behave as machines. Let us examine some aspects of the neo-Darwinian theory a bit closer.

Flaws in the neo-Darwinian Theory of Evolution

The Origins of Life

Charles Darwin, himself, never claimed his theory could explain the origin of life, but he implied as much. In some of his private correspondence, he speculated that perhaps in "some warm little pond" that contained a certain chemical composition, combined with heat and electricity, living creatures might be formed.[5] In any case, the random, chance circumstances of his theory banished any supernatural being from biology, and therefore strongly implied a chance beginning of life on Earth.

The neo-Darwinists have embraced the idea of a chance beginning of life from non-organic matter. For example, G. G. Simpson, the well-known paleontologist (those that study ancient life), who was largely responsible for the incorporation of paleontology into the neo-Darwinian theory, writes concerning the origin of life:

> There is....no reason to postulate a miracle. Nor is it necessary to suppose that the origin of the new process of reproduction and mutation was anything but materialistic.[6]

The oldest known rocks of our Earth are dated at about 3.8 billion years; from this date the Earth is estimated to be about 4.5 billion years old. The Earth and the remainder of our solar system, is believed by scientists to have been formed by the condensation of a huge cloud of interstellar gas and dust about 4.5 billion years ago. Nobody knows what the atmosphere of the early Earth consisted of, but there can not have been any oxygen, since oxygen comes from the photosynthesis of organisms which didn't exist on Earth at that time.

In the 1920s two scientists, Alexander Oparin of the Soviet Union and J. B. S. Haldane of England, independently presented a hypothesis for the origin of life from inorganic material by chance circumstances. Briefly, Oparin and Haldane speculated that the early Earth had an atmosphere consisting of hydrogen, water, methane and ammonia. This early atmosphere was exposed to energy from solar radiation and lightning, they hypothesized, that led to the formation of organic compounds. These organic compounds formed what was called a primordial or prebiotic soup in the oceans. The prebiotic soup built up and eventually a primitive organism formed from these organic compounds that could replicate itself, and life began.

In 1953, two biochemists, Stanley Miller and Harold Urey, designed an experiment that was supposed to approximate the conditions of the Earth when life was formed. They sent an electric charge through the mixture of gases believed to be present in the atmosphere of the early Earth. The Miller-Urey experiment produced some amino acids, two of which were among the twenty amino acids used by organisms to synthesize proteins. That's all. The Miller-Urey experiment produced nothing that was nearly as complex as a living organism and only two of the twenty amino acids used to "construct" a protein of a living organism.

However, this experiment created quite a sensation at that time, both in the scientific community and in the popular press. Surely, the formation of living organisms from non-living matter would take place in some laboratory soon, thought most scientists and much of the general population. Many scientists and non-scientists alike assumed the Miller-Urey experiment had all but proven that living organisms developed by chance from inorganic matter billions of years ago.

This enthusiasm has waned considerably over the years. Hundreds of experiments similar to the Miller-Urey experiment conducted since then have yielded organic compounds that are little or no closer to the extremely complex compounds found in a living organism, than the original experiment.

The first evidence of life on Earth is found in microscopic fossils found in rocks of Australia and South Africa that are about 3.5 billion years of age. They are fossils of bacteria, microscopic one-celled organisms, that appear to be similar to the one-celled bacteria of today.

A one-celled bacterium today has its own DNA, RNA, and the capability of producing about 100 proteins. Our very brief review of molecular biology demonstrates the complexity of just one living cell and of protein synthesis. We saw that a protein usually consists of at least 200 amino acids that must be strung together in an exact arrangement in order for the protein to properly carry out its function. Clearly, a bacterium that can produce 100 proteins, enough to sustain and replicate itself, is not a simple organism. Nucleic acids such as DNA and RNA, as well as protein molecules consist of hundreds to billions of atoms and have very complex three dimensional structures. And, we didn't even review other complex molecules used by organisms. Biochemist Robert Shapiro comments:

> The important construction materials used in a bacterium (or us, if we set aside special equipment such as bones and teeth) are proteins, nucleic acids, polysaccharides, and lipids...None have been detected, in any amount, in Miller-Urey chemistry.[7]

Later he states that the very best Miller-Urey chemistry, " ...does not take us very far along the path to a living organism."[8]

There is no evidence in the fossil record of a gradual evolution from something simpler than a one-celled organism to a one-celled organism. Further, there is no evidence of a prebiotic soup. One's mind is boggled at the thought of a functioning one-celled organism with DNA and/or RNA that codes for proteins that allow it to survive and reproduce, originating by chance circumstances. To quote Dr. Shapiro again, "...virtually all scientists today believe that living cells cannot commonly be generated from their chemical ingredients by random process."[9]

The key word in the above quote is "commonly." Despite the immense odds and evidence to the contrary, many scientists still believe that life of some sort was generated from their chemical ingredients by random processes. As Shapiro points out, the fact that this hypothesis is generally accepted in the scientific and non-scientific communities, despite contrary evidence, is an indication that this hypothesis is now indistinguishable from a myth.[10] Some scientists have formulated alternate hypotheses for the prebiotic soup hypothesis, unfortunately these ideas are based upon calculation and conjecture, not scientific experiment. Other scientists (including the above-quoted Dr. Shapiro) believe a scientific materialist explanation will eventually be developed, or they argue that "...it happened despite the odds, because we are here."

Confronted with these difficulties surrounding the question of the origin-of-life, a few scientists have turned to outer space. Ideas about life from space, known as panspermia, have been around since the early years of this century. A recent version of panspermia comes from the team of the well-known British astronomer, F. Hoyle, and a mathematician, N. C. Wickramasinghe, in their book *Life Cloud*.[11] In their early version of panspermia, Hoyle and Wickramasinghe see organic molecules forming in interstellar clouds. These biochemical materials become attached to passing comets, some of which collide with the early Earth. In this way comets bring the ingredients of a prebiotic soup to the Earth's waters, where eventually a life-form emerged.

Another version of panspermia comes from Francis Crick, who, as we observed, won the Nobel Prize with James Watson for deciphering the structure of DNA. Crick offers the following hypothesis, known as directed panspermia, in his book, *Life Itself*.[12] Crick suggests that life may have arisen on another planet much older than the Earth where conditions for the formation of life were more favorable. Life evolved until humanoids, much like ourselves, developed a technologically high civilization. Eventually their planet became less and less habitable, and their civilization began to perish. After trying unsuccessfully to save their planet and civilization, these humanoids sent out spaceships from their dying planet with species of frozen bacteria. One of these spaceships found our Earth and spread bacteria over the surface of the planet. Some survived and thus life on Earth began.

Later, Hoyle and Wickramasinghe developed their own hypothesis of directed panspermia. In their book, *Evolution from Space* (1981), they postulate that an early form of life evolved elsewhere, and had a silicon base, similar to computer chips. These intelligent forms of life then developed a carbon-based life form. Thereafter, they proceeded to seed our Earth with the carbon-based life.[13]

While many scientists believe in some form of panspermia, the authors of directed panspermia hypotheses have very little, if any support from the scientific community. Shapiro and other scientists contend that any seeding of life on Earth from any life form beyond our planet, even if not from an omnipotent Supreme Being, falls into the realm of religion, and is not science. Furthermore, modern, materialistic science rejects the possibility of any non-physical existence and believes that life arose only here on Earth. Scientists contend that if there is life elsewhere in the universe, we haven't contacted it yet. In contrast to the vast majority of scientists, Hoyle and Wickramasinghe do believe in a spiritual hierarchy and are very serious about their own hypothesis of directed panspermia. Crick apparently has serious doubts about his hypothesis. Shapiro reports that Crick told him that he believes his hypothesis is premature, and presented it only as a suggestion in order to increase public awareness of the difficulties surrounding the origin-of-life question.[14]

While these difficulties are acknowledged by scientists that have seriously considered the origin-of-life question, all but a handful of scientists believe the question will eventually be solved by materialistic science. Maybe it will, but certainly those who do have doubts can be excused.

At the beginning of my university human evolution courses, I said no more than a few words about the beginning of life. I pointed out the results of the Miller-Urey experiment and stated that I assume a scientific solution to this question will be found. Looking back, I feel that I at least should have been more forthright in informing my students that, at present, science has no satisfactory answer to the question of the origin-of-life. I was so immersed in the neo-Darwinian theory of evolution, I had no doubts that a scientific solution would be found. But now, I believe any thinking person should have grave doubts that the neo-Darwinian theory of evolution, or scientific materialism in general, can ever explain the origin-of-life on Earth.

Decelerating Evolution

Stephen J. Gould, a prominent paleontologist and neo-Darwinist, has indicated evolution is a decelerating process.[15] This presents another problem for neo-Darwinism. While several new species have evolved in the past few million years, no new higher categories of organisms have appeared for millions of years.

Simple one-celled organisms, such as bacteria and algae, were all the life forms that existed for almost the first two billion years of life on this planet. Then, quite suddenly, simple, soft multicelled animals, such as marine worms and jellyfish, appeared in the oceans about 670 million years ago. Then, even more suddenly and dramatically 100 million years later (at the start of what is known as the Cambrian period), marine animals with hard bodies "exploded" onto the fossil record. Not only did animals with shells, scales, spines and other skeletal parts appear suddenly about 570 million years ago, but all of the basic body plans of animals, or phyla, appeared at this time. A phylum is one of the highest biological classification categories (see Table 1), and is defined as a distinctive body plan of its members. For example, we humans are classified in the phylum Chordata, along with all animals with a vertebral column (or back bone), from fish to cats, as well as animals known as invertebrate chordates. Lobsters are classified in another phylum, oysters in another, sponges in another, and so on. Evidence obtained by comparing the ribosomal RNA of various species of different phyla suggests that most phyla did appear simultaneously, as the fossil evidence indicates.[16] Many paleontologists refer to this event as the "Cambrian explosion". This remarkable "explosion" is made even more remarkable by the fact that no new phyla, or basic animal body plans, have appeared since. There are twenty-six known phyla and they all appeared at the beginning of the Cambrian period.

J. S. Levinton of the Department of Ecology and Evolution at the State University of New York at Stony Brook states in a *Scientific American* article, that "...(e)volutionary biologists are still trying to determine why no new body plans have appeared during the past half a billion years."[17] Later, Levinton speculates:

> One idea worth entertaining is that evolution occurs more slowly today than it did when the Earth was young. If evolution has slowed for unknown, peculiar reasons, then perhaps too little time has elapsed for new body plans to evolve.[18]

Neo-Darwinists are forced to speculate like this when considering the Cambrian explosion.

Robert Augros and George Stanciu argue in their book, *The New Biology: Discovering Wisdom in Nature*, that the fact that no new phyla have arisen since the Cambrian era, argues against neo-Darwinism.[19] If changes in organisms occur by the accumulation of small random changes that are screened through natural selection as neo-Darwinists argue, why have no new basic body plans or phyla evolved in almost 600 million years, they ask?

Similarly, we belong to the class Mammalia, while other vertebrates are classified in other classes, such as Pisces (fish), Amphibia, Reptilia and Aves (birds). The last vertebrate class, Aves, emerged about 170 million years ago, but most classes of animals stopped emerging about 300 to 400 million years ago. Also, we belong to the order of Primates (placental mammals), and there are several other orders of placental mammals (see Chapter 2). Most, if not all of these orders of placental mammals appeared about 65 million years ago. As with phyla, and many classes, some orders have become extinct over the years, but no new ones have appeared recently. As a result there are more species in the groups that have survived, while formerly there were fewer species in more groups. Augros and Stanciu state:

> This pattern of shift from few species in many groups to many species in fewer groups, flatly contradicts Darwinian gradualism; for if evolution proceeded by species accumulating small variations, we should see over long periods new orders, classes, and phyla emerging with increasing frequency. But just the opposite occurs in the fossils. Darwin's model is backward.[20]

Major Transitions of Organisms

The most glaring shortcoming of the neo-Darwinian theory of evolution, aside from its inability to provide a satisfactory hypothesis for the beginning of life on this planet, is the lack of transitional fossils and a logical explanation of the major transitions of organisms recorded in the fossil record. Darwin himself and neo-Darwinists have claimed that evolution proceeds gradually in small steps, as we have observed. The fossil record provides no unambiguous examples of a gradual transition of one organism evolving into another organism, whether on the lowest level of biological classification, the species level, or in the higher levels, such as class to class (eg., reptiles to mammals). Steven Stanley, a prominent paleontologist, states:

> "The known fossil record fails to document a single example of phyletic (gradual) evolution ac-

complishing a major morphologic transition and hence offers no evidence that the gradualist model can be valid."[21]

The hard truth is that the core of neo-Darwinism, the idea that evolutionary changes occurred in small changes, is not documented by the fossil record. When a new species, or a new class, or new order, or new genus, of organism appears on the fossil record, the new organism appears fully formed.

Kingdom	Animal
Phylum	Chordata
Class	Mammalia
Order	Primates
Superfamily	Hominoidea
Family	Hominidae
Genus	Homo
Species	Sapiens

Table 1 A Biological Classification of Humans

We humans are classified in the phylum Chordata. Cambrian chordates did not have bony parts in their body, but had a rather soft, but strong notochord where the backbones exist in modern vertebrates. But early chordates were bilaterally symmetrical and had their sensory organs (eyes, ears etc.) at one end of the organism, as do modern chordates. It wasn't until about 500 million years ago (in the Ordovician epoch or period) that traces of cartilage are found in primitive armored and jawless fish. The oldest jawed fish with paired fins and a bony vertebral column appeared about 400 million years ago.

Roughly 370 million years ago, the first four-limbed land animals, the amphibians, appeared on Earth. Amphibians are the most primitive of the four-footed (tetrapod) land animals, as most of them that live today depend on water a great deal during their life cycles. The eggs of an amphibian are small and usually laid in the water, as with their fish ancestors. The egg has little yolk and does not have the protective membranes or shells found in the eggs of the later, more advanced reptiles and birds. Amphibians spend the first few days and weeks of their lives as immature aquatic animals. Many readers have watched the tadpoles of frogs as they develop into a frog.

An estimated 330 million years ago, the first land vertebrates appeared on the fossil record, the reptiles. Many modern reptiles, such as crocodiles and some snakes have a rather considerable dependence on the water, but many reptiles depend little on water in their lifetimes, and all reptiles lay their eggs on land. In order for the reptiles to "make it" on land, the structure of their eggs had to change considerably from those of the amphibians. The developing amphibian receives its oxygen and most of its food from the water in which the eggs are laid. On the other hand, the developing reptile obtains its food from a large amount of nourishing yellow yolk (see Figure 1-1). As a substitute for the amphibian's pond, the reptile embryo is protected from desiccation by a liquid-filled sac (the amnion) that develops around the embryo. From the back-end of the embryo's body there grows a tube and sac (the allantois) in which the waste of the embryo is deposited. The outer structure of the reptile egg is stiffened and protected by a firm shell which is completely lacking in an amphibian shell. The reptile eggshell is porous and beneath it is a portion connected to the allantois wherein lies a membrane rich in blood vessels, the chorion, that acts as a lung, taking in oxygen and giving off carbon dioxide. The reptile egg is a precursor to the reproductive system seen in mammals. From this brief description it is obvious that there is an extraordinary difference between the eggs and early developmental patterns of amphibians and reptiles. Neo-Darwinists claim that reptiles evolved from amphibians by gradual, random changes, screened by natural selection.

Another example of a major transition that stretches the credibility of the neo-Darwinian theory of evolution is the example of birds, believed to have evolved from reptiles, or dinosaurs, about 170 million years ago. The skeletons of birds are similar to those of reptiles and many fossil dinosaurs, and some paleontologists have even called birds "glorified reptiles" or "glorified dinosaurs." The main difference between a reptile and a bird, Alfred Romer, the well-known former professor of paleontology at the University of Chicago, tells us, are feathers and the fact that birds can fly.[22] But what a difference!

The oldest known fossil of a bird, the famous, most often quoted as a possible example of a transitional fossil, *Archaeopteryx*, is about 150 million years old. *Archaeopteryx* does exhibit some anatomical features that resemble reptiles and dinosaurs, especially in the structure of the pelvis. But *Archaeopteryx* is a bird, complete with feathers. Feathers are astonishingly, intricate and beautiful configurations. Feathers are very unique structures specialized for flying. They are quite unlike the scales found on reptiles, as anybody who has ever seen a reptile can attest. According to neo-Darwinian theory, feathers gradually evolved from scales through many small, chance mutations. Barbara Stahl in her book, *Vertebrate History: Problems in Evolution*, states with regard to feathers, "...how they arose initially, presumably from reptile scales, defies analysis".[23]

Figure 1-1 *Amniotic egg of reptiles. This egg represents a dramatic advance in the evolution of land vertebrates. Unlike the simple eggs laid by amphibians in water, the amniotic eggs, always laid on land, allowed vertebrates to give up dependency on water. Neo-Darwinian theory indicates that the evolution of the amniotic egg from the much simpler amphibian egg was the result of many small, chance mutations of the genetic code which were maintained through natural selection. Drawn after Nelson and Jurmain, 1988.*

There are other major differences between birds and reptiles: the lungs of birds are quite different from those of reptiles,[24] and the fact that birds are able to maintain a constant body temperature (warm-blooded), whereas reptiles are cold-blooded.

There are many more examples of more subtle evolutionary changes that are hardly understandable using the current neo-Darwinian theory. One such example is found in the evolution of mammals. The first mammals appear on the fossil record about 200 million years ago. As the name indicates, mammals feed their young from mammary glands that the reptiles have not developed. This reflects a major behavioral difference between reptiles and mammals. Mammals take care of their young from infancy through adolescent stages of development, while reptiles generally do not aid their offspring in their "struggle for survival" in any way. To date, the only reptiles that have been observed to care for their young are crocodiles, cobras and pythons.

Mammals have different types of teeth, a more upright way of standing and moving, and a different way of reproducing than reptiles. As we've seen, reptiles lay eggs, but mammals give birth to live young. Most mammals have hair, whereas reptiles have scales. Hence, we have a problem here that is similar to the scales to feather problem; how did hair evolve from the scales of a reptile?

Mammals have much larger brains, relative to body size, than reptiles. Mammals have relatively much longer periods of immaturity and development than reptiles. Since mammals have larger brains and are generally taken care of by at least one parent, these longer periods of development give the mammal young the opportunity to learn much more about their environment and engage in more complex behavior than reptiles.

Today mammals, even primitive mammals such as opossums, have much larger brains, relative to their body sizes, than do reptiles. A group of reptiles that are extinct, called mammal-like reptiles, that lived between 330 and 200 million years ago, are believed to be the ancestors to mammals. Dr. T. S. Kemp, a paleontologist specialist on mammal-like reptiles, writes:

> The evolution of mammal-like reptiles from their earliest appearance on the fossil record until the time of the first mammals occupied 130 million years...This sequence illustrates features of the evolutionary transition from one vertebrate class, the reptiles, to another, the mammals, and indeed this is the only such major transition in the animal kingdom that is anything like well documented by an actual fossil record.[25]

Yet, the skull of the first mammal on the fossil record, the approximately 200 million year old *Triconodon*, indicates that the first mammals had brain sizes comparable to the primitive mammals of today, larger than that of any reptile species, living or fossil.[26]

The early mammals were small rat-sized animals and remained so for more than 130 million years while large and small dinosaurs dominated the land masses during what is known as the age of dinosaurs. As we saw above, all of the modern orders of placental mammals appeared quite suddenly about 65 million years ago, after the dinosaurs became extinct.

Another example of the sudden appearance of a new life-form is seen in flowering plants. Flowering plants (angiosperms) appeared abruptly and quickly, and spread throughout the world about 100 million years ago, without any kind of gradual evolution from the forms of plants that existed before. Darwin, himself, called the sudden appearance and spread of flowering plants "...an abominable mystery."[27]

Darwin was bothered by the lack of intermediate forms, but in his various editions of *The Origin of Species* blamed this serious lack of support for his theory on the imperfection of the fossil record. He pointed out that only a small portion of the world was explored geologically at the time of his writing. He argued that the fossil record was not at all well-sampled. So Darwin used a poorly sampled fossil record to blunt criticism of the fact that no intermediate fossils had been found. As Michael Denton points out in his book, *Evolution: A Theory in Crisis*, using a poorly sampled fossil record to blunt criticism was a bit weak, even in Darwin's time. Denton illustrates that, even though 99.9 percent of all fossil discoveries and their study has been done since the publication of *The Origin of Species* in 1859, almost all fossil species discovered have been either closely related to forms known in Darwin's time.[28] No undoubted transitional fossil has been found more than 130 years after Darwin introduced his theory.

There are a handful of living forms that neo-Darwinists claim are themselves examples of transitional forms. For example, the lungfish has an intestine, fins, and gills like a fish, but has a reproductive cycle (larval stage) and a heart and lungs like an amphibian. Therefore, neo-Darwinists argue, the lungfish represents a transitional form between fish and amphibians. However, the characters themselves of the lungfish, such as the heart, are not transitional between two types. Monotreme mammals of Australia, of which there are only two species living, are another example of "living transitional forms." In the reproductive systems of monotremes, they lay eggs and seem almost wholly reptilian. Yet, in the morphology of their inner ear and the fact that they have hair and mammary glands, they seem mammalian. Again, none of their characteristics by themselves are transitional between reptiles and mammals. Denton states concerning these forms:

> There is no question that those forms are somewhat anomalous in terms of typology, each exhibiting a curious combination of the diagnostic characteristics of two otherwise quite distinct types. But, they provide little evidence for believing that one type of organism was ever gradually converted into another.[29]

On at least one occasion neo-Darwinists have had the rare opportunity to study the soft parts of a living fossil. An ancient fish, known as a coelacanth (of the order Crossopterygii), was recently dragged up from the ocean's depths. Coelacanth fish were believed to be close to forms that were believed to be ancestral to amphibians (the rhipidistians). Amphibians, we recall, were the first animals to use four legs (tetrapods), and use part of the land of the world located near water. The internal organs, the digestive track and the reproductive system of an amphibian are quite different from those of a fish. Rhipidistians had for decades been considered excellent ancestral candidates of the amphibians and had been biologically classified as intermediate between fish and the first amphibians. The coelacanths contain bones in their fins which are believed to provide the bases of the first legs used to move about on land. Based upon these bony fins, the pattern of their skull bones, the structure of their teeth and other skeletal features, the rhipidistians were considered intermediate between fish and amphibians.

In 1938, fishermen near South Africa caught a living relative of the coelacanth, known as *Latimeria chalumnae*. It had been believed that the coelacanth had been extinct for 100 million years, but here was a live one. As the *Latimeria* is closely related to the coelacanth, biological scientists at last had an opportunity to study a live animal that should have shown some transitional features. But a study of the soft anatomy of the *Latimeria* showed it was physiologically and anatomically a fish and had no features that might be considered intermediate between fish and amphibians. The ear, the digestive system and all of the soft parts were characteristically fish-like.[30]

This problem of the lack of data to support the neo-Darwinian theory of gradual evolution from one type of organism to another was confronted by a couple of paleontologists in the 1970s, Niles Eldridge of the Smithsonian Institute and S. J. Gould of Harvard University.[31] They proposed a theory that attempted to explain how the sudden evolutionary changes occurred, that are documented on the fossil record. Their theory has been called *punctuated equilibria*. A major part of their theory consists of the concept of stasis, a situation where a life form remains pretty much the same (does not evolve) for long periods of time, sometimes for millions of years. For example, the shark of today is similar to the shark of 100 million years ago. The above mentioned coelacanth fish is another example of an animal that has remained in a state of stasis for millions of years. The authors of this theory postulated that after long periods of stasis, a new species evolves in a small peripheral area of the geographical range of the species in a short time, geologically speaking (e.g., 50,000 years). This new species that has evolved in a peripheral population of the total area in which the species lives, may or may not have a competitive advantage over the original species. If the new species does have a competitive advantage (e.g., members of the new species might be able to digest a nutritious food source that members of the original, or parent population could not) it might replace the original species through ecological competition. The fact that the new species evolves in such a short time in a small area explains the lack of transitional fossils, according to Eldridge and Gould. It must be pointed out that Eldridge and Gould's explanations of how peripheral populations might evolve to another species and why there are no transitional fossils, is based upon the speculation of the authors. They are certainly highly educated, highly respected scientists, so this may be called educated speculation, but it is speculation nonetheless.

Punctuated equilibria theory has been heatedly debated since it was introduced. Some neo-Darwinists have embraced it proclaiming it to be revolutionary, while others have argued against *punctuated equilibria* being incorporated into neo-Darwinism as a primary or even a secondary mode of evolution. Still others argue that *punctuated equilibria* can easily be incorporated into traditional neo-Darwinism. For example, the prominent neo-Darwinist, Richard Dawkins states, "The debate about this interesting theory is a technical, parochial affair if ever there was one. It lies firmly within the Neo-Darwinian synthesis. It is no more revolutionary than any other argument that has enriched the synthetic theory."[32]

Nevertheless, the prevalence of stasis and saltationalism (sudden changes) observed in the fossil record was not part of the neo-Darwinian theory. According to Darwin and early neo-Darwinists, natural selection was continuously selecting from the variations of individuals within a species to cause continuous change to eventually form new species, new genera, new classes, and so on. This is clearly not the case. Even though most neo-Darwinists, such as Dawkins, see the debate over *punctuated equilibria* theory as a "parochial affair" in which the theory itself is easily incorporated into neo-Darwinism, many non neo-Darwinists see it as emphasizing some of the weaknesses in the neo-Darwinian theory. Denton writes, "Overt saltationalism, that is proposing that new types or organisms arise suddenly, is an obvious way of avoiding the problem."[33]

Evolution by Mass Extinction

Dinosaurs have undergone an upgrading in recent years as paleontologists have learned more about them. Dinosaurs were apparently not the huge, slow-moving, dumb animals depicted during the nineteenth and early part of this century. Dinosaurs possibly were warm-blooded, may have taken care of their young and otherwise been quite different from the reptiles we know. Dinosaurs thrived for millions of years and some of them seemed to have been fairly intelligent.[34]

However, the dinosaurs became extinct about 65 million years ago. Not only did the dinosaurs become extinct at this time, but up to eighty-five percent of all species of life forms that existed at that time became extinct in what must have been a world-wide catastrophe of some sort.

Many scientists now believe a comet or meteorite struck the Earth 65 million years ago and caused this world-wide catastrophe. The collision of the Earth and a large extraterrestrial body, many scientist believe, caused the spewing of millions of tons of debris into the atmosphere, and little or no sunlight was able to reach the Earth's surface, causing the mass extinction of the biological species that existed at that time. Not all scientists familiar with the data agree that this is what happened, but data are accumulating that support this possible extraterrestrial cause of mass extinction. Recently, a huge crater-like structure some 180 kilometers (112 miles) in diameter has been discovered on the Yucatan Peninsula of Mexico that dates to about 65 million years ago; this may have been the crater that remained after the extraterrestrial body struck the Earth some 65 million years ago.[35] At any rate, the heretofore successfully surviving dinosaurs, along with roughly 85 percent of all species of animals became extinct, but small mammals survived and soon began to flourish and develop into the many forms we see today.

Ancient life scientists know of at least six other mass extinctions where hundreds of thousands of species, 60-85 percent of all then existing species, became extinct. The youngest of these mass extinctions is the one we are considering now; the oldest happened about 500 million years ago, in the Cambrian period. University of California at Davis professor, Philip Singer, has recently found evidence in a survey of Cambrian species, of a mass extinction that wiped out more than 80 percent of the then existing species.[36]

The occurrence of mass extinctions in the Earth's geological past is another problem of the neo-Darwinian theory of evolution. Instead of being out-competed by forms that possessed "more favorable variations", as Darwin theorized, many, if not most species and higher categories that became extinct, vanished from the life-history of the Earth in some form of mass extinction.

"Struggle" for Survival

Some of the basic assumptions that Darwin made in formulating his theory of natural selection have come to be questioned in light of recent research. As we observed, Darwin argued in *The Origin of Species* that organisms reproduce at such a high rate that there is a deadly struggle among the offspring for survival. Darwin stated that, "Every single organic being may be said to be striving to the utmost to increase its numbers."[37] Elsewhere he writes, "...a struggle for existence inevitably follows from the high rate at which all organic beings tend to increase."[38] He argued that if it were not for struggle, resulting in the death of many offspring, the Earth would be overwhelmed by populations of animals.

However, observations of populations of animals living in their natural habitats do not confirm Darwin's assumption of organisms trying to reproduce as much as they can. Field studies have demonstrated that natural populations have internal mechanisms that regulate population growth. For example, among several species of mammals, the age at which an individual animal first reproduces varies according to the population density in which it lives. If the density of populations of white-tailed deer or bighorn sheep are high, for example, an individual animal in these high density populations will usually first reproduce at a much older age than if the populations were of low density.[39]

Most bird and mammal populations that have been studied have many non-breeding adults, especially males. In most monkey and ape populations, for example, males establish a dominance hierarchy. One or a few dominant males in a group do most of the mating, while the subordinate males, often younger males, either do not mate at all, or much less frequently than dominant males.

Many animal populations regulate the number of offspring they have each season according to environmental conditions. In abundant years they produce more offspring and in lean years they produce less. Birds and mammals vary their clutch size (number of eggs laid) and litter sizes according to the availability of food.[40]

Some fish and amphibians lay thousands, even millions of eggs, as Darwin observed. But animals that lay a large number of eggs do not protect them or care for their offspring. Most of these eggs are simply food for predators, and only a handful of them actually hatch.

Since Darwin's day, biology and neo-Darwinism have maintained that competition, the struggle of existence, dominates relationships between species. However, competition is rarely observed among wild animals of a different species. Similar species living in the same area avoid competition by dividing their habitat into ecological niches. For example the ungulates of the African savannah, such as gazelles, wildebeests and zebras, each eat different parts of the grasses available and do not compete with each other. While in Zaire observing rain forest monkeys, I frequently observed two or three different species of monkeys feeding in the same tree on the same fruit, located only a few feet from each other. Individuals of species, not only didn't compete, but they didn't interact with each other at all, except for the infants which sometimes played with each other. This has been observed by several researchers of African forest monkeys.

Moreover, many animals in nature spend a good deal of their time sleeping. The monkeys I studied slept all night and, at times, more than 50 percent of the day.[41] I usually took a book with me to read while the monkeys slept. All of those that have studied animals in the wild have reported long periods of resting and sleeping by wild animals. Certainly there is little evidence for Darwin's postulated "struggle for existence." Ecologist, Paul Colinvaux, after a review of the subject of struggle in nature states that, "...peaceful coexistence, not struggle, is the rule..."[42]

A Theory with no Clothes

The major shortcomings of the neo-Darwinian theory reviewed here are:

1. It provides no satisfactory explanation for the origin of life.

2. The observed fact that the evolution of higher categories (phyla, classes etc.) of animals has decelerated, if not ceased altogether, contradicting a basic premise of the theory, that small changes in organisms should continuously over time lead to the formation of higher categories.

3. There is no evidence of the gradual evolutionary mode postulated by the theory, and no satisfactory explanation of the stasis and evolutionary "jumps" observed in the fossil record.

4. Many, if not most species have not become extinct because they were "outcompeted" by other species that evolved superior characteristics, as neo-Darwinian theory claims, but because of mass extinction probably caused by a collision of the Earth and an extraterrestrial object.

5. One of the basic assumption of Darwin and neo-Darwinism, the "struggle for existence", is not borne out by observations of wild-living animals.

6. The theory does not provide a satisfactory explanation for the major vertebrate transitions observed in the fossil record, where not only do totally new and unique morphological characters appear, but also the new class of vertebrates exhibits a new, unique reproductive system and life style.

Several other anomalies of the current biological neo-Darwinistic paradigm could be mentioned, but I refer the interested reader to Denton's book and that of Augros and Stanciu, for a much more detailed account.

From this brief review it should be clear that there are many holes in neo-Darwinism. Yet, except for a handful of voices in the wilderness, neo-Darwinists do not doubt the theory that is today the centerpiece and foundation of biology, is essentially correct. As we have seen, even the anomalous data and theories, such as the *punctuated equilibria* theory, are subsumed into the neo-Darwinian theory of evolution, at least to the satisfaction of a vast majority of neo-Darwinists. Every one of the numerous scientific biological journals assumes the neo-Darwinist theory of evolution is correct, and this tends to reinforce its credibility. The biological and other scientific journals exude confidence. For example, Dawkins states:

> Charles Darwin showed how it is possible for blind physical forces to mimic the effects of conscious design and by operating as a cumulative filter of chance variations, to lead eventually to organized and adaptive complexity, to mosquitoes and mammoths, to humans and therefore, indirectly, to books and computers.[43]

Such confidence, even arrogance, is simply not merited by the data. The fossil data especially do not support the inception and evolution of life through natural selection by gradual, random processes. Denton states that:

> Ultimately, the Darwinian theory of evolution is no more nor less than the great cosmogenic myth of the twentieth century. Like the Genesis-based cosmology which it replaced, and like the creation myths of ancient man, it satisfies the same deep psychological need for an all-embracing explanation for the origin of the world which has motivated all the cosmogenic myth makers of the past, from the shamans of primitive peoples to the ideologues of the medieval church.[44]

Neo-Darwinism is like a theory with little or no clothes - or without data to support it. Readers may wonder why the theory is so widely respected in Academia! Part of the reason for this can be understood by observing how scientists overwhelmingly support the paradigm current in their field. Like all humans, scientists get carried away by the belief systems that are most prevalent and accepted in their circles. In 1962 Thomas Kuhn demonstrated that scientists often support a theory beyond what the theory may merit, simply because it is part of the dominant paradigm.[45] Direct and indirect pressures exist to demand that "members" - in the case we are considering, evolutionary biologists - tow the line and follow the dominant paradigm, even if there are anomalous data that do not fit it. It is currently professional suicide to question the basic assumptions of the neo-Darwinian theory. Nevertheless, anomalies - pieces of data that don't fit the current paradigm - accumulate. As anomalies accumulate, Kuhn points out, eventually a crisis results that will precipitate a completely new paradigm. Right now, it is crisis time for the neo-Darwinian theory - the old paradigm - whether neo-Darwinists presently realize it or not.

A University of California law professor, Phillip Johnson, has written a book, *Darwin on Trial*, where he argues that current neo-Darwinism is actually based on a philosophical doctrine called naturalism rather than scientific data.[46] Denton states, in the climax of his devastating exposure of the holes in neo-Darwinism, that:

> The cultural importance of evolution theory is ... immeasurable, forming as it does the centerpiece, the crowning achievement, of the naturalistic view of the world, the final triumph of the secular thesis which since the end of the middle ages has displaced the old naive cosmology of Genesis from the Western mind.[47]

Neo-Darwinism and Microevolution

I do not wish to close this chapter leaving the readers with the impression that neo-Darwinism is a theory that is totally without merit. The theory explains some aspects of what is known as microevolution rather well.

Natural selection has been observed, on a small scale. The best observed example of natural selection at work is the case of the peppered moth of England, today covered in almost every introductory evolution text.[48] The peppered moth of Great Britain, occurs in two colors, the "peppered" color and a black color. Species of organisms frequently exhibit different forms, such as different colors. This is known as polymorphism. The grayish, peppered color of this moth is practically invisible to its predators on many lichen-covered tree trunks of England. However, as the industrial revolution progressed in England, pollution caused the lichen to disappear and tree trunks near industrial centers reassumed their naturally dark color deepened by the black color of the pollutants strewn into the air by factories. Under these conditions the "peppered" color moth became very visible to birds and other predators, and they were eaten much more frequently than the black-colored moths. So the black colored variation of the peppered moth, which had been rare, became common in industrial areas, while the "peppered" color variety decreased in these areas.

This clear case of natural selection in action does not represent a major change in the organism, as the peppered moth did not become another species. Rather, a change in the frequency of the two major variations of the moth occurred as the environment changed. Examples such as this are termed microevolution. Large changes such as evolution of one class to another (e.g., reptiles to mammals) are known as macroevolution, which, as we've seen, is hardly explained by the neo-Darwinian theory.

Many of the examples Darwin used in his arguments in *The Origin of Species* are examples of microevolution. He used many cases of human breeding of domestic animals to illustrate how evolution occurred. The dog, for example, has been selectively bred by humans for thousands of years. Yet, despite the numerous colors and sizes, a dog is still a dog. Similarly, fruit flies have been used in genetic experiments by geneticists since the early part of this century. Fruit flies have been bred that have different colors of eyes, without wings, etc. But after thousands of generations of selective breeding, fruit flies are still fruit flies. In fact, no breeding experiment has produced a new species of animal. Plants, being simpler organisms, are another story.

New species of plants are produced by doubling their chromosome number. This will be discussed when we cover the "domestication" of plants.

The German zoologist, Bernhard Rensch, does not believe that macroevolution can be explained by the same neo-Darwinian process that seems to work well for microevolution.[49] He provides a list of prominent biologists that have similar doubts. These individuals are not kooks or cranks, but highly respected individuals in their subfields of biology.

Movements Toward a New Theory of Evolution

It would seem, at best, that the current theory needs to be modified, if not completely overhauled. At present, there is no alternative to neo-Darwinism. A few who have studied this problem (including myself) wonder if Lamarck may have been partly right. Earlier we saw that Lamarckism, the idea that characteristics acquired during one's lifetime can be passed on to their progeny, was abandoned around the turn of the century after experiments failed to support it. In the aforementioned experiment of August Weismann, who cut-off the tails of hundreds of mice over five generations, but whose progeny grew tails that were not even shorter than a normal mouse tail, might seem conclusive. However, there was one major drawback to this and other similar experiments; the mice themselves were not participating in this experiment. They had no conscious or unconscious desire to shorten or lose their tails. The current neo-Darwinian theory states that communication of the DNA operates only one way, from the genes to the body. Above, we saw that the genes act as a blueprint for the development of the body of the organism from the time of fertilization. Maybe there is some unknown mechanism whereby the living organisms can participate in their own evolution by communicating with the genes.

Several neo-Darwinists have suggested that the key to major changes lies with the regulatory genes.[50] Regulatory genes constitute only a small percentage of the total genome, the genetic material, of an organism. As we have seen, most of the genome is made up of structural genes, the genes that code for the basic structures of an organism, such as the skin, internal organs, hair, etc. Regulatory genes, on the other hand, are in control of the overall development of the organism. During the development of an embryo, for example, regulatory genes are believed to regulate whether new cells will be bone cells, liver cells, connecting tissue cells, or whatever. Perhaps, small adjustments to a small part of the genome, the regulatory genes, might cause major changes in organisms as seen in the fossil record.

As mentioned earlier, Augros and Stanciu believe the neo-Darwinist paradigm is turned around. They point out that Darwinism emphasized competition, inefficiency and gradualism. They argue that the natural world is characterized by non-competition, efficiency, and saltationalism.[50] They speculate that the jumps observed in evolution are made by the organism somehow rearranging its excess DNA. All more complex organisms have excess DNA, often called "junk DNA" because the reason for its existence is not known. Neo-Darwinists believe that "junk DNA" is leftover DNA from earlier evolutionary history which is no longer needed by the organism. Augros and Stanciu speculate that the rearrangement of "junk DNA" takes place in the regulatory genes and that this is the key to macroevolution, or the jumps of evolution seen in the fossil record.[51] Later we will see that the rearrangement of our "junk DNA" may be a key to our spiritual evolution.

It must be pointed out that Augros and Stanciu, and others who speculate on possible evolutionary mechanisms other than those of the neo-Darwinian theory stand virtually alone. A vast majority of neo-Darwinists would argue that there is nothing wrong with the current theory, and that claims to the contrary are nonsense.

In contrast, my main point here is that neo-Darwinists cannot present a satisfactory, believable theory for the origin and evolution of life on Earth. Let us look more specifically at the Darwinian-anthropological version of human evolution, where we will find still more shortcomings.

---

[1] Cravens, H. 1978. *The Triumph of Evolution*. University of Pennsylvania Press, Philadelphia.

[2] Horn, A. D. 1987. The socioecology of the black mangabey, *Cercocebus aterrimus*, near Lake Tumba, Zaire. American Journal of Primatology, *12*: 165-180.

[3] Dawkins, R. 1976. *The Selfish Gene*. Oxford University Press, New York.

[4] Wilson, E. O. 1978. *On Human Nature*. Harvard University Press, Cambridge, p. 195.

[5] Darwin, F. (ed.), 1888. *The Life and Letters of Charles Darwin, 3 Volumes*. John Murray, London, vol. 3, p. 18.

[6] Simpson, G. G. 1949. *The Meaning of Evolution*. Yale University Press, New Haven, p. 15.

[7] Shapiro, R.1986. *Origins: A Skeptic's Guide to the Creation of Life*. Summit Books, NY, p.104.

[8] *Ibid.*, p. 116

[9] *Ibid.*, p. 118.
[10] *Ibid.*, p. 112.
[11] Hoyle, F., and C. Wickramasinghe 1978. *Life Cloud*. J. M. Dent and Sons, London.
[12] Crick, F. *Life Itself*. Simon and Schuster, New York.
[13] Hoyle, F., and C. Wickramasinghe 1981. *Evolution from Space*. J. M. Dent and Sons, London.
[14] Shapiro, R., *op. cit.*, p. 227.
[15] Gould, S. J. 1983. "Nature's great era of experiments". Natural History 92: pp. 12-21.
[16] Levinton, J. S. 1992. "The big bang of animal evolution." Scientific American 267, pp. 84-91.
[17] *Ibid.*, p. 87.
[18] *Ibid.*
[19] Augros, R. and G. Stanciu 1988. *The New Biology: Discovering the Wisdom in Nature*. Shambhala, Boston. pp. 167-169.
[20] *Ibid.*, p.169.
[21] Stanley, S. 1979. *Macroevolution*. W. H. Freeman and Co., San Francisco, p. 39.
[22] Romer, A. S. 1933. *Man and the Vertebrates*. University of Chicago Press, Chicago, p. 90.
[23] Stahl, B. J. 1974. *Vertebrate History: Problems in Evolution*. McGraw-Hill Book Co., NY, p. 349.
[24] Denton, M. 1986. *Evolution: A Theory in Crisis*. Adler and Adler, Bethesda, Maryland, pp. 160-161. Denton also elaborates in detail on the structure of bird feathers.
[25] Kemp, T. S. 1982. *Mammal-like Reptiles and the Origin of Mammals*. Academic Press, London, p. 296.
[26] Jerison, J. H. 1973. *Evolution of the Brain and Intelligence*. Academic Press, New York, p. 213.
[27] Darwin, F. and A. C. Seward (eds.) 1903. *More Letters of Charles Darwin*, vol. II. Murray, London, pp. 20-21.
[28] Denton, M. 1986. *op. cit.*, pp. 160-161.
[29] *Ibid.*, p. 110.
[30] *Ibid.*, pp. 178-180.
[31] Eldridge, N., and S. J. Gould 1972. "Punctuated equilibria: an alternative to phyletic gradualism." In: *Models in Paleobiology*, T. J. M. Schopf (ed.), Freeman, Cooper and Co., San Francisco, pp. 82-115. And, Gould, S. J., and N. Eldridge 1977. "Punctuated equilibria: the tempo and mode of evolution reconsidered." Paleobiology *3*: pp. 115-151.
[32] Dawkins, R. 1985. "What's all the fuss about?" Nature *316*, p. 683.
[33] Denton, M. 1986. *op. cit.*, p. 192.
[34] Bakker, R. T. 1986. *The Dinosaur Heresies*. William Morrow and Co. Inc.,

New York.
[35] Monastersky, R. 1992. "Giant Crater linked to mass extinction." Science News *142*, p. 100.
[36] Anonymous 1992. "Tragedy found in Cambrian carnival." Science News 142, p. 109.
[37] Darwin, C. 1872. *The Origin of Species*, *6th ed.* Reprinted, (1958) Mentor, New York, p. 77.
[38] *Ibid.*, p. 75.
[39] Fowler, C. 1981. "Comparative population dynamics in large animals". In: *Dynamics of Large Mammal Populations*. C. Fowler and T. Smith (eds.). Wiley, New York, pp. 444-445.
[40] Lack, D. 1954. *The Natural Regulation of Animal Numbers*. Oxford University Press, Oxford.
[41] Horn, A. D. 1987. *op. cit.*
[42] Colinvaux, P. 1978. *Why Big Fierce Animals are Rare: An Ecologists Perspective*. Princeton University Press, Princeton, p. 149.
[43] Dawkins, R. 1985. "What's all the fuss about?" Nature *316*, p. 683.
[44] Denton, M. 1986. *op. cit.*, p. 358.
[45] Kuhn, T. S. 1970. *The Structure of Scientific Revolutions, 2nd ed.* University of Chicago Press, Chicago.
[46] Johnson, P. E. 1991. *Darwin on Trial*. Regnery Gateway, Washington D. C.
[47] Denton, M. 1986. *op. cit.*, pp. 357-358.
[48] Kettlewell, H. B. D. 1959. "Darwin's missing evidence." Scientific American *200*, pp. 48-53.
[49] Rensch, B.1959. *Evolution Above the Species Level*. Columbia University Press, New York, p. 57.
[50] Wilson, A. C., L. R. Maxon, and V. M. Sarich 1974. "Two types of molecular evolution: evidence from studies of interspecific hybridization." Proceedings of the National Academy of Sciences *71*, pp. 2843-2847.
[51] Augros, R., and G. Stanciu 1988. *op. cit.*, p. 17

"Lo, soul! seest thou not God's purpose from the first?
The Earth to be spanned, connected by network,
The people to become brothers and sisters,
The races, neighbors, to marry and be given in marriage,
The oceans to be crossed, the distant brought near,
The lands to be welded together."
Walt Whitman, *Passage to India*

Chapter 2

The Darwinian-anthropological Myth of Human Evolution

Early Primate Evolution

Primates are placental mammals which are the most advanced mammals on Earth. The origin of primates is quite obscure, but the first primates appear on the fossil record at about the same time as almost all of the dozen or so placental mammal orders appear, shortly after the catastrophe that left the dinosaurs and most life forms of the Mesozoic era extinct. The first primates appear about 60 million years ago (mya), in the first epoch or period, the Paleocene (65-53.5 mya), of the Age of Mammals, the Cenozoic or Tertiary era (65 mya-10,000 years ago). Almost all of the dozen or so orders of placental mammals appeared in the Paleocene period. All of the modern horses, elephants, deer, rabbits, bears, camels, and cats are related to some of the mammals that quite suddenly appear on the fossil record in the Paleocene period.

French paleontologists have recently found the earliest known fossil primates, some isolated primate teeth in fossil beds of Morocco.[1] These first primates were very small, mouse-sized (50 - 100 grams) prosimians.[2] Prosimians of today are small, primitive primates that live in the forests of south Asia and Africa. Living prosimians that are as small as these first primates, live mostly by eating insects. So it seems likely that the first primates on Earth were insect eaters, still a major part of the diets of most non-human primates.

The Darwinian-anthropological Myth of Human Evolution

In the next geological period of the Age of Mammals, the Eocene period (53-37 mya), prosimians were common in North America and Europe, and less common in Asia. They were most probably in Africa during this period, but the fossil record there is very poor during this period. These prosimians varied in size from that of a small mouse to that of a cat. They adapted to living in the trees, and, judging from their fossil teeth, they ate insects, flowers, fruit, leaves and probably small reptiles, as eaten by prosimians of today.

For the first 20 million years the primates of Earth were small prosimians. About forty mya the first of what we call higher primates began to appear. All of the monkeys and apes of today, as well as ourselves are higher primates. Higher primates differ from prosimians by a few skeletal features, and have larger and more complex brains which correspond to much more complex behavior patterns. The first higher primates which exhibited some of the higher skeletal features, such as eye orbits that are completely enclosed in bone, still resembled their prosimian ancestors, but quite soon became a distinct group.

The early development of higher primates took place in Africa. The only record we have today of the first 15 to 20 million years of the development of higher primates is from a fossil site in Egypt (The Fayum). Apes, not monkeys, were the first higher primates to develop, although we would hardly recognize these animals as apes. First, apes are not monkeys, as the two groups have many distinct characteristics and separate lines of development and evolution. An easy way to differentiate living apes from monkeys, is that monkeys have tails and apes do not. These thirty-five million year old apes, however, had tails and otherwise didn't resemble living apes, except in the morphological patterns of their molar teeth. Paleoprimatologists who have studied these primitive fossil apes have determined that they did resemble the smaller living apes, in that they lived in the trees, ate nuts and fruit (and presumably insects) and lived in social groups.

Meanwhile, the prosimian groups that had thrived on the continents of North America and Europe for millions of years during the Paleocene and Eocene periods, disappeared. One of the Eocene groups of prosimians (the Adapids) gave rise to most of the modern prosimians (the pottos, lemurs, and lorises of Africa and south Asia). However, the fifty to forty million year old prosimians were much more primitive than the living prosimians and the intermediate steps between them are obscure or mostly missing from the fossil record.

The other large group of Eocene prosimians (the Omomyids) is presumed by most paleoprimatologists to be ancestral to the type of prosimian, the tarsier, that gave rise to higher primates. But this relationship, too, is very tenuous. In reality, it seems that the two major groups of Eocene prosimians may have left no descendants at all. Certainly, the fossil record does not make it obvious that these Eocene prosimians left any descendants. Dr. John Fleagle, professor at the State University of New York at Stony Brook and leading expert in the field states, "(n)one of the North American and European prosimians from the Eocene seem very closely related to living Strepsirhines, Tarsius or anthropoids."[3] To those of you who are not familiar with the scientific jargon, what Dr. Fleagle is stating is that Eocene primates of North America and Europe do not seem closely related to any living primate group!

In the same Fuyum deposits of Africa where the first primitive apes are found, are also found tarsier prosimians. Four species of tarsiers live today on a few islands off southeast Asia. Tarsiers are quite small (~120 grams) and exhibit a number of primitive and more advanced characteristics - characteristics common to both prosimians and simians, or higher primates. Tarsiers are believed to be ancestral to higher primates, and most authorities classify them in the same suborder (Haplorhini) as the higher primates. For example, the tarsier has its eyes almost completely enclosed with bone, a characteristic of higher primates and not of other fossil or living prosimians. The placenta of the tarsiers is like that of higher primates, not other prosimians.[4] Furthermore, biomolecular comparisons of living primates show the tarsiers to be more closely related to higher primates than to other prosimians.[5]

Meanwhile, the prosimians that had thrived on the northern continents disappeared completely. The northern continents had a semi-tropical climate when many prosimian species had thrived there forty to sixty mya, but the Earth's climate was becoming cooler, gradually assuming climatic patterns similar to what we see today, where non-human primates are confined, with rare exceptions, to the tropical and semi-tropical climates of Africa, South Asia and the New World.

The first primates of the New World begin to appear in the fossil records of South America a few million years after the first higher primates had appeared in Africa. In fact, the first South American fossil primates closely resemble a group of the first-known higher primates (the parapithecids) of the Fayum fossils of Egypt. Most paleoprimatologists today believe this group (the parapithecids) of Old World was ancestral to the New World primates, and that members of this group island-hopped across the Atlantic Ocean to establish themselves in South America some thirty million years ago. The New World primates evolved into a variety of beautiful arboreal animals, which today range in size from that of a mouse to that of a medium-sized dog. While they did not biologically figure in the ancestry of humans, studies of their ecology and behavior have contributed much to our knowledge of primates and mammals.

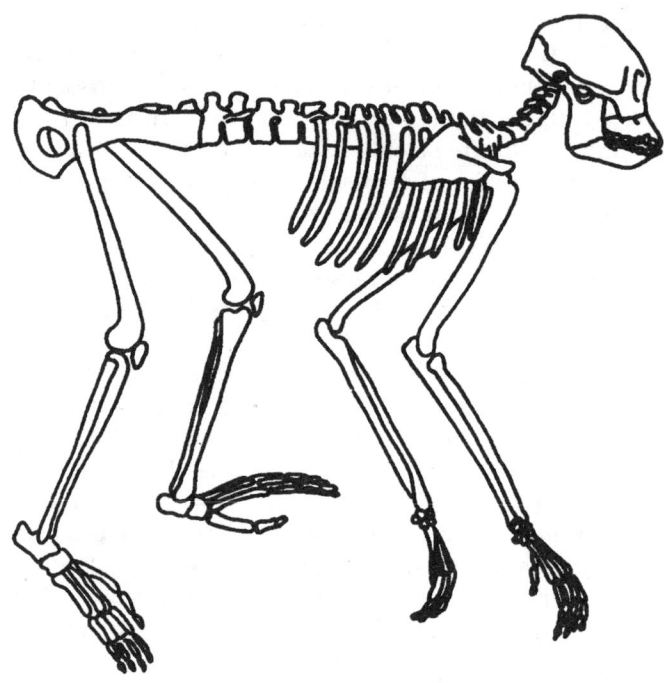

Figure 2-1 *Proconsul africanus. One of several species of primitive apes living in Africa during the early Miocene period (~20m.y.a.). Drawn after France and Horn, 1992.*

Back in Africa the primitive apes thrived. In the early part of the Miocene period, roughly 20 million years ago, there were many primitive ape species (15 to 25 species) living in the forests of Africa. Again, these primitive apes did not have the same features as modern apes, but their teeth leave no doubt that they were apes. Figure 2-1 shows one of the best preserved of the primitive apes of this period. This primate had some ape-like characteristics, such as its teeth and the lack of a tail, but otherwise did not have many characteristics possessed by modern apes, such as long forelimbs and short hind limbs. But they made their living in a way similar to modern apes. They lived in the trees, although the larger ones (~200 lbs.) may have spent a lot of time on the ground, like living chimpanzees and gorillas. They ate fruit, insects, some leaves and flowers and probably small lizards and other small vertebrates.

At about this time (20 mya), Old World monkeys appear on the fossil record for the first time. Old World monkeys have a tail and very distinctive molar teeth called bilophodont molars. Fossil specimens of Old World monkeys appear without any obvious fossil precedents. There is no doubt that these first Old World monkeys are derived from primitive apes, but, at present, there is no indication that bilophodont molars and monkeys evolved gradually, in small steps, from apes. This is another one of those famous "jumps" in the fossil record. Hence, another very successful, very beautiful group of primates was established rather suddenly. Monkeys today are the predominant non-human higher primates in Africa and south Asia. Today, unlike twenty million years ago, there are many species of monkeys and only a few species of apes in the Old World. But, there is no doubt we humans were derived from apes, not monkeys.

Apes began to spread from Africa to Europe and south Asia around fifteen mya. Recently a partial ape skull was found in Pakistan that is about thirteen million years old, that resembles the living Asian great ape, the orangutan. This fossil ape, classified in the genus *Sivapithecus* (named after the great Hindu god Shiva), has facial and palatal features that closely resemble those of the orangutan. Most authorities believe *Sivapithecus* was ancestral to the orangutan. If the interpretations of this fossil face turn out to be mostly correct, *Sivapithecus* represents the only fossil ape to which a direct relationship to a living ape can be demonstrated. There are some problems, however. To date, most of the several *Sivapithecus* fossils discovered are teeth and jaw fragments, none of which suggest as close a relationship of *Sivapithecus* and the orangutan as does the partial fossil skull from Pakistan mentioned earlier. Furthermore, a recent discovery of three upper arm fragments attributed to *Sivapithecus* calls into question the proposed close relationship of *Sivapithecus* and the orangutan. These arm fragments do not fit the pattern expected by paleoprimatologists of a potential thirteen million year old ancestor to the orangutan.[6]

The Australopithecines

Fossil apes become more and more scarce as we approach the present. The biggest disappointment in the fossil study of human origins is the lack of discoveries of African fossil ape ancestors to the first members of the human fossil family, the hominids. The bipedal hominids emerged upon the east African scene between three and four million years ago. The first hominid species, known as

Australopithecus afarensis, exhibits many ape-like characteristics, as well as human-like characteristics. *A. afarensis* was 3½ to 5 feet tall, walked on two legs in a manner similar to you and I, and had brains that were about one third the size of ours. The *Australopithecus* species (of which four to six species are known) clearly led to the first fossil hominids we classify in our biological genus, *Homo*, at about two million years ago. In other words, there is no doubt today that the australopithecines are our distant ancestors (see Figure 2-2).

Figure 2.2 *One of the more robust species of several species of Australopithecus of Africa.*

Naturally, paleoanthropologists want to know all they can about the ape ancestors to Australopithecus, but they have been unable to find them. There are few fossil apes in Africa of the correct age, five to ten million years, and the few that have been discovered, are mostly isolated teeth and jaw fragments. They do not tell us anything about the stages of evolution and development of the australopithecines.

Since the first members of the australopithecines, humankind's distant ancestors, came from east Africa, they must have been derived from African apes. In fact, several biochemical comparisons between humans and apes demonstrate that humans are clearly more closely related to living African apes (chimpanzees and the gorilla) than to Asian apes (gibbons and the orangutan). So there must have been an ape that lived in Africa between five and ten million years ago that was the ancestor to both the first australopithecine and the African apes, paleoanthropologists ask? Paleoanthropologists would love to know what this common ape-human ancestor would have looked like and how it lived. Did it live in the forest, or in more open country? Was it bipedal, some, or even most of the time? Did it resemble more a living or something more similar to the first australopithecine?

In any case, the small-brained, bipedal australopithecines lived in the savannah habitats in eastern and southern Africa for about two million years. Evidence indicates that they lived in small groups, as do almost all higher primates today. They had large teeth and jaws and must have eaten many tough foods, such as nuts and seeds. Like the baboons that live on the same savannah habitats today, they probably ate mostly of the products of plants, such as fruit, flowers, and grasses, plus insects and a few small vertebrates when they could catch them. There is no evidence that they used fire or any types of tools. However, field studies have illustrated that chimpanzees occasionally use sticks and unmodified stones as tools, and so probably did the australopithecines.

Homo habilis

At roughly 2 to 2 1/2 million years ago, the next major step of human development took place in Africa. Again, rather suddenly after almost two million years of evolution of the small-brained (to us) australopithecines, larger-brained hominids began to appear which paleoanthropologists call *early Homo*, or *Homo habilis*. Specimens of early *Homo* had brain capacities of around 600 to 800 cubic centimeters (cc) while the australopithecines had cranial capacities of 400 to 500 cc. This is still small when compared to the average cranial capacity of modern *Homo sapiens* of around 1,300 to 1,400 cc. But, the jump in brain size of early *Homo* over its australopithecine ancestors is quite significant. The dental patterns of early *Homo* were much more like ours than those of the australopithecines. Also, the first crude stone tools, called Olduvai tools, are found in east and south Africa at the time of the appearance of early *Homo*. These stone tools were made by simply pounding off a few flakes at one end of an approximately hand-sized stone, leaving a sharp edge. So this marks the beginning of a material culture that is today so characteristic of humans and their ancestors.

At this point, around two million years ago, the brain size of early *Homo*, our ancestors, was about half the size of modern human brains. In the next two million years, the brain-size of our ancestors increased from an average of around 700 cubic centimeters to the average brain capacity of modern humans, around 1,400 cc. There is nothing like this rapid increase in the fossil record of any other animal. This rapid increase in hominid brain size has prompted many comments from scientists over the years.

Homo erectus

After early *Homo*, or *Homo habilis*, the next species of our human development was *Homo erectus*. *H. erectus* was a species with a brain about two-thirds the size of the modern human brain. The cranium of *H. erectus* had a distinctive shape that is seen on fossil hominid specimens for more than a million years, 1.8 million years ago to as recently as 12,000 years ago (see Figure 2-3). Recently, a nearly complete fossil skeleton of *H. erectus* was found in Kenya. The bones of this fossil skeleton are quite robust, but otherwise there are only minor differences between it and the skeleton of a modern human.

For years the evidence seemed solid that *H. erectus* evolved first in Africa, then, soon after one million years ago, populations migrated out of Africa into Europe and Asia. Recent fossil discoveries and new dates from the island of Java, Indonesia, and the Republic of Georgia give evidence that there were populations of *H. erectus* in Europe and Asia at about the same time as the oldest known *H. erectus* in Africa. These recent discoveries have shaken up paleoanthropologists' view of how the evolution of humans occurred.[7,8,9] Now paleoanthropologists don't know if *H. erectus* evolved first in Africa or Asia; whether *H. erectus* populations of Africa, Europe and Asia represent one or several species; which of the groups (or all of them) may have given rise to *Homo sapiens*; or what was going on at this stage of human evolution. Paleoanthropologist Philip Rightmire of the State University of New York at Binghamton states that,

"Paleoanthropology is always interesting and bound to be surprising. Just when everyone is comfortable with something, a new fossil is found or a new date." [10] Bernard Wood, a paleoanthropologist of the University of Liverpool, states that the new dates from Java mean "the whole pattern of human evolution is much more complicated than a simplistic linear explanation." [11] In any case there was a primitive hominid species throughout the Old World soon after two million years ago.

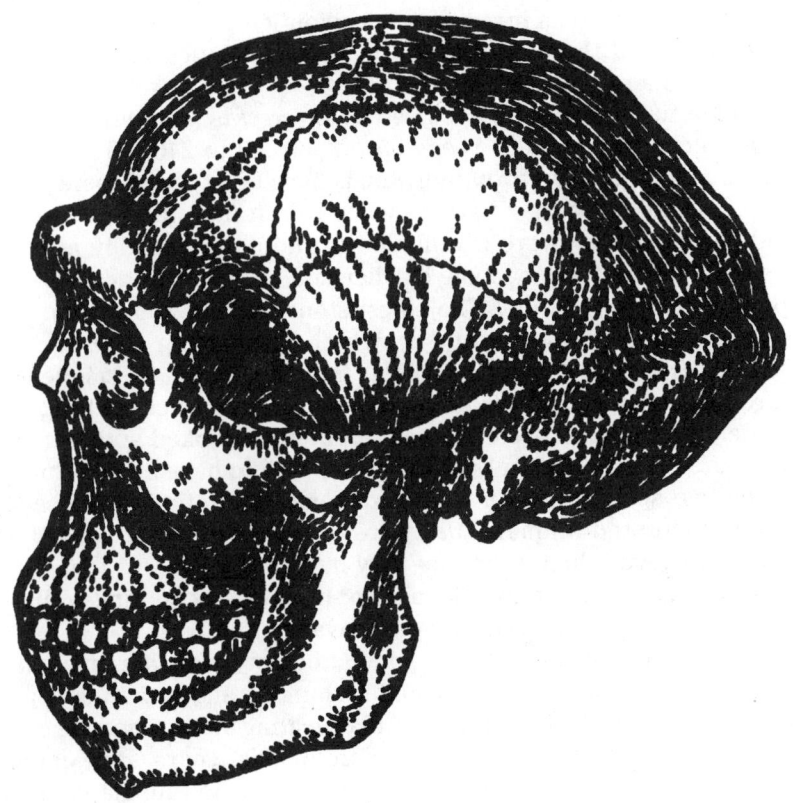

Figure 2-3 Homo erectus cranium from Zoukoutien Cave, China. Drawn after Montagu, 1962.

A new, more sophisticated stone tool appeared with African and European *H. erectus* and, in general, the material culture of *Homo erectus* is more advanced than that of early *Homo*. Even so, the cultures of *H. erectus* populations remained quite crude and primitive, even when compared to modern human populations that are nomadic and live off the wild animals they hunt and the products of wild plants that they gather (hunters and gatherers).

Nevertheless, sites in both France and China show that younger *H. erectus* populations (~500,000 years before the present) lived in small nomadic bands, used fire, lived in caves and sometimes built temporary living shelters out of sticks, leaves, and stones. They relied on hunting, or scavenging, and gathering for subsistence. Let us look in detail at one of the sites, Zoukoutien, near Beijing in China. Zoukoutien (old spelling, Choukoutien) is the most famous *H. erectus* site of all that have been excavated. Recently, a team of more than 100 Chinese scientists investigated Zoukoutien giving us a more complete picture of these ancient *H. erectus* populations than we have heretofore had available.[12]

Zoukoutien caves were occupied by *H. erectus* populations on and off for more than 200,000 years - from 460,000 to 240,000 years ago. Remains of more than forty individuals have been found there. There is no evidence that they buried their dead; in fact, there is disputed evidence that they were cannibalistic, and quite possibly ate their dead. The stone tools become smaller and more sophisticated as we approach the present day. The oldest stone tools at Zoukoutien were large and made of soft stone on which the cutting edge is easily blunted. The younger stone tools at this site are much smaller and more finely made, and the stone is much harder, such as flint. Zoukoutien populations used fire and apparently cooked some of their meat, as some of the deer bones found appear to have been burnt.

The two species of deer found are some of the first evidence that hominids hunted animals that were as large or larger than themselves. However, the notion that they hunted the animals they ate, is debated. Certainly, nothing like a spear-head or any hunting-point has been found. Perhaps, the deer were scavenged.

In any case, these *H. erectus* populations probably ate much more plant material than meat, as do modern hunting and gathering populations.[13] Most all of the remains of plant eating would disappear in a handful of years, let alone more than 200,000 years. But, copious remains of hackberry seeds have been unearthed in the cave and many wild fruits, tubers, grass seeds, nuts that surround Zoukoutien today, which must have been eaten by *H. erectus* populations that lived there.

So, we have a picture of *H. erectus* living in these caves some 300,000 years ago, at least part of the year. They ate whatever wild plant food they could gather, hunted or scavenged meat, they may or may not have been cannibals, and there is even some doubt that they had control over fire. This is not a very flattering picture of a species that would give rise to a species that would begin to establish some

The Rise of Modern Homo sapiens

quite sophisticated civilizations in just a few hundred thousand years.

The Rise of Modern Homo sapiens

There is much confusion and debate among paleoanthropologists concerning the last and most important step in human biological evolution, the transition from *Homo erectus* to modern *Homo sapiens*. In fact, there are currently two mutually exclusive hypotheses to explain this transition. One hypothesis, usually known as the gradualistic hypothesis, has modern *Homo sapiens* evolving gradually in a few locations from local populations of *Homo erectus*. Following this hypothesis, modern *Homo sapiens* evolved gradually from *H. erectus* and other archaic populations in Africa, Europe and east Asia.[14] The other hypothesis, sometimes called the migration hypothesis, has modern *H. sapiens* evolving from a *H. erectus* population in Africa, then the moderns migrated out of Africa into other areas of the Old World replacing the more primitive types where they found them.[15]

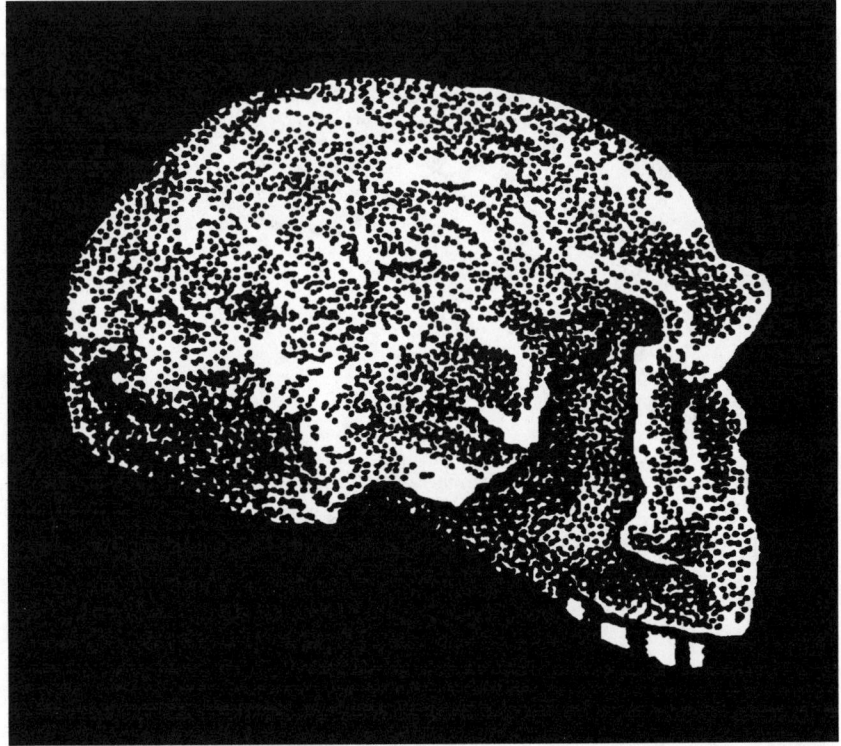

Figure 2-4 *Steinheim skull.*

The fossil evidence does not confirm or absolutely rule out either hypothesis. First, there are a few partial fossil skulls which are dated between 300,000 and 200,000 years before the present that show some *Homo erectus* traits and some more modern characteristics. For example, a fossil skull found near Steinheim, Germany dated at around 250,000 years ago, has a small cranial capacity similar to *H. erectus*, but a rounded back of the skull unlike its older relatives (see Figure 2-4). Partial skulls from near Swanscombe, England, Fontechevade, France, Shaanxi Province, China (Dali) also show a mosaic of *erectus* and more modern traits.

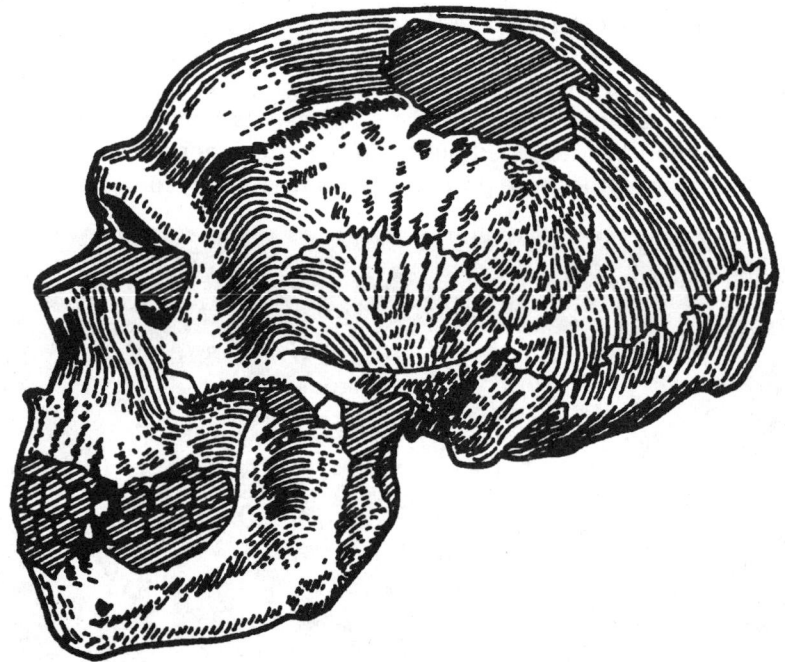

Figure 2-5 Neanderthal skull from La Chapelle-aux-Saints, France.

Then there are fossil hominids found in Africa, Europe and the Middle East dated from 300,000 to 35,000 years before the present that had brains almost as large or larger than modern humans, but otherwise show many primitive features. These types of specimens are usually called archaic *H. sapiens* by paleoanthropologists, although many experts would prefer to divide these fossils into several species.

The most famous and well-known population of archaic *H. sapiens*, is the Neanderthal population of Europe and the Middle East, dating from about 120,000 to 35,000 years ago (see Figure 2-5). Neanderthals had a brain that was sometimes larger than that of modern humans, but it was shaped differently, being long and low. Otherwise the skulls of Neanderthals show many primitive features, such as heavy eyebrow ridges, large teeth and a sloping chin. They were about 5 to 5 1/2 feet tall and were very robust.

Studies have been done on the flat cranium bases of Neanderthals, and it was determined that they had much shorter vocal tracts than modern humans and most likely could not pronounce the vowels we do.[16] This controversial conclusion indicates that Neander-thals and all that came before either could not speak, or spoke very slowly.

Neanderthal culture was a bit more complex than the *H. erectus* culture from Zoukoutien. For example, there have been more than a dozen Neanderthal burials found, although why they were buried is a matter of speculation. Neanderthal's stone tools are a bit better and more varied than that seen in *H. erectus*, but the cultural improvements of Neanderthals were very slight compared to what modern humans would do later. One of the leading experts on Neanderthals, professor Erick Trinkhaus of the University of New Mexico, reports that Neanderthal sites are monotonous, and that their groupings consisted of only a dozen or so people that lead a hard life.[17] At about 35,000 years ago Neanderthals were either replaced or evolved into modern humans in Europe, depending on what theory a paleoanthropologist believes. At present, most paleoanthropologists see little evidence in the fossils or cultural remains in Europe to support a gradual evolutionary transition from Neanderthals to modern humans, but there is enough ambiguity in the data that those supporting a gradual transition can (and do) make a case.

As was the case for the *Homo erectus* stage of human evolution, for years the evidence was that the earliest fossil hominids that seemed fairly close to modern humans were found in Africa, dated at a little more than 100,000 years before the present. A new, recent date from China, however, places near modern-looking humans there at least 200,000 years before the present.[18] If this new date holds up, it will add to the confusion and debate among paleoanthropologists concerning the course of human evolution. The fossils found in Africa dated at around 100,000 years before the present have many modern human characteristics, such as high foreheads and strong chins, but also retain more primitive features, such as heavy eye-brow ridges. Recently, molecular scientists began to use mitochondria DNA (mtDNA) in their research to help decipher the origins of modern humans. Mitochondria are minute granular bodies in all cells of an individual's body. They function to metabolize food and have their own DNA, or genetic material, quite apart from the nuclear DNA of the cells. Inheritance of mtDNA is through the females only, since the male sperm does not carry any mitochondria. Molecular scientists of the University of California at Berkeley compared the mtDNA of most human races and concluded that the African blacks have the oldest pattern seen in mtDNA of modern humans. They concluded, using a molecular clock, that modern humans first evolved in Africa some 200,000 to 400,000 year ago.[19, 20, 21] Researchers are leaning toward the 200,000 years ago date since it seems to fit better with the fossil evidence.

This mtDNA study received much publicity in the popular press, and convinced many paleoanthropologists that the evolution of modern humans first took place in Africa. This supported the migration hypothesis of the origin of modern *H. sapiens*. However, there have been recent studies of mitochondrial DNA that do not support this "African Eve" hypothesis. A. R. Templeton, a molecular biologist of Washington University in St. Louis, and others report that the existing mitochondria DNA data indicates that modern humans have both African and non-African roots.[22] So, there is not a clear, unambiguous picture of the emergence of modern humans from molecular biologists to help clarify the confusing fossil evidence of human evolution.

To confuse matters worse, there is evidence that modern human populations lived side by side with Neanderthal populations for years in what is now Israel around 90,000 years before the present. Since moderns were supposed to have evolved from an archaic population similar to Neanderthals by at least 200,000 years before the present, this creates some problems of interpretation.[23]

Let me leave this confusion and cut directly to the best documented case we have of the biological creation of what we call modern humans, the first writings of humans from the ancient Mesopotamian civilizations.

---

[1] A summary of recently discovered Paleocene primate fossils found in: Gingerich, P. D. 1990. "*African dawn for primates.*" Nature 346: p. 411.

[2] *Ibid.*

[3] Fleagle, J. G. 1988. *Primate Adaptation and Evolution.* Academic Press, San Diego, p. 319

[4] *Ibid.*, p. 102.

[5] Dene, H. T., M. Goodman and W. Prychodko 1976. "Immunodiffusion evidence on the phylogeny of the primates." In: *Molecular Anthropology*, M. Goodman, R. E. Tashian and J. H. Tashian (eds.), Plenum Press, New York.

[6] Pilbeam, D., M. D. Rose, J. C. Barry and S. M. Ibrahim Shah 1990. "New

Sivapithecus humeri from Pakistan and the relationship of *Sivapithecus* and *Pongo*." Nature, *348*: pp. 237-239.

[7] Petit, C. 1994. "New fossil date shakes up ideas on evolution." *San Francisco Chronicle*, February 24, pp. A1 and A11.

[8] Swisher III, C. C., G. H. Curtis, T. Jacob, A. G. Getty, A. Suprijo, Widiasmoro 1994. "Age of the earliest known hominids in Java, Indonesia." Science *263*: pp. 1118-1121.

[9] Gibbons, A. 1994. "Rewriting - and redating - prehistory." Science *263:* pp. 1087-1088.

[10] P. Rightmire quoted in Gibbons, 1994, p. 1088.

[11] B. Wood quoted in Gibbons, 1994, 1087.

[12] For a brief summary of the recent work at Zoukoutien see: Nelson, H. and R. Jurmain 1988. *Introduction to Physical Anthropology, 4th ed.* West Publishing Co., St Paul, pp. 491-499.

[13] *Ibid.*, pp. 340-349.

[14] Wolpoff, M. H. 1989. "Multiregional evolution: the fossil alternative to Eden." In: *The Human Revolution: Behavioral and Biological Perspectives on the Origins of Modern Humans, vol. 1.* P. Mellers and C. B. Stringer(eds.). Edinburgh University Press, Edinburgh.

[15] Stringer, C. and P. Andrews 1988. "Genetics and fossil evidence for the origin of modern humans." Science *239:* pp. 1263-1268.

[16] Lieberman, P. & E. S. Crelin 1971. "On the speech of Neanderthals." Linquistic Inquiry *2*: pp. 203-222.

[17] Trinkaus, E. 1984. "Western Asia." In: *The Origins of Modern Humans: A World Survey of the Fossil Evidence*, F. H. Smith and F. Spencer (eds.). Alan R. Liss, New York, pp. 251-291, and personal communication.

[18] Bower, B. 1994. "Asian hominids make a much earlier entrance." Science News *145*: p. 150.

[19] Cann, R. L., M. Stoneking and A. C. Wilson 1987. "Mitochondrial DNA and Human Evolution." Nature *325*: pp. 31-36

[20] Cann, R. L. 1988. "DNA and human origins." Annual Review of Anthropology *17*: pp. 127-141.

[21] Stoneking, M. and R. L. Cann 1989. "African origin of human mitochondrial DNA." In: *The Human Revolution: Behavioural and Biological Perspectives on the Origins of Modern Humans, vol. 1.* P. Mellers and C. B. Stringer (eds.). Edinburgh University Press, Edinburgh.

[22] Templeton, A. R., S. B. Hedges, S. Kumar, K. Tamura, and M. Stoneking 1992. "Human origins and analysis of mitochondrial DNA sequences." Technical comment. Science *255*: pp. 737-739.

[23] Wolpoff, M. H. 1989. *op. cit.*

"At last Ooota pointed to me and spoke to each person, pronouncing the same word repeatedly. I thought they were trying to say my first name, but then decided they were instead going to call me by my last name. It wasn't either. The word they used that night, and the name I continued to carry for the journey, was MUTANT."
Marlo Morgan, *Mutant Message Downunder*

Chapter 3

The Ancient Mesopotamian Account of the Creation of Humans

Renaissance and post-Renaissance Europeans thought that Greece and Rome were where civilization first began. Gradually the magnificence and antiquity of ancient Egypt became known, but by the beginning of the twentieth century the remains of an even older civilization had been discovered in the area which is now southern Iraq.

Early travelers and explorers had reported huge mounds - "tells" in Hebrew and Arabic - in the land between the two rivers, the Tigris and Euphrates, which were said to be the remains of ancient cities. The excavation of these ancient cities began in the 1840s. One of the very first people to dig in this area, an Englishman named Henry Layard, dug up an Assyrian military center named Kalhu, which was north of the modern city of Baghdad. In Kalhu, Layard found an obelisk set up by the King Shalmaneser II, on which the king listed those that were paying him tribute. Among those appearing on the list was the King of Israel. This find was the first of many that helped confirm the historical veracity of at least some of the Old Testament of the Bible.

In the last half of the nineteenth century many cities of ancient Babylonian and Assyrian empires were excavated by teams of archeologists of mostly European countries. Soon almost all of the Babylonian and Assyrian cities which are referred to in the Old Testament were found. The Babylonian empire waxed and waned from about

The Ancient Mesopotamian Account of the Creation of Humans

2,000 B.C. to 500 B.C. The Assyrian empire had roughly the same time span, although the height of Assyrian power occurred after about 900 B.C.

Soon the royal city of Akkad, south of Babylon, was discovered along with other remains of the Akkadian empire (2,350 - 2,200 B.C.). However, the writings of all of these ancient civilizations contained words of an even older language. The Akkadians referred to their predecessors as Shumerians, and the Bible, written more than 1,000 years after the Akkadian empire disappeared, often has references to the land of Shin'ar. What is now called the Sumerian civilization was discovered by archeologists in the last decade of the 19th century.[1] The Sumerian civilization is the first human civilization of which we have archeological remains (see Figure 3-1).

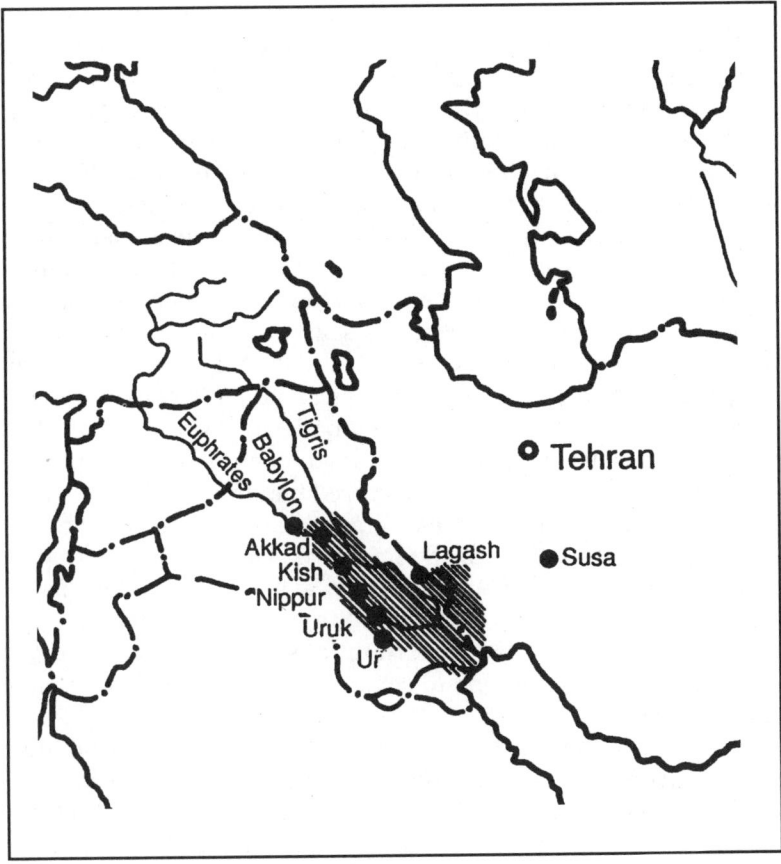

Figure 3-1 Map of ancient Sumer showing some of the major cities, plus two capital cities, Babylon and Akkad, of later Mesopotamian empires.

51

The Ancient Mesopotamian Account of the Creation of Humans

The exact dates of the beginnings of this civilization are not clear, as there are few carbon fourteen dates available from the ancient Sumerian archeological sites, but the initial Sumerian cities and civilization were founded some time in the fourth millennium B.C., shortly before the Egyptian civilization. The Sumerians had a rich civilization that included long irrigation canals, monumental architecture in their cities, foreign trade, and much more, as we shall see. The Sumerian civilization appeared quite suddenly, as if initiated by another culture. We'll discuss the rise of the Sumerian and other early civilizations in some detail in later chapters, but for now we are interested in the Sumerian account of the creation of humans.

Two of the most important inventions of the Ancient Sumerians were the cylinder seal and writing. The cylinder seals were small cylinders made of ordinary or semiprecious stone, such as lapis lazuli, ranging in length from about one to three inches and as thick as a thumb or less. Designs were engraved on the surface of the seals so that these designs would appear on clay when the cylinders were rolled over wet clay. The cylinders first appeared in the beginning of what the archeologists call the Uruk Period (ca. 3,500 B.C). Uruk was one of the five major cities in the Sumerian civilization. The early seals were beautifully carved and are considered miniature masterpieces. Cylinder seals became a tradition of all of the ancient civilizations of this region, the Sumerian, Akkadian, the Babylonian, and Assyrian civilizations, collectively referred to as Mesopotamian civilizations. Mesopotamian civilizations spanned a time period of about 4,000 years.[2] The seals often depicted scenes of daily life or scenes of what archeologists consider mythological subjects performing ceremonies that are not known or understood.[3] As we'll see, it is almost certain cylinder seals do not depict mythological scenes, but actual historical scenes of events that occurred hundreds and thousands of years before the cylinder scenes were made. It is probable that these seals are the first attempt of the primitive primate humans to depict and preserve their own ancient history, as best they could, that had come down to them through their oral traditions. In non-literate societies, even those of today, oral traditions of epics and history are passed along from generation to generation by specialists who memorize the stories.

The first primitive writing also appeared in the Uruk period toward the end of the fourth millennium. Not every Sumerian was reading and writing when writing developed, as the Sumerian writings were prepared by a small group of professional scribes. In fact, this was the general case for the first 6,000 years of civilization, as only a small percentage of humans in any society could read and write until the

The Ancient Mesopotamian Account of the Creation of Humans

twentieth century.

The Sumerian scribes shaped clay into a tablet usually only a few inches square, large enough to hold in one hand. Then the writings were etched into the wet clay tablet with the end of a reed stalk or another sharp object. The tablet was then either baked in an oven or in the hot Sumerian sun. The first written pictographs were much modified over the next few hundred years until their evolution was complete in the mid-third of the third millennium. The Sumerian writing and that of subsequent Mesopotamian civilizations is called cuneiform writing after the Latin word *cuneus*, which means wedge or nail (see Figure 3-2).[4]

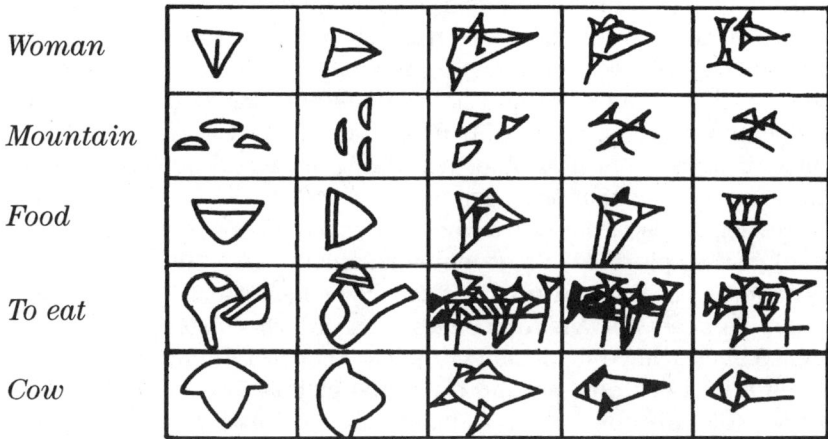

Figure 3-2 *The evolution of a few words of Sumerian cuneiform writing.*

The early Sumerian writing was strictly for administrative purposes. There are accounts of deeds of ownership, accounts of transactions, receipts, lists of goods and workers etc. Certainly these accounting lists were not developed to create myths. Nothing close to what we would call an historical document appears until several hundreds of years after writing began at around 2,600 B.C.[5] Again, it appears that these historical accounts are the writings of the oral traditions that were passed from generation to generation. Hardly any of these historical documents of Sumer and succeeding Mesopotamian civilizations are considered history by most modern historians and archeologists. Archeologists Lamberg-Karlovsky and Sabloff consider most of these documents as "pure fable", as the recorded tales include many fantastic stories, such as accounts of gods "flying about" and individual kings who ruled for hundreds and thousands of years.[6]

The Ancient Mesopotamian Account of the Creation of Humans

By now, a little more than one hundred years after scholars established that Sumer existed and that it was the first human civilization, hundreds of these tablets have been found containing these epic tales of the early history of man, stretching back in time hundreds and thousands of years before Sumer was established. Many of the original Sumerian accounts that have not been discovered, have been found among the remains of subsequent Mesopotamian civilizations, the Akkadians, the Babylonians and the Assyrians. One of the greatest discoveries of these tablets deserves note: that is the discovery of the library of Ashurbanipal in the remains of Nineveh, north of Baghdad. Ashurbanipal (668-626 B.C.) was an Assyrian king apparently of high culture. Several works of art, in addition to his library, were discovered in the remains of his palace at Nineveh. Ashurbanipal's library contained more than 25,000 tablets which were arranged by subject. Ashurbanipal apparently collected every text he could lay his hands on, and had copies and translations made of tablets that were ancient by his time. Many tablets were identified by his scribes as "copies of older texts" of Sumer and other Mesopotamian civilizations that preceded them.

Academic historians and archeologists have translated and studied many of the epic tales of Mesopotamia, and relegated them to mythology and "pure fable". Fortunately, today we have more than the translations of academicians to depend on in order to understand the immense significance of these epic tales. Zecharia Sitchin has written seven books centered around the initial civilizations of Earth, translating not only the Mesopotamian tablets, but Egyptian, ancient Hebrew and other ancient sources.[7, 8, 9, 10, 11, 12] Sitchin claims that many ancient Sumerian, ancient Hebrew and other ancient writings have been mistranslated. According to Sitchin, when he was a young man attending a class in Tel-Aviv, Israel, and studying the Old Testament in its original Hebrew, he asked the teacher why the ancient Hebrew word Nefilim was commonly translated as "giants" (such as in the familiar King James version of the Bible) when, in fact, Nefilim meant "Those who had come down." His teacher reprimanded him, telling him not to question the Bible. Of course, he had not questioned the Bible, but its interpretation. This incident prompted Sitchin to a lifelong project of studying and interpreting the Bible and other ancient writings, learning Sumerian and many other ancient and modern languages.[13] Sitchin has established that what most scholars consider mythology (or "pure fable"), are instead the repository of ancient memories, and that the Mesopotamian writings, as well as the Bible, deal with history and should be read as historic/scientific documents. Sitchin first recounted the Mesopotamian story of the creation of humans in *The Twelfth Planet* in 1976.[7] He reviews this story in later books, especially in *Genesis Revisited*.[11] The creation-of-humans story is contained mostly in what modern-day Mesopotamian scholars call *The Atra-Hasis Epic* supplemented by several other texts discovered in various states of preservation. The Atra-Hasis text recounts how a group of ETs, called the Anunnaki (Those Who from Heaven to Earth Came) by the Sumerians, came to Earth about 450,000 years ago to mine gold. The Anunnaki came to Earth from the "12th planet" called Nibiru. The Sumerians knew of the nine planets of which we moderns know, and they counted the sun and the Earth's moon, as well as Nibiru, of which modern astronomers do not know, making a total of twelve celestial bodies in our solar system. The orbit of Nibiru brought it close to Earth every 3,600 years; Sitchin shows its orbit to be similar to that of a comet.[14] We will have more to say about Nibiru later, as there is evidence that the passing of Nibiru near the Earth has caused many catastrophes in the past.

It should be pointed out here that it is hardly surprising that our modern astronomers might not know about this "12th planet", if it exists. The three planets we know of in our solar system that are furthest from the sun, Uranus, Neptune, and Pluto, cannot be seen with the unaided eye, and were only "discovered" in the past 200 years, 1781, 1846, and 1930 respectively, with the aid of telescopes. What is surprising, and even startling, is the fact that the ancient Sumeri-ans knew about these planets. How the Sumerians, who were just learning how to write, knew about these distant planets has never been addressed by academic scholars (historians, archeologists, linguists) to my knowledge. Sitchin demonstrates that the Sumerians not only knew of these distant planets in our solar system, but knew some of the details about them that have only been discovered by twentieth century science with the aid of unmanned satellites that have passed near them.[15] For example, scientists and the public discovered in August, 1989, as the unmanned satellite, *Voyager 2*, passed by Neptune sending back pictures, that this distant planet contained much water, and was not strictly gaseous as had been thought. Using ancient Sumerian texts, Sitchin had written about the waters of Neptune even before Voyager 2 was launched in 1977.[16] Of course, the Sumerians learned about such details of the heavens from "Those Who from Heaven to Earth Came", their Anunnaki gods.

The Anunnaki operated throughout a strict hierarchy in which Anu was ruler of Nibiru and became the number one "god" of the Sumerian hierarchy of "gods." Sitchin makes it clear that neither the Sumerians nor the later Akkadians referred to these visitors to Earth as "gods." It was through the later paganism of Greece and Rome that the notion that they were divine beings, or gods, seeped into our language and thinking. The Sumerians called them DIN.GIR, or "the Righteous Ones of the Rocket Ships." The Akkadians calledthem the Ilu, or the "Lofty Ones" (from which the biblical EI stems).[17]

Anu, the chief Anunnaki of the "12th planet" Nibiru, left his son Enlil in charge of the affairs here on Earth. Enlil was the son of Anu and Anu's half-sister and therefore the heir of Anu, as the Anunnaki apparently passed their inheritance through the son of a half-sister. Anu's firstborn son, Enki, a master scientist/engineer, was also given much power on Earth. Enki and Enlil were often involved in some sort of conflict with each other. This rivalry eventually included their own offspring and extended for millennia. Generally speaking, Enki almost always took the side of humans in his conflicts with Enlil and other Anunnaki. Enlil, it seems, did not much care for humans.

The Anunnaki had been mining gold on Earth for more than 100,000 years when the rank-and-file Anunnaki, who were doing the back-breaking work in the mines, mutinied about 300,000 years ago.[18] Enlil, their commander-in-chief wanted to punish them severely and he called an Assembly of the Great Anunnaki, which included his father Anu. Anu was more sympathetic to the plight of the Anunnaki miners. He saw that the work of the mutineers was very hard and that their distress was considerable. He wondered out loud in this Assembly of the Great Anunnaki, if there wasn't another way to obtain gold. At this point, Enki suggested that a Primitive Worker, an *Adamu*, be created to take over the difficult work. Enki pointed out that a primitive humanoid (what we call *Homo erectus*, or a closely related hominid) was quite prevalent in the AB.ZU (Africa) where he worked. Why don't we "Bind upon it the image of the gods"[19] in order to give it the intelligence and ability to carry out the mining? The Assembly of the Great Anunnaki liked this idea and approved it. Thus, the decision to develop humans to work as slaves was made.

Enki was put in charge of the creation of an *Adamu*, a primitive worker, that would assume the difficult work of mining for the Anunnaki. Enki was already located in the AB.ZU, the land of the mines, which Sitchin believes was located in southern Africa where much gold has been mined in modern times. Sitchin quotes the journal of South Africa's leading mining corporation, the Anglo-American Corporation, which states that many ancient mining shafts, with depths of up to fifty feet have been found. Three of these ancient shafts have been carbon 14 dated at 35,000, 46,000, and 60,000 B.C., and the archeologists hired to find these ancient mining shafts believe that mining was being done in southern Africa prior to 100,000 B.C.[20] Of course, there is no evidence that stone-age humans could sink

The Ancient Mesopotamian Account of the Creation of Humans

shafts into the Earth in search of precious metals, and why would they even want or need to mine gold?

Enki requested help in the form of Ninti, a daughter of Anu and very high up in the Anunnaki hierarchy, who was the chief medical officer. Ninti was the half-sister of both Enki and Enlil and it seems likely that Enki was seeking an heir, as well as an assistant in his attempts to create an *Adamu*.[21]

The binding of the image of the gods on the primitive creature - probably something close to what we have called *Homo erectus* - was, of course, accomplished through some sort of genetic engineering. Piecing this creation story from several Mesopotamian tablets in various states of preservation, Sitchin has concluded that the *Adamu* was the first test-tube baby, so to speak.[22] On one tablet, quoted by Sitchin, Enki instructs Ninti to mix clay of the core (or basement) of the Earth with that of a young Anunnaki. In *The 12th Planet* and again in *Genesis Revisited*, Sitchin points out that the Sumerian and Akkadian terms that are usually translated as "clay" or "mixed" stemmed from the Sumerian TI.IT which meant "that which is with life" with derivative meanings of "mud" and "clay" and "egg".[23] So, evidently Enki took an "egg", or ovum, from a *H. erectus* female, and fertilized it with the sperm of a young Anunnaki in vitro. The fertilized egg was then implanted into the womb of Ninti herself. Ninti's name meant "Lady Life" and she was later nicknamed Mammi, the source of the universal Mamma/Mother.[24] Ninti gave birth to the first *Adamu* ten months after the egg was implanted in what was apparently a very difficult birth. One clay tablet found quite damaged has indicated that there was a lot of cutting during the birth, probably what would occur in a Cesarean birth. But finally the infant came forth and Ninti became very excited. She held the newborn baby up for all to see and shouted, "I have created! My hands have made it!"[25]

This dramatic event is pictured in Figure 3-3. This image is from an Assyrian cylinder seal and is almost certainly a copy of a much more ancient Sumerian cylinder seal. Since this and other illustrations appeared about 1,000 years before the Sumerian historical epics appeared in writing, it is not known exactly how these illustrations were used in Sumerian society. Perhaps, this and other illustrations accompanied the telling of these epic tales orally.

In any case, we see Ninti sitting near the tree of life holding the rather large newborn infant that already has a pony-tail, the first *Adamu*, facing Enki who is holding some sort of vessel. Behind Enki is a worker and more vessels or flasks. This would appear to be a laboratory. The place where the first *Adamu* was created was a special structure whose Akkadian name was Bit Shimti. This comes from the

57

The Ancient Mesopotamian Account of the Creation of Humans

Sumerian word SHI.IM.TI, which literally meant "house where the wind of life is breathed in".[26] The texts state how cleanliness of the participants and purifications were necessary before this procedure of creation would work.

Figure 3-3 An Assyrian cylinder seal illustration of the birth of the first human. From The 12th Planet, by Zecharia Sitchin, Avon Press.

After the first *Adamu* was born and it was determined that the procedure worked, more Anunnaki "goddesses" were found and the procedure was repeated. We read on one tablet that Enki and Ninti used fourteen birth-goddesses to create seven males and seven females. The procedure must have been repeated many times, since it was a complicated procedure requiring many "birth goddesses." The "primitive workers" were hybrids of something close to what we call *Homo erectus* and the Anunnaki. Nowhere in Sitchin's books do we find anything close to a biological description of the Anunnaki, but they must have been close enough to the primitive hominids and/or they knew of some special genetic engineering tricks to allow a hybrid to be created in the first place. But the primitive workers were not able to reproduce themselves.

Unfortunately, the Mesopotamian tablets relating the rendering of the "primitive workers" fertile are lost. Early Assyrian tablets exist (~850 B.C.) that relate to the "primitive workers" becoming fertile, but the one that carries most of this story is too damaged to reveal what the original Sumerian text had to say.[27] So, Sitchin turns to the creation story as contained in the book of Genesis in the Bible. Many academic scholars have pointed out that much of Genesis is a strongly edited version of Mesopotamian texts. Genesis has been monotheised and Hebraized (for the most part), but otherwise it has been

The Ancient Mesopotamian Account of the Creation of Humans

clear to Mesopotamian scholars since late in the nineteenth century that much of Genesis stems from Mesopotamian texts that are as much as 2,000 years more ancient than the Old Testament.

In any case, from the King James version of the Bible, we read:

> And the Lord God caused a deep sleep to fall upon Adam and he slept; and he took one of his ribs, and closed up the flesh instead thereof; And the rib, which the Lord God had taken from man, made he a woman and brought her unto the man. And Adam said, This is now bone of my bones, and flesh of my flesh: she shall be called Woman because she was taken out of Man. Therefore, shall a man leave his father and his mother, and shall cleave unto his wife: and they shall be one flesh.[28]

The great Sumerologist, Samuel N. Kramer, pointed out near the middle of this century that the tale of Eve's origin from Adam's rib probably stemmed from the double meaning of the Sumerian word TI, which means both "rib" and "life."[29] Sitchin points out that there are many other multiple meanings to many Sumerian words and an apparent penchant of the Sumerian scribes to play with these multiple meanings. Sitchin believes the Biblical tale of the creation of Eve represents a second genetic manipulation of the *Adamu* in order to make the "Primitive Worker" fertile. Bone marrow, probably from a rib, was taken from a male *Adamu*, and implanted into a female in a genetic manipulation that is unknown to the modern genetic manipulators, to allow the *Adamu* to procreate.[30] The more procreation, the more slaves.

And the new slaves, the "Primitive Workers", were in high demand by the Anunnaki. Early on, the new "primitive workers" were confined to the mining areas of the Abzu, Enki's stronghold in Africa. But the Anunnaki from E.DIN. ("the abode of the righteous ones"), which was located in the area of what later became Sumer, wanted "Primitive Workers" to do the manual labor for them. E.DIN was Enlil's domain and contained a spaceport that shipped the gold to a permanent spacecraft that orbited the Earth, whence the final transfer to their home planet was made. The tension between the two half-brothers, Enlil and Enki, heightened until Enlil led a raid on the Abzu and forcibly removed a large number of Primitive Workers to E.DIN.[31]

There can be little doubt that the Sumerian E.DIN provided the basis for the biblical Eden:

> And the Lord God planted a garden eastward in Eden; and there he put the man whom he had formed.[32]

So, despite the missing texts, we have remarkable documentation of our own "creation". Sitchin states:

> Bearing in mind that these ancient texts come to us across a bridge of time extending back for millennia, one must admire the ancient scribes who recorded, copied, and translated the earliest texts - as often as not, probably, without really knowing what this or that expression or technical term originally meant, but always adhering tenaciously to the traditions that required a most meticulous and precise rendition of the copied texts.[33]

The Sumerian and later Mesopotamian scribes have actually provided us with a record of our creation that took place hundreds of thousands of years ago. The ancient Sumerian scribes couldn't have possibly known about the technicalities of the process of "creating a human", but they described it as best they could. For example, how could they know what a gene was, or how they could be manipulated and re-combined to create new organisms? And yet they described the mixing of clay using a word that also means egg. As we've seen, scientists of the modern era only came up with the concept of a gene at the beginning of this century. DNA, the substance of the genetic material, was only discovered in 1946, with the structure of the DNA only being understood in 1953. Our understanding of genetics has expanded rapidly since then, and the first test-tube baby was born in 1978. Since 1978, Sitchin has claimed that Adam was the world's first test-tube baby.[34] Certainly, the process of creating humans from the primitive hominids and the Anunnaki would seem to have followed procedures closely resembling modern procedures for "creating" a test-tube baby.

Sitchin makes it clear throughout his books that the Anunnaki treated their created slaves poorly, much like we treat domestic animals we are simply exploiting - like cattle. Slavery in human societies was common from the first known civilizations until quite recently. Perhaps it shouldn't surprise us to learn that the Anunnaki were vain, petty, cruel, incestuous, hateful - almost any negative adjective one can think of. The evidence indicates that they worked their slaves very hard and had little compassion for the plight of humans. Yet, the Anunnaki eventually decided to grant humankind their first civilization, the Sumerian civilization, which had its beginnings around 4,000 B.C. We'll continue reviewing the fascinating Mesopotamian historical accounts of the interactions of their "gods", the Anunnaki, and their human creations in later chapters. In *Genesis Revisited*, Sitchin points out how the recent mitochondria DNA data and the African genesis theory of paleoanthropologists support the Anunnaki story of our "creation."[35] From the last chapter, we recall that many paleoanthropologists support the theory that modern humans originated in Africa between 200,000 and 450,000 years ago, and then gradually spread throughout the remainder of the world. Certainly this theory would seem to fit Sitchin's interpretation of the Mesopotamian tablets very closely. However, we saw that the earlier interpretations of the human mtDNA data are now being questioned, and that other fossils have recently been discovered, and the "multi-regional" hypothesis of the origin of modern humans is regaining ground among paleoanthropologists.

Sitchin believes that the ancient Mesopotamian mythic-historical account of the creation of the "Primitive Worker" by the Anunnaki totally explains the appearance of what we call modern *Homo sapiens*. There is evidence, however, that other ET groups beside the Anunnaki of the "12th planet" have been involved in human affairs, as the following human historical account will attest.

A Human Historical Account of a "Sirian Connection"

Aside from the ancient Mesopotamian accounts, one of the strongest exoteric (non-esoteric) pieces of evidence indicating ET visitation and interaction with human affairs comes from the Dogon tribe of West Africa. The Dogon historical account of ET visitation is an oral historical account, as the Dogons are not a literate tribe. As mentioned, oral history is the only way all human groups, including the ancient Sumerians, had to remember their past until they learned how to write. The credibility of Dogon oral history is greatly enhanced by the fact that they have knowledge of the heavens that our scientists have only recently acquired, as we observed with the ancient Mesopotamians.

The Dogons have carried an oral tradition for hundreds and thousands of years of a visit by entities from the vicinity of Sirius to their ancient ancestors. The Sirian visitors were apparently partially aquatic and preferred flopping about in the water to moving about on

The Ancient Mesopotamian Account of the Creation of Humans

land, although they did both. Apparently, a Sirian airship landed in a slight depression of the land surface, drilled a hole to the water level and filled the depression with water. The Sirians evidently tried to teach the Dogons, among other things, how to live in accordance with their own divine nature. The Dogons have worshipped the Sirians for centuries.[36]

R.K.G. Temple has written about the Sirius connection with the Dogons and other ancient cultures, especially that of ancient Egypt, in *The Sirius Mystery*, first published in 1976. What is impressive about the Dogon material is not only the fact that they have an oral tradition - a religion, if you will - of a Sirius connection, but also the fact that they had the knowledge that Sirius is a double star, centuries before this was discovered by modern astronomers.

Sirius, known as the Dog Star, is the brightest star in the heavens and is located about eight and one half light-years from Earth. In 1844, an astronomer (Bessel) announced that the slight irregularities in the movement of Sirius indicated that it had an obscure companion star. This companion star is today classified as a white dwarf star of the tenth magnitude (quite dim). Temple argues that the Dogons' legend of contact with ETs from the vicinity of Sirius must be true since they possessed such detailed information for centuries about the Star, which our own astronomers have only recently discovered, or rather rediscovered. As we saw with the Sumerians regarding their knowledge of water on Neptune, ancient cultures often had more knowledge of the heavens than was possessed by the modern science of astronomy until recently, with the advent of much more powerful telescopes and unmanned satellites.

Sirius played a prominent role in several ancient cultures. The ancient Greeks and Romans considered the rising of Sirius in August as a token of summer heat. The ancient Egyptians were obsessed with Sirius, indicating that they had a Sirius connection. The rising of Sirius was a sign of the beginning of the annual Nile flood to the ancient Egyptians. Temple points out that the chief female deity of the ancient Egyptian pantheon, Isis, was identified with Sirius.[37] According to Temple, the ancient Egyptians had as clear a conception of Sirius as the Dogon tribe demonstrates.[38]

Temple, in his very scholarly book, wishes to assure fellow scholars that he hasn't "gone over the edge," despite his radical hypothesis. He carefully documents all his work and assails others who talk of spacemen and ancient cultures who make statements that cannot be checked. Temple assures his readers that he doesn't believe in UFOs and refers to those of us who know about UFOs and the ET presence on Earth as the "lunatic fringe." [39] While thus assuring his readers,

The Ancient Mesopotamian Account of the Creation of Humans

Temple states:

> ...(I)n considering the very origins of the elements of what we call human civilization on this planet, we should now take fully into account the possibility that primitive Stone Age men were handed civilization on a platter by visiting extraterrestrial beings...[40]

After pointing out some of the similarities of Egyptian and Mesopotamian cultures, such as their identical astronomy, Temple argues that Egypt and Sumer had a common origin with the Sirians. Sitchin, on the other hand, using Mesopotamian historical records, believes the Anunnaki, that biologically created the Sumerians and founded their civilization, founded also the civilization of Egypt, as well as other ancient civilizations. Temple feels that the Sirians came to Earth, started civilization, and then left, leaving monitors to watch our progress. Eventually, Temple believes, they will return. There is no record from the Dogons or Egyptians of genetic manipulation of humans by the Sirians.

---

[1] Lamberg-Karlovsky, C. C. and J. A. Sabloff 1979. *Ancient Civilizations: The Near East and Mesoamerica*. The Benjamin/Cummings Publishing Co., Inc., Menlo Park, pp. 28-29.

[2] *Ibid.*, pp. 147-148.

[3] *Ibid.*, p. 147.

[4] *Ibid.*, p. 150.

[5] *Ibid.*, p. 160.

[6] *Ibid.*

[7] Sitchin, Z. 1976. *The 12th Planet*. Avon Books, New York.

[8] _____ 1980. *The Stairway to Heaven*. Avon Books, New York.

[9] _____ 1985. *The Wars of Gods and Men*. Avon Books, New York.

[10] _____ 1990. *The Lost Realms*. Avon Books, New York.

[11] _____ 1991. *Genesis Revisited: Is Modern Science Catching Up with Ancient Knowledge?* Bear and Company Publishing, Santa Fe, New Mexico.

[12] _____ 1992. *When Time Began*. Avon Books, New York.

[13] Sitchin, Z. 1987. Introduction. *In Breaking the Godspell*, by N. Freer, Falcon Press, Phoenix.

[14] Sitchin, Z. 1980. *op. cit.*, p. 89.

[15] Sitchin, Z. 1991. *op. cit.*, pp. 3-22.

[16] *Ibid.*, pp. 5-9.

[17] Sitchin, Z. 1980. *op. cit.*, p. 86.
[18] Sitchin, Z. 1991. *op. cit.*, p. 158.
[19] *Ibid.*, p. 160.
[20] *Ibid.*, p. 22.
[21] *Ibid.*, pp. 163-164.
[22] *Ibid.*, p. 162.
[23] *Ibid.*, p. 166
[24] *Ibid.*, p. 163.
[25] *Ibid.*, p. 170.
[26] *Ibid.*, p. 185.
[27] *Ibid.*, p. 187.
[28] *The Holy Bible*, King James Version, Oxford University Press, New York, Genesis 2: 21-24.
[29] S. N. Kramer quoted in Sitchin, Z. 1991. *op. cit.*, p. 184.
[30] *Ibid.*, p. 187.
[31] *Ibid.*, pp. 170-171.
[32] *The Holy Bible*, King James Version, op. cit., Genesis 2: 8.
[33] Sitchin, Z. 1991. *op. cit.*, p. 171.
[34] *Ibid.*, p. 162.
[35] *Ibid.*, pp. 192-202.
[36] Temple, R. K. G. 1987. *The Sirius Mystery*. Destiny Books, Rochester, Vermont.
[37] *Ibid.*, p. 179.
[38] *Ibid.*, p. 182.
[39] *Ibid.*, pp. 213-214.
[40] *Ibid.*, p. 199.

Note: All of Sitchin's books are available in paperback editions through Avon Books, and in hardback editions through Bear and Company Publishing.

> *"With Earth's first Clay*
> *They did the last man knead,*
> *And there of the Last Harvest sowed the Seed:*
> *And the first Morning of Creation wrote."*
> Rubaiyat *of* Omar Khayyam

Chapter 4

The Creation of Heaven and Earth and All of Life

An Exercise in Logic

Let us not spend much time here trying to prove that life exists outside of this planet. We have already reviewed strong evidence for the inception of modern humans by an ET species, and there will be much more evidence presented of ET intervention in human affairs throughout this book. For now, let us take a simple journey in logic. Let us begin with time. We've already seen that Earth scientists now estimate the age of the Earth at about 4.5 billion years. Astronomers and Earth scientists estimate our universe to be between eight and twenty billion years old. So to the best of our knowledge, our Earth is anywhere from about one-half to one-quarter as old as the universe in which it is located.

Let us consider the space in our universe. As we know, our Sun is but one medium-sized star located towards the edge of our galaxy that we call the Milky Way. Our Sun is one star out of about 100,000,000,000 (100 billion) stars in our galaxy. There are hundreds and thousands of galaxies visible to our most advanced instruments our scientists use to explore the universe.

Now consider how many habitable planets there may be in the universe. We know that our particular solar system contains at least nine planets. At present it seems there is only one planet in our solar system capable of supporting life forms such as ourselves, our Earth. There is some speculation as to whether Mars may have supported life systems in the past, but Mars as well as the rest of the planets in our solar system are not believed capable of supporting life forms at

The Creation of Heaven and Earth and All of Life

the present time. Our scientists have concluded this despite the fact that we have barely investigated the planets of our solar system. Nevertheless, there is at least one habitable planet in our solar system.

There has been a question among our astronomers as to whether there are planets revolving around other stars in the universe. They haven't been able to positively identify planetary systems of other stars, but they have not ruled out this possibility and are developing new and better technologies that will enable them to detect planetary systems if they are there.[1]

To me, this demonstrates one of the extreme weaknesses of the current scientific approach to the reality of the universe. To an entity who has evolved a logical capability, such as we have, it is logical to assume that there are probably many planets and solar systems similar to our own in the universe. To rely on our primitive astronomical instruments to confirm that there are planets outside our solar system, would seem ludicrous to me. Relying strictly on our primitive science of the physical universe would, logically, so impair a person's thinking about our probable origins, history and existence, that he would be incapable of rendering much reliable information about such subjects.

Considering that this star, our Sun, supports one habitable planet, it is logical to conclude that there are hundreds and thousands of habitable planets in the universe. Nobel laureate, Sir Francis Crick conservatively estimates that there are one million planets in our own galaxy where life could have originated. Further, Crick points out that there are an estimated 100 billion galaxies of various sizes in the universe.[2] There may be billions of planets in the universe where conditions were and are favorable for life.

In view of so many probable planets, let us again look at the question of the origin of life. If the neo-Darwinists of the late twentieth century can theorize that life with its extreme complexity arose from a prebiotic soup on Earth by chance circumstances, why couldn't this have happened on another one of the billions of planets in the universe, billions of years before our planet even existed, since our planet is only one-fourth to one-half as old as the universe itself? As we saw in Chapter 1, Crick and a few other scientists, confronted with the problems of explaining the origins of life, are throwing around just such possibilities.

Now let us consider the question of the enormous distances in our galaxy and universe. The nearest star to our Sun is 4.3 light-years away. Because the distances in space are so enormous, astronomers have devised a measurement called a light-year. One light-year is the amount of distance an object traveling at the speed of light can travel in one Earth year. The speed of light has been calculated at 186,000 miles (300,000 kilometers) per second. In order for an object to travel from Earth to the nearest star in the universe, the object must travel many miles and spend a lot of time doing it. How much more time would it take to travel to more distant stars in our galaxy, or another galaxy? This would take most or all of a human's lifetime, or several human lifetimes, depending on the distance the star traveled to is from Earth. So how, you might ask, would space travel be practical, since distances are so great?

In the first place, the ETs may not have been limited by the eighty-year lifetimes of we humans. Evidence suggests that ETs may live for hundreds, and even thousands of years. If one has lots of time, distances don't mean much. Furthermore, there is evidence that some of the extraterrestrial beings have mastered forms of space-warping, wherein space and time are distorted to enable them to travel light-years in just hours. Robert Lazar, a young physicist who claims to have worked on a secret government space propulsion project in Nevada, makes just such a claim on video tape and in print.[3] There is no reason to logically assume that ETs, with a civilization that is millions of years old, feel restricted by time and distance as we do in our primitive state of knowledge. It is much more logical to assume that the many ETs of this universe have no trouble with time and distance.

I am not saying that it is certain, based on logic, that there are habitable planets out there with civilizations older and far more technologically advanced than we are. I am saying that using the above information, we could logically entertain such a probability. After all, considering our technological advancement over the past few decades, how much further might other civilizations have advanced if they had had a few more hundred, or thousands, or millions of years than us? Given the above information about our place and time in the universe, proclamations by our scientists that we humans are the only intelligent life forms in the universe, (and if intelligent life did develop somewhere else in the universe, time and distance limitations have kept us isolated from such life forms), seem more than slightly absurd and very provincial.

As we've seen, we do have one good historical account that ETs genetically engineered us, and evidence from the Dogon tribe of Africa that their ancient ancestors were visited by ETs from a planet near the double star, Sirius. One might logically ask, however, that if ETs visited and exploited the Earth during humanity's ancient past, why is there so little evidence of this? There is considerably more evidence for ET influence in human affairs, and much of this evidence will be covered later in this book. We will review evidence that regressive ETs have gained control of this Earth thousands of years ago and have sought to control and restrict human spiritual potential by concealing the truth of the ET influences of the biological and cultural development of humankind. In other words, there has been a massive cover-up of the very existence of intelligent, technologically advanced extraterrestrial life in this universe, especially in our "Milky Way" galaxy. Knowledge of the development of ET civilizations in this galaxy and ET influences on this planet would help clear up the inconsistencies of the development of humans on this Earth. Because of the historical conspiracy to keep such information from us, we must turn to esoteric sources in order to help fill in the gaps.

The Spread of Life and Civilization from Planets in the Lyran Ring Nebula

Let's review the observations we've covered thus far that concern human origins. We have briefly reviewed the history of life on this planet, and found glaring weaknesses in the scientific explanations, and religious explanations as well. A few of the glaring weaknesses we observed in the Darwinian-anthropological myth of human origins are, (1) the origin of life itself on the planet Earth, (2) a satisfactory explanation of the transitions of organisms observed on the fossil record, especially the major ones, such as reptiles to birds or mammals, and (3) the rapid development of humankind in the past several thousand years. Neo-Darwinism is the only biological evolution theory we have, and the literal interpretation of the six-day creation myth of Genesis, which does not take into account any human relationship to lower forms of animals, is absurd to thinking people. Then we reviewed the ancient Mesopotamian accounts of the origin of humans through genetic engineering by ETs, the Anunnaki ("Those Who from Heaven to Earth Came"); and strong evidence from a non-literate African tribe of visitation to Earth by ETs other than the Anunnaki.

As mentioned in the introduction, esoteric information will only be used here if it seems to fit what historical data we have. One such esoteric source is found in the book, *The Prism of Lyra: An Exploration of Human Galactic Heritage*, by Lyssa Royal and Keith Priest. The book is the result of insights, deductive reasoning and channeling by Lyssa Royal.[4] The authors state in the preface that Lyssa Royal channels a few entities, and that these have been cross-referenced with other channels. The authors also state that they have consulted a

The Creation of Heaven and Earth and All of Life

number of respected anthropological and metaphysical works.

While I am using this esoteric source here to help us gain some insights into human origins, I am not giving it a blanket endorsement. In the first place, all "ultimate origin" stories, including the one related below, seem quite vague and not entirely satisfactory to me. Also, Royal and Priest handle the grey ETs, the ones responsible for the millions of human abductions and cattle mutilations in the late 20th century, and many other atrocities (see Chapter 12), with kid gloves. Nevertheless, even considering these reservations, *The Prism of Lyra* does corroborate with the historical information we have reviewed concerning the origins and development of life on this planet, including humans.

Royal and Priest themselves do not claim that their book represents the ultimate truth of our galactic origins. They call *The Prism of Lyra* an introductory book, and that is the light in which we should view the book, as well as all books that are currently coming out that are concerned with our origins, including this one. As we will see, this critical source will be supplemented and further corroborated as we go along.

According to *The Prism of Lyra*, all of consciousness and energy was once fused into an integrated Whole, the Whole of All That Is, or the Divine Creative Existence. This Divine Whole segmented and was responsible for the creation and continuing evolution of this universe.[5]

The original Creation in this universe took place near Lyra, which is now a constellation of stars visible in the northern hemisphere of Earth. Within the "time/space fabric" of the Lyran constellation was a "white hole," which Royal and Priest liken to a prism. As a portion of the Whole passed through this "prism," several "frequencies" were created. Consciousness fragmented away from other segmented consciousness. Apparently the purpose of this experience is to first experience aspects of separateness, then bring back what is learned and experienced and then re-integrate into the Whole.[6]

In addition to consciousness, the three-dimensional (third density) universe was also created - the planets, stars, gases, and atoms that make up the physical universe. This third density reality represents only a small part of the energy frequencies that emerged from the segmentation of the Whole.

In the first division at the "prism of Lyra," a group of mostly non-physical beings, known as the Founders, were created. These were what may be called the supervisors of the creation of this galaxy; and they remembered the "blueprints" of creation from the Whole, and began to segment themselves until they began to create individualized consciousness. The first of what we might call humanoids of this galaxy lived and developed civilizations on planets in the constellation of Lyra billions of years before our Earth was forged-formed-created from gases and other material of the universe.

Very briefly, this Lyran theory is presented as an ultimate creation story of our universe. This ultimate creation story, like all ultimate creation stories, leaves me dissatisfied. This accounting seems quite vague to me. Where did the Whole come from? Who or what created the Whole before it fragmented to create our universe? The approach taken here is not to dwell on this frustration concerning the ultimate creation. Instead, we will investigate the development and evolutionary life of this galaxy with special emphasis on this world. We have much, much to learn regarding this world alone. Perhaps, after we absorb much of the truth of our Earthly origins and join our galactic family, we will be able to better understand our ultimate creation.

The first life in this universe and the first humanoids (creatures with two arms, two legs, a torso, and a head) were created and evolved in the Lyran system. Most, if not all of the original humanoids of the Lyran system were primates or of primate stock, and to this day many of the ETs are primates, or have some primate genes. Over the course of the long creation/evolution of life, many non-primate humanoids were also developed. As civilizations developed on the planets of the Lyran Ring Nebula, as the Earth astronomers call it, and the humanoids there advanced technologically, the capability of space travel was developed. Many wars were fought using nuclear weapons and other forms of advanced technology that were even more deadly than what we Earth humans currently have developed. The various civilizations developed different philosophies reflecting the polarities of creation, which we would call negative to positive. Civilizations rose and fell; some were destroyed by others in what appears to be a dreary succession of wars, as the humanoids evolved exploring different polarities of their existence.[7]

Some, wanting to escape the conflicts of the Lyran civilizations, or perhaps simply wanting to establish their own cultures in relative isolation, emigrated from the Lyran system and looked for suitable planets in other star systems. Planets of the star system we know as Sirius, were the first to be settled outside of Lyra. Sirius is 8.7 light-years distant from our Sun, and many Sirians of both positive and negative orientation have had many important interactions with the primitive primate species developing here on Earth.[8]

Next, planets of the Orion Star Constellation were settled. As with all groups of civilizations, both positively oriented and negatively oriented civilizations developed near Orion, but Orion civilizations have a very negative reputation in this Universe. By negative orientation, I mean that most of the individuals and civilizations of Orion sought after service to self, primarily through conquering, dominating and controlling others, rather than loving service to others and the recognition of the unity of all.

According to *The Prism of Lyra*, the Orions have been responsible for spreading war and suffering throughout the universe in their attempts to conquer and dominate.[9]

The next group of ETs to be considered in this drama, come from civilizations on planets of a cluster of seven blue stars known as the Seven Sisters of Pleiades. In the not too distant past - within the last sixty million years - a group of Lyrans decided that they wanted to develop their civilization in isolation, so they came to this Earth to live. While here, they mixed some of the genes of Earth primates with their own in order to become better adapted to environmental conditions here on Earth. Soon other Lyrans came to Earth, producing some of the same conflicts that the "Earth-Lyrans" had left the Lyran system for in the first place. So the "Earth-Lyrans" left the Earth and decided to settle in the Pleiades.

With regard to the Earth itself, The Founders were the overseers of the Earth Inception Project, and facilitated the nonphysical aspects of it. The physical aspects of the Earth Inception Project were orchestrated by various Lyran races of ETs. The Lyran group had instinctive drives for genetic seeding and genetic manipulations derived from their creators, the Founders. They also employed a Sirian group to assist them. Royal and Priest suggest that the conscious or unconscious motivation of the Lyran Inception Group, aside from the pure joy of creating, might have been to accelerate their own consciousness evolution.[10]

In the latter stages of the creation/evolution of life on Earth, the Lyran directors of the Inception Project needed some extraterrestrial genes that might blend with the genes of the most advanced Earth primates, Royal and Priest report. The Pleiadians whose ancestors had been "Earth-Lyrans" were the most suitable, since they already had some Earth primate genes in their genetic make-up. The Lyran directors of our inception here on Earth needed certain aspects of Pleiadian DNA for the Earth primate. Further, the Lyran directors of our inception induced the Pleiadians to become involved with the developing Earth species in order that the Pleiadians might face their own negativity, which they had suppressed for generations in the Pleiades. And so, somewhat reluctantly at first, the Pleiadians agreed to interact with the developing primitive primate species here on Earth. According to Royal and Priest, the ancient Pleiadians interacted with nearly every primitive culture here on Earth, and provided most of the ET genes that made modern man, by combining Pleiadian genes with those of primitive primate humanoids that had developed here on Earth.[11]

When the Pleiadians that began to settle on Earth began to interact with the primitive primates, their culture did not think of itself as godly any more than we humans do. The primitive Earth primates, however, overwhelmed by their advanced technology and the fact that "they came from above", treated them like gods. Apparently, during the early developmental stages of any humanoid species in the universe, it is common for them to give up their personal power to godlike figures[12], as the Pleiadians must have seemed to the first Earth humans. Soon, the Pleiadians began to relish the godlike power they were given by the Earth humans, and began to use fear in order to manipulate early humans. The ancient myths involving warlike, sensual, jealous gods are directly linked to ETs of other systems that were involved with the early cultures of Earth humans.[13]

The Prism of Lyra gives a picture of our Milky Way galaxy being inhabited by billions of three-dimensional humanoids, many of whom are technologically advanced to a degree which we Earth humans can only imagine. We are only able to imagine their technology, because our own technology has advanced so breathtakingly fast in this, our twentieth century.

A Re-examination of the History of Life on Earth

The reader is referred to Chapter 1, where we exposed some of the major weaknesses of the neo-Darwinian theory of evolution. As we saw, the very origin of life on Earth is perhaps the most serious deficiency of this materialist, scientific theory. Judging from the information we have presented thus far, including from *The Prism of Lyra*, it must have been a Lyran group that seeded this planet with life in the form of one-celled organisms, complete with their own DNA and ability to reproduce, about 3.5 million years ago (if our dating techniques are correct). The planet first had to be made suitable for higher forms of physical life by creating an environment conducive to higher life forms. Oxygen would be needed in the oceans and atmosphere in order for life forms to exist on Earth that need oxygen. This was accomplished through millions of years of the existence of the one-celled, blue-green algae which, through photosynthesis, produce food for themselves while releasing oxygen at the same time.

After millions of years, when the planet had evolved to the point where it could support higher forms of life, the Lyran Inception

Group, possibly with the help of Sirian and other ET groups, must have come to Earth periodically to genetically engineer many of the major jumps of evolution the paleontologists observe in the fossil record (Let me remind the reader here that for such beings a million years are, as it were, an insignificant timespan). As with the example given in *The Prism of Lyra* with regard to the Lyran humanoid group that migrated to the Earth and later to the Pleiades, they probably incorporated genes of Earth life forms, which had adapted to environmental conditions on Earth through microevolution, with the genes of extraterrestrial life forms. The terrestrial genes were necessary for the survival of the organism being engineered for life on earth. We can only guess at the evolutionary-creation histories of the genes of the life forms brought to this planet by the Creators.

The sudden appearance of simple, multicelled marine animals on the fossil record about 670 million years ago after almost two billion years of the existence of only one-celled organisms on this planet, as well as the "Cambrian explosion" where marine animals with hard skeletal parts suddenly appeared at 570 million years ago, would seem to be clear examples of genetic manipulation by some ET Inception Group. The reader will recall that not only did animals with hard body parts appear suddenly at the beginning of the Cambrian period, but also all of the basic body plans, or phyla, of all the types of animals that have existed on Earth appeared simultaneously at this time. Based on the fossil record, the Mesopotamian historical records of ET genetic manipulation of an Earth life form (humans), the Dogon oral history of ET visitation, plus the esoteric information provided in *The Prism of Lyra*, the ET (Lyran) Inception Group must have laid the foundation of the development and evolution of all animal life on Earth during the "Cambrian explosion", and 100 million years before that. This scenario would explain why no new phyla, or basic body plans, have appeared since the Cambrian period.

The intervention of ET Inception Groups would also provide an explanation for the huge jumps in the fossil record, such as from one animal class to another. The readers will recall from Chapter 1 that the transitions from one class of animals to another, as recorded on the fossil record, stretches the credibility of the neo-Darwinian theory of evolution beyond the breaking point. The transitions from fish to amphibians (~370 mya), from amphibians to reptiles (~340 mya), from reptiles to mammals (~200 mya), and from reptiles to birds (~170 mya), all involve the appearance of radically different modes of reproduction and entirely new morphological (physical) characteristics. The entirely new mode of reproduction that the reptiles (amniotic eggs) and mammals (live births) exhibit, and the other entirely new characteristics (e.g., from the scales of reptiles to the feathers and hair of the birds and mammals) are better explained by the intervention of ET master genetic engineers, than by the "gradual evolution" of the neo-Darwinian theory. The sudden appearance of flowering plants (angiosperms) about 100 million years ago, and the simultaneous, or near simultaneous appearance of all known orders of placental mammals soon after 65 million years ago, are two more possible, if not probable, examples of the work of ET master genetic engineers.

The mass extinctions of the early history of life on this planet might also be evidence of ET intervention. We recall that paleontologists and Earth scientists (geologists) have found evidence that there have been at least six episodes of mass extinction, the oldest occurring about 500 mya and the youngest about 65 mya, where 60% to 85% of all species that existed at that time, abruptly disappeared. Earth scientists have found evidence that the last mass extinction event 65 million years ago may have been caused by the collision of a large extraterrestrial body with Earth. Perhaps this was a method by which ET Inception Groups directed evolution on Earth. For example, during the last mass extinction event 65 mya at the end of the age of dinosaurs, maybe the ET creators were tired of working with dinosaurs and wanted to begin working on mammalian development, especially that of the primate order. Since they had been capable of space travel for millions of years by then, perhaps they had the capabilities of altering and directing the course of comets and asteroids. So, after making sure that mammals and other living organisms they may have wanted to preserve would survive a global catastrophe, they forcefully directed an extraterrestrial body into the Earth.

There are countless other possible examples of possible and probable ET intervention in the ancient development of life on this planet, but let us examine specifically possible ET intervention in the primate order that led to human beings.

Possible ET Interventions in Primate and Human Evolution

In *The Prism of Lyra*, we read:

> For many thousands of years during the early phases of the Earth Inception Project, the Lyrans watched the developing primate race on Earth with cautious eyes. Occasionally they took samples and made slight alterations to the DNA structures. At critical points in development they began

inserting genetic material from the Pleiadians (and other groups) into these primates...[14]

Let us briefly look here at the stages of primate and human evolution where it is possible that an ET Inception Group may have intervened in the evolution and development of primates and humans. The reader is referred to Chapter 2, where the obvious gaps in the Darwinian-anthropological version of primate and human evolution are presented.

The origin of the orders of placental mammals, incuding primates, at the beginning of the age of mammals, shortly after 65 mya has already been discussed. Primates remained at a prosimian stage for more than twenty million years, until the first primitive apes appear in the fossil beds of Fayum, Egypt, shortly after 40 mya. As we observed, ancient tarsier prosimians would seem to be the precursors of these first apes. The origin of tarsiers, and the relationship of the vast majority of prosimians (adapids and omomyids), that had thrived for millions of years, is unclear and much debated by paleoprimatologists. Perhaps an ET (Lyran) Inception Group intervened here to spur the evolution of more advanced primates.

In any case, primitive apes evolved and thrived for millions of years in Africa, giving rise to Old World monkeys around 18 mya. Eventually, about 15 mya, apes migrated to southern Eurasia. The first member of the human biological family (Hominidae) - the australopithecines, as they have been dubbed - appeared abruptly shortly after four mya in east Africa. Australopithecines are clearly more closely related to apes than monkeys. The experts on primate and human evolution have frequently bemoaned the fact that no good fossil ape ancestral candidates for the australopithecines have been found, despite years of diligent searching. In fact, of the apes that exist today, the small gibbons and large orangutans of southeast Asia, plus the gorillas and chimpanzees of Africa, the only ape with a good possible fossil ape ancestor is the orangutan. It was pointed out that a roughly thirteen million year old fossil ape from Pakistan, known as *Sivapithecus*, has facial features (but little else!) that resemble that of the living orangutan. There are no fossil apes known in Africa between five and ten million years old that make good ancestral candidates of the modern African apes or the australopithecines and humans.

The origins of the australopithecines would seem to be a strong possible point in human evolution where an ET Inception Group might have intervened. Perhaps, the ET Inception Group took genes from something close to *Sivapithecus* and combined them with extraterrestrial genes, and the first members of the Earth human family were created.

This possible scenario begs an explanation of why the biochemical studies have shown that the African apes are closer genetically to living humans than the Asian apes. Various techniques used by molecular biologists over the years consistently show that the chimpanzee nuclear DNA differs from human DNA by only one to two percent, and the gorilla DNA only slightly more than this. The orangutan nuclear DNA exhibits differences from human nuclear DNA that are generally about twice that of the African apes, and the gibbons' DNA is more distant still from that of humans. This data is generally interpreted as indicating that the ancestors of the African apes were also ancestors of *Australopithecus* and humans (Figure 4-1). However, if ET genetic engineers intervened to create the australopithecines, it is possible that the African apes evolved from the australopithecines (Figure 4-2). In other words, following this scenario, when some of the very first australopithecine populations entered woodland and forest habitats, they began to make adaptations for forest living, which included longer arms for tree climbing abilities, and, eventually, quadrupedal walking on their knuckles. This is all speculation, of course, but there is nothing in the fossil record, molecular biology, and the behavior and ecology of living apes that would rule out this scenario. Furthermore, given the evidence presented here of ET intervention in the evolution of humans, plus the data we currently have available, the scenario of the African apes evolving *from* the first members of the human biological family, the australopithecines, is much more plausible than the other way around.

This rather close genetic relationship between apes and humans is somewhat surprising, because the behavioral, ecological, and physical differences between humans and our non-speaking, four-legged cousins suggest that there would be a larger genetic difference. This illustrates, as many geneticists are claiming, that only a small percentage of our organisms' genetic material (regulatory genes?) may be responsible for the development of our physical organisms, as discussed earlier.

The australopithecines thrived for more than three million years, from roughly 4 mya to 1 mya. These small-bodied (three and a half to five feet in height), bipedal primates had very large, robust teeth and small brains, one-third the size of that of modern humans. Paleoanthropologists now recognize several species of *Australopithecus*.

At about two and a half, to two mya, the first member of the genus *Homo* appeared, called *Homo habilis*. *H. habilis* had a brain that

The Creation of Heaven and Earth and All of Life

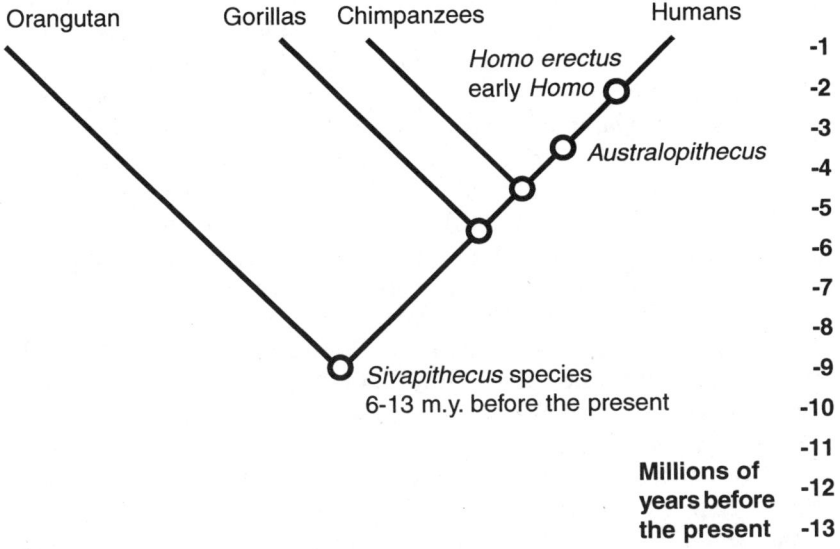

Figure 4-1 Evolution of the orangutan, the gorilla, the chimpanzee and the human according to current neo-Darwinian dogma.

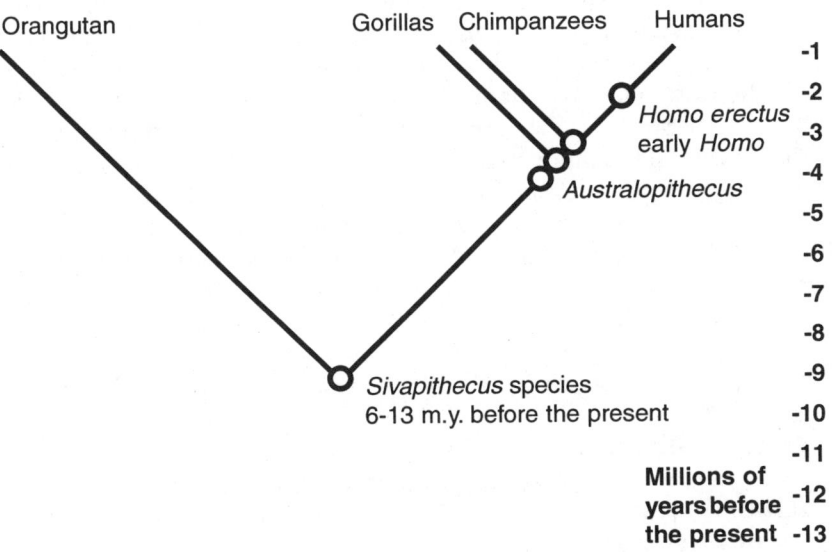

Figure 4-2 Possible evolution of the orangutan, the gorilla, the chimpanzee, and the human. (see text)

was approximately fifty percent larger than that of the australopithecines, with much smaller, more human-looking teeth. There is a strong possibility that an ET Inception Group intervened here.

The appearance of the next species in *Homo* lineage, *Homo erectus*, around 1.8 mya, in east Africa, south Asia, and probably Europe, is another time when an ET inception group may have used their genetic skills in the development of humans, and spread this new species throughout the Old World. The confusion among paleo-anthropologists we reviewed in Chapter 2 concerning the origin, biological classification and distribution of *H. erectus* populations throughout the Old World might be best resolved by assuming that one or more ET Inception Groups were present at this time. *H. erectus* remained quite stable physically from 1.8 mya to 300,000 years ago, and survived in isolated areas of the world until as recently as 12,000 years ago. At 300,000 years before the present, many human fossils begin to appear that are slightly more similar to modern humans (e.g., larger brains) which paleoanthropologists refer to as archaic *Homo sapiens*. We have already reviewed Mesopotamian historical accounts that indicate that at least one ET group (the Anunnaki) genetically engineered a more advanced human species at this time, and esoteric information indicates that several ET groups were involved with genetic manipulations with the primitive hominid lineage. Again, the debate and confusion expressed by paleoanthropologists which we reviewed in Chapter 2, concerning the biological classification of the varied archaic *H. sapiens* fossil specimens found throughout the Old World and the appearance of anatomically modern *Homo sapiens*, is clarified by the presence of ET groups on the Earth at this time. Many of the numerous archaic *H. sapiens* fossils found throughout Africa, Europe and Asia 300,000 to 35,000 years before the present represent genetic manipulations of the hominid lineage by ET groups. The modern species of humans had begun to arrive.

---

[1] Coven, R. 1992. "Hunting planets with a gravitational lens." Science News *141*: p. 327

[2] Crick, F. 1981. *op. cit.*, p. 105.

[3] Lindemann, M. 1991. *UFOs and the Alien Presence: Six Viewpoints.* The 2020 Group Visitors Investigation Project, Santa Barbara, California, p. 102. The Lazar Tape ordering address: 1324 S. Eastern, Las Vegas, Nevada.

[4] Royal, L. and K. Priest 1991. *The Prism of Lyra: An Exploration of Human Galactic Heritage.* Royal Priest Research, Sedona, Arizona, preface.

[5] *Ibid.*, p. 2

[6] *Ibid.*, p. 4
[7] *Ibid.*, pp. 21-27
[8] *Ibid.*, pp. 27-36
[9] *Ibid.*, p. 40
[10] *Ibid.*, p. 70
[11] *Ibid.*, pp. 49-53
[12] *Ibid.*, p. 54
[13] *Ibid.*
[14] *Ibid.*, pp. 71-72
[15] Romer, A. S. 1933. *Man and the Vertebrates*. University of Chicago Press, Chicago, pp. 201-205.
[16] Levinton, J. S. 1992. "The big bang of animal evolution." Scientific American 267: p.84.
[17] See reference number 39 of chapter 3.
[18] Levinton, J. S. 1992. *op. cit.*, p. 87.
[19] Gould, S. J. 1989. *Wonderful Life: The Burges Shale and the Nature of History*. Norton Press, New York.
[20] Levinton, J. S. 1992. *op. cit.*, p.87.
[21] Romer, A. S. 1933. *op. cit.*, p. 90.
[22] Denton, M. 1986. *op. cit.*, pp. 210-213.

*"Men work together
I told him from the heart,
Whether they work together
or apart."*
Robert Frost, *The Tuft of Flowers*

Chapter 5

Ancient Humans, Regressive and Progressive ETs

In this chapter we will review a few more historical accounts of the development of ancient humankind. Again, some of these accounts were written, and other accounts were passed down orally and only recently written down. All demonstrate evidence of much ET involvement in ancient human affairs. We will supplement these human historical accounts with more esoteric accounts of ancient human history and human development that corroborate the human historical accounts in order to get a clearer picture of ancient humankind. This is necessary because, as we shall see, ETs of ancient human history were often regressive in nature and wished only to dominate and control the young, naive humanoid species of Earth, and have subsequently covered up and distorted the truth of human origins.

The Book of Genesis Accounts of the Creation of Humans

We already have referred to the creation story as contained in Genesis, the first book of the Old Testament in the Bible, familiar to generations of Jews and Christians alike. We have already seen that it has been known by Biblical scholars for more than a century that a good part of Genesis is a much abbreviated, heavily edited, Hebraized and monotheised version of ancient Sumerian texts. The Old Testament consists of a body of literature that was written between the twelfth and second centuries B.C. Thus, when the first books of the Bible were written, the Sumerian texts had been written on clay tablets at least 1,300 years before (~ 2,500 B.C.) and the evidence indicates that the stories had been orally passed down for hundreds and thousands of years before that. I'm not saying that Genesis is nothing but an edited, monotheised version of an ancient Sumerian text. Far from it! The Genesis account of our creation might be a conflation of the Sumerian creation story, and at least one other creation story. There is some evidence that the entity commonly known as Jehovah or Yahweh, worshipped as the one and only God almighty by generations of Jews and Christians, had his own creation and subsequent story which has become entwined with the Sumerian story.[1]

When one looks at the creation story in Genesis, one finds not one, but three separate stories that relate to the biological creation of humans. First, in Chapter 1, verse 26 of Genesis of the King James Version, we read, "...and God said, Let us make man in our image, after our likeness....".[2] Sitchin points out that this is one example of where the ancient Hebrew editors must have slipped. "Our image" clearly refers to more than one "god". When the ancient Hebrew editors referred to this and a few other examples of plural gods, Sitchin argues, they were probably copying from a Sumerian/ Mesopotamian text that told of the assembly of the Great Anunnaki when they decided to create a "primitive worker" to replace the Anunnaki workers in the gold mines.[3] In the next verse we read, "So God created man in his own image, in the image of God created he him; male and female created he them."(Genesis 1: 27) Now "God" is the one, masculine God depicted throughout most of the Old Testament.

We move to the second chapter of Genesis to read another version of our creation:

> And the Lord God formed man of the dust of the ground, and breathed into his nostrils the breath of life; and man became a living soul.[4]

As mentioned, Sitchin points out the word "clay" or "dust" came from the Sumerian word TI.IT, which means literally "that which is with life." Later this word assumed derivative meanings of "clay" and "mud" and "egg."[5]

Later, in Genesis, God turned his attention to creating a woman:

> And the Lord God caused a deep sleep to fall upon Adam, and he slept: and he took one of his ribs, and closed up the flesh instead thereof. And the rib, which the Lord God had taken from man, made he a woman, and brought her unto the man.[6]

As discussed in Chapter 3, Sitchin believes that the above quoted passage represents an unknown genetic manipulation to make humans fertile because the first created humans, being a hybrid between two species (the Anunnaki and *Homo erectus*), was infertile. From the Bible only, all we can say is that in the first version of the creation of humans, God created both male and female and in the second version, God created a man, then a woman out of a rib of the man.

In Genesis 6, we find out about the third strong genetic influence on humankind:

> And it came to pass, when men began to multiply on the face of the Earth, and daughters were born unto them. That the sons of God saw the daughters of men that they were fair; and they took them wives of all which they chose... There were giants in the Earth in those days; and also after that when the sons of God came in unto the daughters of men and they bare children to them, the same became mighty men which were of old, men of renown.[7]

Pleiadian Accounts of the Origins of Humans

The Hebrew word Nefilim, which is translated as "giants" in the above passage, actually and literally means, "Those who had come down", as we have seen from Sitchin's writings. Clearly, the Nefilim means ETs that have come down to Earth, as does the phrase "sons of god." We don't know the details of these ET visitors, such as how long they were here, and with how many of the daughters of men they "came in unto", but clearly if a number of ETs left many babies with the daughters of men at sometime in the distant past, they had to leave a significant genetic impact on humans. Possibly we all have several strands of DNA that were formed on another planet or planets in our DNA as a result of the "sons of god" having sexual intercourse with the daughters of men.

Sitchin equates the Nefilim of the Bible with the Anunnaki of the ancient Sumerians. However, we have already seen evidence that several ET groups visited the Earth in the ancient past. The term Nefilim may well have been a catch-all word that meant any of the various species of ETs that came down from the heavens. In this regard let us look at what information has been given to us by the Pleiadians with regard to our creation. If you'll recall from *The Prism of Lyra*, the Pleiadians are supposed to be our genetic cousins in this galaxy.

There is evidence that the Pleiadians have been trying to contact us for a number of years and have given us much information about the early history of humanity and this galaxy.

The Pleiades is an open cluster of stars in the vicinity of the constellation Taurus. The cluster consists of several hundred stars that are about 325 to 410 light-years from Earth. Six of the stars are easily visible to the naked eyes. These "six" stars have been known since ancient times as the Seven Sisters. The ancient Greeks named the Seven Sisters after the seven daughters of Atlas and the nymph Pleione. A faint, seventh star is visible with a telescope that is associated with six visible stars. It is speculated that in ancient times this very faint star may have been much brighter and visible, thus accounting for the many references to the seven stars. Another speculation might be that perhaps the Pleiadian "gods" simply told the ancient humans of the seventh star that was closely associated with the six visible stars of the Pleiades star cluster.

It is not known how many habitable planets there are in the Pleiades, but there may be several. Many who call themselves Pleiadians have contacted humans, either directly or through a channel. Let us review two esoteric Pleiadian sources that tend to corroborate with what ancient human historical data we have.

There have been many UFO contactees since the modern UFO phenomenon started in the late 1940s. Many of the messages these contactees bring to us from our supposed "space brothers" make little sense, or contradict each other. But, there are many of these "messages" that seem to be good information from beings of ET civilizations that are trying to help us evolve consciously and spiritually.

The materials provided to a Swiss farmer of little education, Mr. Eduard "Billy" Meier, by those who claimed to be Pleiadians and friends, are possibly some of the best information we have. The information provided by the "Billy" Meier phenomenon checks out with other information we have from both exoteric and esoteric sources, concerning the beginnings of humans on this Earth, as well as the knowledge of the spiritual truths they incorporate into the fabric of the material passed on to Mr. Meier that concerns our history and current affairs. In short, the whole, amazing Billy Meier phenomenon, including the bits of information the Pleiadians give us about our ancient history fits in well with, and greatly supplements our work here.

The Billy Meier phenomenon itself takes a book to cover adequately. Gary Kinder, a writer and investigative journalist, has provided such a book.[8] Billy Meier, Kinder reports, has been contacted directly with an actual physical presence by the Pleiadians more than one hundred times. Usually these contacts involved Meier actually entering a Pleiadian ship (a UFO to us), where conversations between Mr. Meier and a Pleiadian, or one of their allies took place. The Pleiadians also communicated with Mr. Meier through what we would call mental telepathy, a form of communicating that seems to be common in this galaxy. The messages the Pleiadians have imparted through Meier are so radical relative to the commonly held belief systems of this world, like this book, the Pleiadians knew most of the people of the world would need some "proof" that these ET contacts were genuine. Therefore, they allowed Meier to photograph their "flying saucers" on several occasions so that Meier would have more credibility. The result is that we have more than five hundred excellent color photos of such craft, probably the best ever taken.

When the photographs and messages began to appear in the late 1970s and early 1980s, Meier was pronounced a hoax by most UFO organizations. The reasoning of the UFO organizations seems to have been, "It just can't be!," suddenly this Swiss farmer turns up with breathtaking, clear, color photos of UFOs and messages that are far-fetched, even for the UFO community. By now, several people have investigated Mr. Meier and his photos and have found no evidence of fraud. The photos appear to be genuine photos of mostly hovering craft, with the Swiss countryside providing the background.

Wendelle Stevens, a UFO researcher and former U.S. Air Force officer, conducted a lengthy investigation of Billy Meier, and became convinced that his contacts with the Pleiadians were genuine and valuable. Mr. Stevens has provided the best documentation published to date of the Billy Meier phenomenon, and the messages the Pleiadians and their allies have given humankind through Mr. Meier, in four books.[9, 10, 11, 12]

This group of Pleiadians (through Meier and Stevens) also speak of planets in the Lyra star system as being the first planets in this Milky Way galaxy to have third density (three-dimensional) intelligent life forms, as we saw in *The Prism of Lyra*. The "Swiss Pleiadians" state that these ancient Lyrans are their ancestors who fled the wars of ancient Lyra. Lyrans wishing to escape the destruction and ruin of Lyran civilization fled to other planets, both nearby planets, such as Vega, as well as planets in other star systems, such as the Pleiades and Hyades.[13] The Hyades is, like the Pleiades, a star cluster in the constellation Taurus and is located some thirty light-years from Earth.

The "Swiss Pleiadians" tell of many civilizations rising and falling on Earth for millions of years, usually destroyed by nuclear wars. They talk of the legendary civilizations of Mu and Atlantis as being only the most recent of ancient civilizations that have been destroyed. There have been many ETs, besides those of Pleiadian descent, that have come to Earth, participated in one of Earth's ancient civilizations, then fled as fighting and destruction broke out again. The Earth, they say, was first settled by Lyrans, then Lyran descendants, such as those from Vega and the Pleiades, all of whom were warlike when they first came to Earth. However, many of the descendants of these early war-like entities have now reached a higher level of spiritual evolution and no longer indulge in war.[14] The Pleiadians and others are here trying to undo some of the karma (you reap what you sow) and somewhat negative effects left by their less spiritually developed ancestors.[15] The "Swiss Pleiadians" talk of many of their genetic ancestors whose technological and (in a smaller measure) spiritual knowledge was considerably more advanced than that of Earth humans, and, therefore, were treated as "gods" and referred to as YHWHs. These YHWHs often abused their positions of power and behaved in a very un-godlike, barbaric manner.[16]

The "Swiss Pleiadians" tend to support the transmissions contained in *The Prism of Lyra*, that they are our genetic ancestors. They talk of some of their ancestors copulating with Earth creatures including *Australopithecus*, *Homo erectus* and Neanderthal man.[17] In each case, and several similar cases, genes that the ancient Pleiadians left in Earth gene pools of human-like (hominid) creatures contributed to the rapid biological evolution of hominids, including modern humans, of the past few thousand years. The "Swiss Pleiadians" tell of more recent meetings of Pleiadians and the "daughters of men" some 11,000 to 13,000 years ago.[18] They caught wild Earth creatures "descendants of former human beings from cosmic space"[19] - descendants of survivors of earlier disasters:

> Wild and beautiful female beings were tamed and mated with by the sub-leaders of the Pleiadian YHWH, "ARUS" who called themselves "Sons of Heaven"...Semjase, the highest leader of the sub-leaders, mated with an EVA, a female being, who was still mostly human-like and also rather beautiful...The descendant of this act was of male sex and a human being of good form. Semjase called him "A-DAM," which was a word meaning "Earth human being." A similar breeding produced a female, and in later years they were mated to each other.[20]

Ancient Humans, Regressive and Progressive ETs

This is a very interesting version of the Adam and Eve story. In any case, the Pleiadian contacts with Billy Meier tend to support the basic information imparted to us through ancient writings, such as the Bible, that:

> ...the sons of God saw the daughters of men that they were fair; and they took them wives of all which they chose. (Genesis 6:2)

Indeed, the Pleiadians are said to look like northern Europeans, and their descendants elsewhere in the world.[21] Meier and others report that they could pass unnoticed on the streets, except that they are very healthy looking. So our genetic heritage would seem to be most closely linked to the Pleiadians among all the numerous ET or alien groups that have visited Earth. But, what about the genetic engineering in the Sumerian records depicting the Anunnaki and their probable genetic manipulation of primitive hominids, *Homo erectus* and their like? To begin to answer this we must turn to other Pleiadian interdimensional transmissions.

There are apparently several groups of Pleiadians in contact with humans. Wendelle Stevens, the editor of much of the Billy Meier-Pleiadian communications now knows of at least five other groups, mostly in South American countries, who have had contact with what they consider to be human-looking extraterrestrials who say they are from one of the civilized planets in the Pleiades. The "Swiss Pleiadians" themselves told Meier that they were contacting at least two other groups and had used at least one other contactee before they began to use Mr. Meier.[22] Stevens provides a long and interesting list of the characteristics of the six Pleiadian human contact groups.[23] The characteristics these groups have in common include:

1. All the Pleiadians told their contactees that humankind is descended from their own ancient ancestors, "....from whom we had inherited our unusual aggressiveness."

2. All the Pleiadians said that they "...participated in some of the great events in our religious works and our mythologies, sometimes being adored and worshipped as Gods, which they said they were not."

3. All the Pleiadians had lifetimes that were at least ten times longer than ours.

4. All had partially organic computers and elaborate space-traveling

capabilities such as the ability to travel faster than light.[24]

Stevens lists several other amazing characteristics of the Pleiadians and their contactees, including the fact that none of the contactee groups had ever heard of each other until 1979.

Entities calling themselves Pleiadians also contact humans through channelers who often transmit before rather large audiences. Such an event occurred on the nights of November 15th and 16th, 1990, in Terman Auditorium, Stanford University, Palo Alto, California when Barbara Marciniak channeled messages from a Pleiadian group. Barbara Marciniak has been transmitting the Pleiadians since 1988. This particular group of around one hundred Pleiadian entities now calls themselves Pleiadian Plus, as these Pleiadians have joined forces with other extraterrestrials.[25] Pleiadian Plus is not going to come down with their mighty ships and save us, but they are giving us information, they claim, that might help us save ourselves, both in a physical and spiritual sense. In fact, most of the messages of the Pleiadian Plus group is considered spiritual, as are the messages of the "Swiss Pleiadians." Again, the material transmitted through Barbara Marciniak by the Pleiadian Plus group is considered very good because it fits well with the other exoteric Earth human sources of information we have covered and will cover. Pleiadian Plus is a group of ETs that have our best interests in mind, they claim, and are here to help us by informing us of our subjected position of being manipulated and controlled by ETs that are not so positively oriented. Since our beginnings and through the thousands of years of our brutal history, they claim, they have been helping us learn how to escape our position of being "victims" by developing our spiritual existences, and encouraging us to join the "Family of Light" by understanding what they call the Prime Creator.

Pleiadian Plus, too, talks of very ancient, but highly evolved civilizations as having existed for more than 500,000 years on Earth.[26] They also talk of many conflicts and wars between the ETs that settled here, mostly over ownership of this beautiful piece of property called Earth, "...truly a prime hunk of real estate". Members of what they call the Family of Light were the original owners and developers of the planet Earth. But about three hundred thousand years ago, this Earth was raided by forces that were not of the Light and took the Earth from the original owners, members of the "Family of Light." They talk of much of the Earth being rent asunder by nuclear wars, and the defeat of the "Family of Light." What does this say about the idea or the concept of The Light, they ask rhetorically, if it can be replaced and defeated by forces that are not of The Light? We will try to provide a satisfactory answer to this critical, even painful question before this book is finished.

The Reptilian Nature of the Anunnaki

Pleiadian Plus informs us that the new owners, the creator gods, as they call them, had much different agendas than their predecessors. These creator gods, our creator gods to which the Sumerian tablets and the Bible and other ancient writings refer, live off fear and chaos. Pleiadian Plus informs us that the electromagnetic energies of humanity's consciousness can act as food, and so it is quite possible that fear can act as food for some entities.[27] Also, Pleiadian Plus says that our creator gods do not exist and operate under the same laws that we do, and that they live, at least some of them, hundreds of thousands of years. The "Swiss Pleiadians" also told Billy Meier about ETs that lived for thousands of years.[28] Furthermore, Pleiadian Plus states that our creator gods were reptilian in nature and have nicknamed them the "lizzies." They call them lizzies in order to inject some humor into the circumstance - the discovery of the true nature of our "creator gods" - that might be terrifying to some.

The reptilian nature of the creator gods of the ancient Sumerians, the Anunnaki, is documented by obscure human historical sources, summarized in a book recommended by Pleiadian Plus, *Flying Serpents and Dragons: The Story of Mankind's Reptilian Past*, by R. A. Boulay.[29] Boulay uses mostly ancient Hebrew writings that were purposely omitted from the Bible by the ancient Hebrews, he believes, because they contained information that was embarrassing. Boulay lists such works as the Book of Jubilees, three books of Enoch, the recently discovered Dead Sea Scrolls, the teachings of Gnostic Jewish sects that existed around the time of Jesus, the Haggadah, or the oral tradition of the Jews, and other works.[30]

Boulay quotes from the Haggadah, "...the source of Jewish legend and oral tradition", a story of Adam and Eve:

> The first result was that Adam and Eve became naked. Before, their bodies had been overlaid with a *horny skin*, and enveloped with the cloud of glory. No sooner had they violated the command given them than the cloud of glory and *the horny skin dropped from them, and they stood there in their nakedness and ashamed* (his emphasis).[31]

Boulay also quotes a Gnostic version of Adam and Eve:

> Now Eve believed the words of the serpent. She looked at the tree. She took some of its fruit and ate, and she gave to her husband also, and he ate too. Then, their mind opened. For when they ate, the light of knowledge shone for them; they knew they were naked with regard to knowledge. *When they saw their makers, they loathed them since they were beastly forms. They understood very much* (his emphasis).[32]

The Gnostics and their writings were eliminated from the Church by the early Christians. We can begin to see why this was probably so. The Gnostics possessed too much of the Truth, so, for better control, most of their works had to be eliminated. Fortunately, for us, they were not completely destroyed. This Adam and Eve story indicates that light and knowledge in humans began when Eve listened to the serpent and ate from the tree of knowledge. And this story tends to confirm our information that we humans might have been genetically manipulated and "created" by "beastly forms."

We can see also why the story from the Haggadah was not included in the Old Testament, as well. Anything that reported Adam and Eve, our ancestors, had horny skin, could not be brought before the human consciousness if a cover-up of our origins was to be maintained. But, it is now, in the late twentieth century, being brought into our consciousness. Our ancestors, our creator gods, may well have been lizzies! Boulay uses these ancient legends, the Mesopotamian writings, and much more to argue that indeed the "creators" of humans were lizzies who came to this Earth hundreds and thousands of years ago in order to live and exploit the resources of this Earth. Boulay emphasizes the beastly manner in which these lizzies often behaved and how they controlled and exploited humans, which, as we've briefly seen, is also covered in Sitchin's works.

Also, lizzies have been seen by a few people that are living today, mostly by people who have been abducted and taken aboard a UFO, or to an underground ET facility. The three-dimensional existence of these creatures will be documented in later chapters. As for the staggering life spans of these lizzies, these are documented in the Sumerian King List.[33] The Kings listed that reigned before the Great Flood, were Anunnaki that reigned over the area which became Sumer after the flood. Eight Kings reigned for a total of two hundred and forty thousand years in this ancient kingdom, which Sitchin identifies as

E.DIN. Furthermore, as most Westerners know, the Bible documents ancient humans as having lived for hundreds of years. For example, Noah, who survived the Great Flood, lived nine hundred and fifty years (Genesis 9:29). Boulay cites evidence that as humans became less reptilian, less like their creators as time passed, the human life span began to decrease.[34] Correspondingly, humans had shorter and shorter life spans as time approached the human historical period - the beginnings of historical Sumer.

The Sumerian King List has received mixed reviews from scholars. Many understandably dismiss it, because it states that certain kings reigned thousands of years. But, archeological evidence has confirmed the reigns of later kings of Sumerian city-states of the fourth and third millennia B.C. And, certainly by this time (4,000 B.C. to 2,000 B.C.), the reigns of kings had become as short as those of historical humans you and I know. The ancient Sumerians, themselves, recorded that their gods were immortal. As for fear and chaos being some sort of food or nourishment, I cannot document or understand this concept. But we all know that this world we live in is filled with fear and chaos, and has been since civilization, as we know it, began.

Humans Controlled

The Pleiadian Plus group indicates through Marciniak, that our world, including ourselves, has been influenced and controlled since our creator gods gained control of it three hundred thousand years ago in a manner that is beyond our understanding. Our creator gods being experts in genetic manipulation worked in their laboratories and created versions of humans with different DNA. They disassembled the original DNA of humans they found on Earth and took out - unplugged - that which might help us to become more enlightened beings. All entities of this universe operate at certain frequencies, the Pleiadian Plus group informs us, and our creator gods created us to operate at frequencies whereby we could be controlled, yet remain operable in order to mine their gold, dig their ditches, build their temples or any other dirty work of theirs that needed to be done. They made it very difficult for frequencies of light - of information of the truth - to penetrate our being.[35] This is why we are still so ignorant about spiritual matters after thousands of years of being able to carry a higher consciousness. The creator gods did not want slaves that could spiritually evolve and, thereby, escape their control. But, they could not design a "primitive worker" that was somewhat efficient, that did not carry a higher consciousness, a soul, and some small degree of spiritual awareness.

To gain more insight into our creation to be controlled, spiritually ignorant beings, and bring in another Earthly, exoteric reference, we turn to some knowledge passed on by the ancient Maya civilization. The Mayans had a rich civilization which flourished in what is now central America and southern Mexico, from a few hundred years B.C. to A.D. 950, when it rather suddenly collapsed for unknown reasons. There is evidence that the Mayans, as with other New World civilizations, may have had an existence thousands of years previous to the known historical civilization.[36] As is well known, when the Spaniards invaded the Americas in the sixteenth century, they deliberately destroyed nearly everything they could of ancient American civilizations, except their gold and silver, because the Spaniard considered the native Americans heathens. As we will see, there is much evidence, despite the general lack of written records, that creator gods similar (if not the same) to those of the Sumerians, controlled the peoples of the ancient American civilizations, including the Mayans. Today, there is only one book available that purports to be a record of ancient Mayan beliefs and legends, and even it is not a record written during the height of Mayan historical civilization. This book is known as the *Popul Vuh* and was written by an unknown Mayan in the sixteenth century. The *Popul Vuh* was translated from Mayan to Spanish by a Spanish priest, Father Ximenez, and first published in Vienna in 1857.[37] There are many Christian ideas incorporated into the *Popul Vuh*, either by the original Mayan author or Father Ximenez, or by both. Despite these obvious distortions, it is all we have of the ancient Mayans, except archeological excavations of their religious centers, and it does preserve a creation story of humans that is probably close to the original Mayan story of human creation, because it is so similar to Sumerian and the other creation stories of other cultures.

According to the *Popul Vuh*, the Mayan gods created humankind to be servants and slaves for them, as in the Sumerian creation epics. This ancient Mayan history book tells how the Mayan gods first made "figures of wood" to be their servants. These creatures could walk and talk like humans, and even reproduce, but they were not good servants to the gods. The reason these "figures of wood" were poor servants was because they didn't have souls.[38]

We will recall that something similar is passed on to us in the allegorical story of Adam and Eve in Genesis. Adam and Eve ate fruit from the tree of knowledge at the encouragement of the serpent. Because of this, the "Lord" became angry and kicked them out of Eden and condemned them and all of humankind to a hard, difficult life. Adam and Eve, ate of the tree of knowledge and thereby learned the difference between good and evil, a basic criterion for the learning and discernment required of knowledge and spiritual evolution. For this, the "Lord" becomes angry, almost certainly because he knew that if humankind did evolve spiritually, they would escape the enslavement of the "gods." The *Popul Vuh* tells us of the saga where the "gods" of the Mayans destroyed their "figures of wood" and then created humans that were more intelligent than they wanted:

> They (the first humans) were endowed with intelligence; they saw instantly they could see far, they succeeded in seeing, they succeeded in knowing all that there is in the world. When they looked, instantly they saw all around them, and they contemplated in turn the arch of heaven and the round face of the Earth... Great was their wisdom: their sight reached to the forests, the rocks, the lakes, the seas, the mountains, and the valleys. In truth, they were admirable men...[39]

This indicates that the Mayan "gods" had gone too far in their genetic engineering to create a "workable" human being. This story clearly implies that the spiritual beings that animated human bodies they "created" were too highly evolved for the Mayan gods.

The gods asked themselves:

> What shall we do with them now? Let their sight reach only to that which is near; let them see only a little of the face of the Earth! It is not well what they say. Perchance, are they not by nature simple creatures of our making? Must they also be gods?...[40]

These are quite shocking words! Again the *Popul Vuh*, like the Bible and the Sumerian "myths", is certainly a flawed document in many ways, but like the Bible, and the Sumerian writings, there are almost certainly many historical and even spiritual truths in this document. And here, as we saw expressed even more clearly in the Sumerian clay tablets, "gods" created humans to be stupid, yet capable of carrying out hard labor so that they-the "gods"-could control and enslave their creations.

Most of the messages imparted by the group calling themselves Pleiadian Plus, are about what we would call spiritual intelligence and spiritual evolution. Pleiadian Plus, like the "Swiss Pleiadians", talks of enormous changes that are going to take place on this planet during the next few years. They talk of a grand plan, part of which is designed to place our planet Earth back in the hands of the "Family of Light," from whom it was rudely seized by the "Family of Darkness" some three hundred thousand years ago, ("....Did you think they were going to give up so easily?..."). The Original Planners of this Earth, who apparently were here until ousted three hundred thousand years ago, have apparently called upon many members of the "Family of Light" to incarnate (as in reincarnation) to assist in bringing light information to this Earth. And Pleiadian Plus talks of a multitude of members of the "Family of Light" who are incarnate and are also trying to help this world to become once again a dwelling place of the "Family of Light." Again, let me remind the reader, this may all be disinformation spread by the dark forces that are in control of this planet to continue to confuse and hoodwink we humans.

Nevertheless, here we have this spiritual specter - or spiritual intelligence splashed across our consciousness. This is no accident. As we have seen, much exoteric and esoteric information indicates that one cannot intelligently discuss our biological origins, our history, the UFO-ET factor, or anything about our existence, unless one has some understanding of spiritual realities. And, the spiritual and biological realities are, that we humans have been created and genetically manipulated by entities who do not have our best interests in mind. In fact, our "creator gods" have wanted only to enslave us, to control us, and keep us ignorant of our true spiritual natures and choices. But, there are those presences of the "Family of Light" who are available to genuinely help us.

Let us return to ancient history and observe how the regressive ETs of the ancient Sumerians, their creator gods, failed to care for their creations (humans) during one of the largest catastrophes that has ever occurred on the Earth, the Great Flood.

[1] See Chapter 16 for more information on "the Jehovah entity."
[2] *The Holy Bible*, King James Version, Genesis 1: 26.
[3] Sitchin, Z. 1976. *op. cit.*, p. 338.
[4] *The Holy Bible*, King James Version, Genesis 2: 7.
[5] Sitchin, Z. 1991. *op. cit.*, p. 166.
[6] *The Holy Bible*, King James Version, Genesis 2: 21-22.
[7] *Ibid.*, Genesis 6: 1,2,4.
[8] Kinder, G. 1987. *Light Years: An Investigation into the Extraterrestrial Experiences of Eduard Meier*. Atlantic Monthly Press, New York.
[9] Stevens, W. C. 1982. *UFO Contact from the Pleiades: A Preliminary Investigation Report*. UFO Photo Archives, Tucson, Arizona.

[10] _____ 1988. *Message from the Pleiades: The Contact Notes of Eduard Billy Meier.* UFO Photo Archives, Tucson, Arizona

[11] _____ 1989. *UFO Contact from the Pleiades: A Supplementary Investigation Report.* UFO Photo Archives, Tucson, Arizona

[12] _____ 1990. *Message from the Pleiades: The Contact Notes of Eduard Billy Meier* 2. UFO Photo Archives, Tucson.

[13] Stevens, W. C. 1982. *op. cit.*, p. 226.

[14] *Ibid.*, p. 235.

[15] *Ibid.*

[16] Stevens, W. C. 1988. *op. cit.*, p. 129.

[17] *Ibid.*, p. 84.

[18] *Ibid.*, pp. 128-129.

[19] *Ibid.*, p. 129.

[20] *Ibid.*

[21] Stevens, W. C. 1989. *op. cit.*, p. 60.

[22] *Ibid.*, pp. 162-163.

[23] *Ibid.*, pp. 164-165.

[24] *Ibid.*, p. 164.

[25] Marciniak, B. 1992a. "The Harmonics of Frequency Modulation and the Human DNA", Part I: The Pleiadians, through Barbara J. Marciniak at Stanford University, Palo Alto, California, November 15, 1990. Sedona *2*, No. 2, pp. 17-33.

[26] *Ibid.*, p. 18.

[27] *Ibid.*

[28] Stevens, W. C. 1988. *op. cit.*, p. 128.

[29] Boulay, R. A. 1990. *Flying Serpents and Dragons: The Story of Mankind's Reptilian Past.* Galaxy Books, P.O. Box 8542, Clearwater, Florida 34618.

[30] *Ibid.*, p. 61.

[31] *Ibid.*, p. 1.

[32] *Ibid.*

[33] Woolley, C. L. 1965. *The Sumerians.* W. W. Norton & Company, Inc., New York, pp. 21-26.

[34] Boulay, R. A. 1990. *op. cit.*

[35] Marciniak, B. 1992a. *op. cit.*, p. 19.

[36] Sitchin, Z. 1990. *op. cit.*

[37] Goetz, D. and S. G. Morley 1950. *Popul Vuh, The Sacred Book of the Ancient Quiche Maya.* University of Oklahoma Press, Norman.

[38] *Ibid.*, p. 89.

[39] *Ibid.*, p. 168.

[40] *Ibid.*, p. 169.

> *"Great Spirits have always encountered violent opposition from mediocre minds."*
> Albert Einstein

Chapter 6

The Anunnaki, Humankind and the Great Deluge

Most readers of western cultures know of the story of the Great Flood from Genesis in the Bible. We have reviewed evidence, however, that this version of the Flood and all of the first few chapters of Genesis are much edited, "monotheised", and "Hebraized" versions of much older epics of Mesopotamian civilizations. The first Mesopotamian versions of the flood, the Sumerian version, was written on clay tablets at about 2,500 B.C., but it told a story that occurred hundreds and thousands of years previously that had been handed down through oral traditions.

Most scholars believe the story of a worldwide Flood was simply a fable. Up to eleven feet of flood-deposited mud, however, has been found in the archeological excavations of the cities of ancient Sumer, but scholars believe that these very large mud deposits were left by local floods of the annual floods common in the basin of the Tigris and Euphrates rivers. Certainly, scholars do not believe that the mud deposits found under the ruins of ancient Sumerian cities are deposits of a flood that inundated most of the world and caused the near-elimination of humankind and much life on Earth.

Here, however, we are following Sitchin and others who contend that the ancient Mesopotamian epic stories are based on some event or events that actually did happen in humankind's ancient past. And, a Great Flood was a prominent event in the history of the ancient Mesopotamians and other ancient cultures throughout the world.

The previously mentioned Sumerian King List chronologically lists all of the kings, the cities, and events of Sumer.[1] The King List divides the "Sumerian" history into two parts, that which happened before the Deluge and that which occurred after the Deluge. The period

The Anunnaki, Humankind and the Great Deluge

before the Deluge was very long, lasting hundreds of thousands of years. The King List states that "Kingship" was first lowered from heaven to Eridu, an antediluvian city of Sumer. After several other antediluvian cities and kings are listed, the text states that "then the Flood swept over the Earth."[2] After the Flood, the list of kings continues. Kingship was again lowered from heaven, this time to the historical Sumerian city of Kish, and the kings names of the cities of Sumer and their reigns can be verified through other historical data. Of course, the kings and cities of Sumer that are said to have occurred before the Flood cannot presently be verified.

Much of the Sumerian version of the Flood has been inscribed on broken clay tablets, of which much is missing. Many missing pieces of the story of the Flood can be gleaned from the broken tablets of later Mesopotamian civilizations, which must ultimately be copies of Sumerian versions with some embellishments. While the Sumerian version was written downaround 2,500 B.C., the Akkadian version must have been written roughly between 2,300 B.C. and 2,200 B.C.; the Babylonian version must have been recorded sometime between 1,900 B.C. and 1,500 B.C.; and the Assyrian version after that time. The version of the Flood of Genesis must have been written about 1,000 B.C.

With all of these accounts and more, the Babylonian version of the Flood is the most complete. Sitchin frequently refers to *Atra-Hasis: The Babylonian Story of the Flood*, by scholars, W. G. Lambert and A. R. Millard, a 1970 compilation of Mesopotamian accounts of the Flood with emphasis on the Babylonian version.[3] The hero of the Babylonian version is called *Atra-Hasis*. In the earlier Akkadian version, the "Noah" is called Utnapishtim. In one Akkadian version of the Flood (Epic of Gilgamesh), Utnapishtim is called by Enki "the exceedingly wise." In the Akkadian language, *Atra-Hasis* means "the exceedingly wise." Hence, the connection between the Akkadian version of the Flood to the Babylonian one seems quite solid. There can be no doubt that the Babylonian version of the Flood was taken from earlier Mesopotamian versions.[4]

The *Atra-Hasis* epic tells of the creation of humans as "Primitive Workers" by the Anunnaki (including Enki) and how humans began to procreate and multiply. Before long, humankind began to upset Enlil, the chief administrator of the Anunnaki on Earth, and the chief god of the ancient Sumerians besides Enlil's father, Anu:

> The land extended, the people multiplied
> In the land like wild bulls they lay.
> The god got disturbed by their conjugations (couplings?)

> The god Enlil heard their pronouncements,
> and said to the great gods:
> "Oppressive have become the pronouncements of
> Mankind:
> Their conjugations deprive me of sleep."[5]

Enlil was so upset by humans disturbing his sleep that he tried to decimate humankind through pestilence and sickness. This was followed by sickness, disease and pestilence suffered by humans and their livestock. Atra-Hasis, the Babylonian Noah, appealed to Enki for help (called Ea in this passage):

> "Ea, O Lord, Mankind groans;
> the anger of the gods consumes the land.
> Yet, it is thou who hast created us!
> Let there cease the aches, the dizziness,
> the chills, the fever!"[6]

What a poignant plea! *Atra-Hasis* seems to say that, after all, "....it is thou who hast created us!"; why all of the pain and suffering for humankind? It is not clear what Enki did here, because the appropriate tablets are broken and lost.

Apparently, Enki did do something to relieve the chills and fever of humans because, a bit later in the epic, Enlil complains that humans are even more numerous than before. Enlil then plots to have humankind diminished by famine:

> Let the rains of the rain god be withheld from
> above
> Below, let the waters not rise from their
> sources
> Let the wind blow and parch the ground;
> Let the clouds thicken, but hold back the downpour.[7]

Enlil employed many of the other Anunnaki to make sure that humans did not get any food and could thereby be "diminished." This included Enki, the god of water, who was supposed to make sure that humans didn't get food from the sea.

So, for six *Sha-at-tams* humans suffered through a terrible famine. *Sha-at-tam* is usually translated as one year, but Sitchin believes it may have been a much longer period.[8] In any case, human life steadily deteriorated during the famine:

The Anunnaki, Humankind and the Great Deluge

> When the sixth *Sha-at-tam* arrived
> they prepared the daughter for a meal;
> the child they prepared for food...
> One house devoured the other.[9]

The "human containment plan" cooked up by Enlil and the rest of the Great Anunnaki had driven humankind to cannibalism. We briefly reviewed in the last chapter human historical data[10] and an esoteric source[11] that indicate that the Anunnaki of the ancient Sumerians were humanoids of a reptilian nature, called lizzies or reptoids, who had little compassion for their human creations. Certainly, these "human containment plans" of the Anunnaki in the mythic-historical accounts of pre-Deluge Sumer demonstrate a shocking lack of compassion for humans by their "gods." It is important to note that while the translations of Mesopotamian texts by Sitchin used here clearly show the lack of compassion demonstrated by the Anunnaki toward their human subjects, Sitchin nowhere mentions the possibility that the Anunnaki were lizzies, but rather expresses the belief that the Anunnaki were a lot like humans. Enki alone, among the Anunnaki "gods," showed compassion for humans throughout the ancient Mesopotamian texts. These texts state, as we have seen, that Enki was an Anunnaki, the half brother of Enlil. Later, however, we'll review evidence that Enki was a Sirian ET in the "employment" of the Anunnaki.

In the "human containment plan" we are considering here, Enki declared that he had opposed the acts of the other gods, and apparently did take some action to provide humankind with some food. There is much of the Mesopotamian texts that are lost here, but a bit later we find that Enlil is furious. He called for an Assembly of the Gods and sent his sergeant of arms to bring Enki. Enlil accused Enki of breaking the pact made by, "All of us, Great Anunnaki,..." to starve the people. There ensued several heated exchanges between Enlil and Enki. It is worth noting here that one of the reasons for believing Enki was something other than a "Great Anunnaki," is the frequent insubordination toward the Anunnaki gods he displays throughout the Mesopotamian historical epics. Enlil said before the Assembly of the Gods that there were many times that Enki "broke the rule."[12]

Enlil then comes up with the "final solution" to the human problem. He announced that a "killing flood" would soon wipe out all of humankind. It is clear from the Mesopotamian texts that the Anunnaki knew that this catastrophic flood was coming. It makes sense that an intelligent, space-traveling race like the Anunnaki would be monitoring the environmental variables of the Earth and of this solar system. And, it also makes sense that a long-lived race like the Anunnaki would be aware of the cycles of the universe, such as movements of comets and other extraterrestrial bodies. It is not unreasonable to assume that the Anunnaki could foresee a natural disaster that would have devastating effects for all living entities on the surface of this planet. Enlil announced to the Assembly of the gods that a killer flood was in the offing. He insisted that the news of this approaching calamity be kept from humans. He called upon all in the Assembly to swear themselves to secrecy:

> Enlil opened his mouth to speak and addressed the Assembly of all the Gods: "Come, all of us, and take an oath regarding the Killing Flood!" Anu swore first. Enlil swore; his sons swore with him.[13]

But, Enki balked at taking the oath at first and asked, "Why will you bind me with an oath? Am I to raise my hands against my own humans?"[14]

Enki, however, devised a way to get around the oath; he talked to a wall, or a reed screen that concealed his human servant. So Enki addressed the Sumerian, "Noah," Ziusudra, or the Akkadian, "Noah," Utnapishtim, or the Babylonian, "Noah," Atra Hasis, through a reed screen. In an Akkadian version of the Flood contained in the "Epic of Gilgamesh," Enki summoned Utnapishtim, the ruler of Shuruppak, an antediluvian city:

> Man of Shuruppak, son of Ubar-Tutu
> Tear down the house, a ship!
> Give up possessions, seek thou life!
> Forswear belongings, keep soul alive!
> Aboard ship take thou the seed of all living things;
> That ship thou shalt build -
> Her dimensions shall be to measure.[15]

Unlike the Biblical Noah, people did not ridicule and make fun of Utnapishtim while he was building the ship. At Enki's advice, Utnapishtim told the curious that he was no longer allowed to live in the land of Enlil and was building a ship to sail the Lower World to dwell with his Lord Enki (Sitchin believes that Enki's Lower World was in Southern Africa). Not only did people not make fun of him, but many helped him as he served them meat and wine.

On the seventh day, the ship was completed. With difficulty he

The Anunnaki, Humankind and the Great Deluge

launched the ship into the Euphrates River. Utnapishtim put his family and kin on the ship plus "whatever I had of all the living creatures." He also put aboard the craftsmen who helped him build the ship, as well as a boatman.

According to the Akkadian text, Enki had instructed Utnapishtim to build a boat that was to be roofed over and below and sealed with pitch. There were to be no decks and no openings whatsoever in the boat. Sitchin believes that the boat was a submarine.[16] Boulay, after examining Sumerian and Akkadian texts, concludes the boat was saucer shaped.[17] In any case, we can be certain that the boat was not the huge wooden boat envisioned by European medieval followers of the Genesis account of the Flood.

Utnapishtim, himself, waited outside the ship to look for a signal that Enki had instructed him to watch for:

> When Shamash
> who orders a trembling at dusk
> will shower down a rain of eruptions
> board thou the ship,
> batten up the entrance![18]

Sitchin has established that Shamash, one of the gods, was in charge of the antediluvian spaceport of the Anunnaki at the city of Sippur. Sippur was located a little over one hundred miles north of Shuruppak where Utnapishtim lived. There is no doubt in Sitchin's mind that Enki told Utnapishtim to watch for space-ship launchings at Sippur. The launchings were to take place at dusk so Sitchin has no doubt that the launchings of rocket ships could be seen in Shuruppak with their "rain of eruptions" that would "shower down."[19]

Why were rocket ships being launched? So that the "Great" Anunnaki could escape the catastrophic Deluge. They orbited the Earth in space shuttles while humankind was destroyed - almost.

When Utnapishtim saw the light of the rocket ships and felt the trembling, he boarded the ship and "battened down the whole ship" and turned it over to the boatman he had prudently slipped aboard. Sitchin quotes evidence that Enki always planned that Ziusudra - Utnapishtim - Atra-Hasis should navigate the boat toward Armenia - toward Mount Ararat which the Mesopotamians called Mount Salvation.[20] Otherwise, why would Utnapishtim have carried a boatman?

The storm came at dawn from the south and it must have been awesome. There was thunder, huge black clouds, darkness and "the wide land was shattered like a pot." The storm lasted for six days and nights, according to the Akkadian records,

> Gathering speed as it blew,
> submerging the mountains,
> overtaking the people like a battle....
> When the seventh day arrived,
> the flood-carrying south-storm
> subsided in the battle
> which it had fought like an army.
> The sea grew quiet,
> the tempest was still,
> the flood ceased.
> I looked at the weather.
> Stillness had set in.
> And all of Mankind had returned to clay.[21]

Soon after the storm had subsided and Utnapishtim opened the hatch of the boat:

> There emerged a mountain region;
> on the Mount of Salvation the ship came to a halt;
> Mount Nisir ("salvation") held the ship fast,
> allowing no motion.[22]

For six days, Utnapishtim watched from Mount Salvation (Mt. Ararat). Finally, after he had released a raven that did not return (It had found a resting place), he released all the other animals with him.

Meanwhile, the Anunnaki watched the destruction of humankind from their orbiting space ships. They were all apparently crowded into a few space ships for the duration of the Flood where the accommodations were not the best. The Mesopotamian texts report that they did feel some grief and compassion for humankind. For example, Ninhursag, known to the Mesopotamians as the Mother Goddess because she had given birth to the first *Adamu*, the first "primitive worker":

> ...saw and she wept...
> her lips were covered with feverishness...
> "My creatures have become like flies -
> they filled the rivers like dragonflies,
> their fatherhood was taken by the rolling sea."[23]

Elsewhere:

> The Anunnaki, great gods,

> were sitting in thirst, in hunger....
> Ninti wept and spent her emotion;
> she wept and eased her feelings.
> The gods wept with her for the land.
> She was overcome with grief,
> she thirsted for beer.
> Where she sat, the gods sat weeping;
> crouching like sheep at a trough.
> Their lips were feverish of thirst,
> they were suffering cramp from hunger.[24]

Following the argument that the Anunnaki were lizzies, humanoids with much reptilian genes, this might be a case where these creatures did show compassion for humankind, as it was being destroyed. There is no record of a human accompanying the gods to their Earth-orbiting sojourn, so the events in the space ships must have been told to the human survivors and the offspring of Atra-Hasis - Utnapishtim and his friends he had with him on the ship; so perhaps the grief expressed by the Anunnaki in the Mesopotamian texts was exaggerated. After all, following Enlil, they had all taken an oath not to tell humankind of the impending disaster, so that humankind would be "eliminated." Perhaps too much should not be attributed to this apparent show of compassion shown by the Anunnaki for the plight of humankind.

In any case, it seems clear that they were thirsty (for beer?) and hungry. Apparently stuffing all the Anunnaki into space ships in order to avoid the Flood, taxed their immediate resources.

But, after six days Utnapishtim emerged from his ship on Mt. Salvation, built an altar, and offered a sacrifice. The hungry Anunnaki were soon at Utnapishtim's side:

> the gods smelled the savor,
> the gods smelled the sweet savor,
> The gods crowded like flies about a sacrificer.[25]

Enlil finally arrived at the scene and became furious when he saw that some humans had survived the Flood and wanted to know how this had happened. Ninurtu, Enlil's son, pointed his finger at Enki. "Who, other than Ea, can devise plans? It is Ea alone who knows every matter."[26] Enki (Ea) denied that he had broken the oath; he had allowed only one "exceedingly wise" human to find out for himself what the gods' secret was. Furthermore, it was pointed out to Enlil that the gods would have a harder time existing on the Earth without humans; had not Utnapishtim given them food when they were hungry? Enki argued that a human that is so "exceedingly wise" as Utnapishtim deserves the attention and consideration of the gods. Finally, Enlil's anger abated as he was influenced by Enki's argument. The "Epic of Gilgamesh" relates that Enlil finally blessed Utnapishtim and his wife, and that, ultimately, Utnapishtim and his wife were taken above and granted eternal life, like the gods.[27]

Sitchin believes the melting of the Antarctic ice cap, the ending of the last ice age, and the passing of the 12th planet close to Earth, were major factors in causing this Flood. Sitchin quotes Dr. John T. Hollin of the University of Maine who contended that the Antarctic ice sheet periodically breaks loose and slips into the sea causing an enormous tidal wave.[28] According to Hollin, as the Antarctic ice sheet becomes thicker, the pressure and friction of the ice cap causes it to develop a slippery, slushy layer at the base of the enormous ice cap. This slushy layer acts as a lubricant and allows a huge portion of the ice cap to slip into the ocean. Hollin calculated that if only half of the present ice sheet were to slip into the ocean, the tidal wave that would follow would raise the sea-level around the globe about sixty meters, more than enough to inundate all coastal cites and lowlands of the world.

In this regard, it is interesting to note what scientists have been discovering about the Antarctic ice cap in recent years. Traditionally the Antarctic ice cap has been regarded by Earth scientists as a stable feature that has persisted for up to fourteen million years. Recent research, however, suggests that the ice cap has passed through unstable periods when large parts of it disappeared into the sea and then reformed. For example, during a project that drilled through the ice sheet, scientists found fossils of a tiny one-celled, sea dwelling algae that evolved two million years ago. The presence of this fossil under the ice, which is generally about one mile thick, indicates that at the time the algae lived, open ocean existed in West Antarctica. Reed Scherer of Ohio State University reports that the West Antarctica ice sheet has formed and collapsed several times.[29]

The ice sheet in the eastern part of Antarctica is much larger and seems more stable than the ice sheet in West Antarctica. However, fossils found in the 1980's indicate that trees and lakes existed in eastern Antarctica as recently as three million years ago.

Recently, scientists have discovered a volcano under the West Antarctica ice cap, which when it erupts, might possibly cause some of the ice sheet to melt.[30] It seems possible that the Antarctica ice sheet, once considered very stable, but recently shown by Earth scientists to be unstable, may have had quite a lot to do with past catastrophic floods on the Earth. Geologists calculate that if the Antarctic ice sheet were to melt, rise in sea levels would amount to a catastrophic sixty meters (approximately 200 feet).[31]

Sitchin's contention that the destruction of the Antarctic ice cap corresponds to the end of the last ice-age, would certainly seem reasonable. Geologists have gathered much paleoclimatological data, such as the fossil data of Antarctica mentioned above, and determined that there have been several glacial periods during the past few million years. These glacial periods were characterized by huge continental glaciers over one mile (~1.6 km) thick that covered the northern areas of North America and Europe. Sea levels were lower, as much water was "locked up" in the continental glaciers. In addition to being cooler, the world's climate was dryer than at present because so much of the world's water was in the form of ice. The last glacial period, known as the Würm glacial period, began about 75,000 years ago and ended around 10,000 years ago. However, this period was not just one long cold spell with a beginning and then an end, as Sitchin implies. Instead, there were several fluctuations from being very cold to being warmer and then back to cold. After a glacial maximum at around 20,000 years ago, the climate of the Earth fluctuated relatively frequently until around 13,000 years ago, to then return to a very cold period for a couple of thousand years. Around 10,000 years ago the world entered a warm interglacial period in which we live today.[32]

The last and key climatic variable that Sitchin believes had a strong bearing on the occurrence of the Flood is the close passage of the "12th planet", called Nibiru by the Sumerians. According to the Sumerian texts, Nibiru is the home planet of the Anunnaki. To summarize the Sitchin scenario, Nibiru's gravitation as it passed close to the Earth caused a large portion of the Antarctic ice sheet, made unstable by rising temperatures at the end of the last glacial period, to slide into the ocean. The sudden slipping of the Antarctic ice sheet into the ocean was the major factor, Sitchin believes, in a world-wide flood.

As mentioned, Sitchin prefers a date of about 11,000 BC. or 13,000 years ago as the date of the Deluge. Boulay, who doesn't worry about any correlation between the Deluge and the "12th planet" argues for a date of around 4,000 BC., only a few hundred years before the beginnings of the Sumerian civilization.[33] And, there are probably about as many estimates of the date of the Deluge as there are people who have tried to estimate it.

Some light can be shed on the Great Flood from an esoteric source we have already discussed. The "Swiss Pleiadians" told Billy Meier in 1975 that the Deluge took place 10,079 years ago, which places the date at 8,104 BC.³⁴ This, as we've seen, puts the occurrence of the Deluge right at the end of the Würm glacial period, as determined recently by Earth scientists. The "Swiss Pleiadians" told Meier that the Flood was caused by a comet they called the "Destroyer," which has an orbit in this solar system of 575 1/2 years. The Pleiadians said the comet displaced the Earth from its orbit and changed the period of revolution of the Earth around the Sun to more than forty hours. For some time the Sun did not rise in the East as it does now. The "Swiss Pleiadians" stated that the passing of this comet near the Earth at that time, 8,104 BC., very nearly destroyed the Earth.

Our reputed galactic cousins claimed that the "Destroyer" comet has passed close to the Earth twice since that date, 3,453 years before 1975, or 1,478 BC., and in AD. 1680, but not as close as in 8,104 BC. Consequently, the damage done was not nearly as great, especially in 1680. However, the date 1,478 BC. is close to the estimated date of the end of the Middle Kingdom of Egypt, where there is some geological and archeological evidence of major catastrophes, such as earthquakes, volcanoes and floodings of coastal cities and bizarre "behavior" of the Sun.³⁵ The exact dates of ancient Egyptian history are in question, and there are reputed to be many chronologies of ancient Egypt as there are Egyptologists.³⁶ Thus, the information provided by the "Swiss Pleiadians" correlates with Sitchin's scenario, except that he named the wrong "heavenly body," and his estimated date was too old by about 3,000 years.

While the death and destruction of such a world-wide Flood in coastal and lowland areas would have been almost unimaginable, paleontological, anthropological and geological evidence does not suggest that flood waters were so high as to cover every mountain top. Most species of animals and plants and many human populations did not suddenly become extinct around 8,000 BC. People, animals and plants who lived in areas of high altitudes were able to survive the Deluge, and of course, Ziusudra/Utnapishtim/Atra-Hasis/Noah and other people that may have been warned and were able to build a ship and/or escape to high ground.

We will discuss the effects of the Great Flood more as we cover the "domestication" of plants and animals, the beginnings of human food production, in the next chapter.

---

¹ Woolley, C. L. 1929. *op. cit.*
² Sitchin, Z. 1990. *op. cit.*, p. 204.
³ Lambert, W. G. and A. R. Millard 1969. *Atra-Hasis, the Babylonian Story of the Flood with the Sumerian Flood Story* by M. Civil. Claredon Press,

Oxford.
[4] Sitchin, Z. 1976. *op. cit.*, p. 390.
[5] *Atra-Hasis, op. cit.*, quoted by Sitchin, 1976, p. 390.
[6] *Atra-Hasis, op. cit.*, quoted by Sitchin, 1976, p. 391.
[7] *Ibid.*
[8] Sitchin, Z. 1976. *op. cit.*, p. 392.
[9] *Ibid.*, p. 394.
[10] Boulay, R. A. 1990. *op. cit.*
[11] Marciniak, B. 1992a. *op. cit*
[12] Sitchin, Z. 1976. *op. cit.*, p. 394.
[13] *Atra-Hasis, op. cit.*, quoted by Sitchin 1976, p. 394.
[14] *Ibid.*, p. 395.
[15] Akkadian "Epic of Gilgamesh", quoted by Sitchin 1976, p. 381.
[16] *Ibid.*, p. 396.
[17] Boulay, R. A. 1990. *op. cit.*, pp. 195-201.
[18] Akkadian "Epic of Gilgamesh", quoted by Sitchin, 1976, p. 397.
[19] *Ibid.*
[20] Sitchin, Z. 1976. *op. cit.*, pp. 399-400.
[21] Akkadian "Epic of Gilgamesh", quoted by Sitchin, 1976, p. 383.
[22] *Ibid.*, p. 384.
[23] *Atra-Hasis, op. cit.*, quoted by Sitchin 1976, p. 398.
[24] *Ibid.*
[25] Akkadian "Epic of Gilgamesh", quoted by Sitchin, 1976, p. 384.
[26] *Ibid.*
[27] *Ibid.*, pp. 384-385.
[28] Sitchin, Z. 1976. *op. cit.*, p. 402.
[29] Monastersky, R. 1993. "Fire beneath the ice", Science News, *143*, p. 107.
[30] *Ibid.*, pp. 104-107.
[31] *Ibid.*, p. 107.
[32] Nelson, H. and R. Jurmain 1988. *Introduction to Physical Anthropology.* West Publishing Co., St. Paul, Minnesota, pp. 477-479.
[33] Boulay, R. A. 1990. *op. cit.*, pp. 167-168.
[34] Stevens, W., 1988. *op. cit.*, p.61.
[35] Schellhorn, C. C. 1990. *Extraterrestrials in Biblical Prophecy.* Horus House Press, Inc., Madison, Wisconsin. p. 294-296.
[36] *The Columbia Encyclopedia* 1950. Columbia University Press, Morningside Heights, New York. p. 596.

> *"The glory of the farmer is that, in the division of labors, it is his part to create. All trade rests at last on his primitive activity."*
> Emerson, *Society and Solitude*

Chapter 7

The Advent of Human Food Production

We now approach the beginnings of historical human civilization. As we've seen, Darwinism contends that the origin of modern humans is through chance circumstances that characterize the origin of life and subsequent evolutionary processes. This paradigm usually contends either that we are the only intelligent life form in this universe, or that other intelligent life-forms are located at such a great distance from us that we have been unable to make contact thus far. This paradigm also denies that there is any type of existence other than the three-dimensional, physical existence we can detect with our five senses. So, the archeologists contend that humans, after thousands of years of leading a nomadic existence eating wild plants and wild animals, domesticated some plants and animals, became farmers and herdsmen, and founded elaborate, sophisticated civilizations entirely on their own. Furthermore, archeologists contend that humans did this at least twice, if not three times in the past 10,000 to 12,000 years.

Archeological data indicates that there were three centers of "domestication", the Middle East and the Far East in the Old World, and at least one in the New World. In the Middle East, sheep, goats, cattle, and pigs were "domesticated," along with wheat, barley and legumes. In the Far East, wheat, millet, rice and yams were "domesticated" along with chickens, pigs and water buffalo. In the New World "domesticated" animals did not play a role, but maize, beans, squash, peppers, tomatoes and potatoes were "domesticated," I put quotes around "domesticated" indicating that I do not believe the archeological myth that many wild animals and wild plants were in fact domesti-

The Advent of Human Food Production

cated by humans. Some may, in fact, have been domesticated by humans but, as Sumerian and other records indicate, initially these plants and animals were genetically manipulated by ETs and given to humans.

Archeologists realize that they have a tremendous task in trying to explain why humans rather suddenly gave up their nomadic hunting and gathering lifestyle for a life of farming. To begin with, an efficient system of food production had to be in place in order to feed the thousands of people who would live in the towns and cities of the civilization. The Sumerian and other texts of ancient civilizations tell us that the "gods" gave humans the capability of farming and animal husbandry, and thus the foundation of civilizations. We'll investigate this story in some detail in the second part of this chapter, but, now I would like to give a brief account of the current scientific view of the advent of food production, that of the archeologists. Von Däniken, Sitchin and others who have presented the extraterrestrial hypotheses to account for the beginnings of human civilization, have not presented the details of the archeological case for this remarkable event. As a former biological anthropologist, I have given a brief account of the view of my former colleagues concerning the biological advent of modern humans, and I want to do the same for archeologists, as archeology is a major sub-division of modern anthropology, I have known several archeologists and have become good friends with some of them. The archeologists I have known are hard-working, very congenial people, and, generally, have high integrity, although, I feel their views on the subject of the rise of civilization are quite mistaken. Nevertheless, their views and the data they've accumulated should be given some review and treated with respect.

Studies of living groups of people who lead, or have recently lead, a hunting and gathering lifestyle have shown that hunters and gatherers, for the most part, lead a good life. They generally have plenty of food and they do not have to work long hours to get it. In fact, they generally spend considerably fewer hours than farmers procuring enough food to live on. Furthermore, the food that hunters and gatherers do get is often of greater nutritional value than the food eaten by farmers. Of course, the amount of food available world-wide to a hunting and gathering way of life is much less than food provided by farmers and herders. If there were no farming and herding in the world, there would be far fewer people and no civilizations. There would be only a few million people in the world, instead of the billions of people that populate the world today. Furthermore, when people began to shift from eating the wide variety of wild plants and concentrated on one or a few farming crops, they became vulnerable to crop failure, famine and starvation. When a drought or some other ecological disaster struck an area where hunters and gatherers lived, they generally had a wide variety of "second-choice" or "third-choice" foods to fall back on, and would not suffer from famine and starvation.

In this regard archeologist, Kent Flannery, believes that maize was "domesticated" from teosinte, a wild grass found in many parts of highland Mexico. Not all archeologists are sure that maize was "domesticated" from teosinte and classify teosinite as a "third-choice" food. Flannery considers the disadvantages of farming in the first place. He states:

> ...(a) farming may be more work than hunting, judging by the available ethnographic data, and (result in) (b) an unstable man-modified ecosystem with low diversity index results. Since early farming represents a decision to work harder and eat more "third-choice" food, I suspect that people did it because they felt they had to, not because they wanted to farm. Why they felt they had to, we may never know, in spite of the fact that their decision reshaped all the rest of human history.[1]

Archeological literature is full of such statements of puzzlement. Archeologists Dexter Perkins and Patricia Daly state that the changes from a hunting and gathering subsistence to an agrarian subsistence,

> ...took place in such a relatively short time and were of such fundamental importance that the question immediately arises: Why did they occur? Why did a lifestyle that had been so successful for tens of thousands of years give way to one so different. Modern hunting peoples, though living for the most part in marginal areas, are frequently better nourished and always more leisured than their agrarian neighbors... Agrarian peoples not only must work a good deal harder for their sustenance, but are much more precariously balanced in relation to their environment, since they have substantially altered the natural ecology of their surroundings...[2]

The Advent of Human Food Production

Perkins and Daly state that even if archeologists do not understand the causes for the adoption of agrarian subsistence, "...it is necessary for us to at least speculate about possible causes for the change, which is in many ways so extraordinary."[3] Archeologists C. Lamberg-Karlovsky and Jeremy Sabloff point out that in the Near East,

> The evidence is clear that following 8,000 B.C., several villages had developed simple but effective means of food production. Just what the adaptive mechanisms were which led ever increasing numbers of people to adopt the agriculture mode of subsistence remains one of the most intriguing archeological questions...[4]

Lamberg-Karlovsky and Sabloff point out that archeology as a science is less than two hundred years old, and serious investigations of the rise of agriculture have taken place only since World War II. With regard to the rise of agriculture in Mesoamerica, the area from the isthmus between North America and South America and Central Mexico, these same two archeologists state that,

> (u)nfortunately, most attention has been paid to the highly visible remains of the great civilizations of pre-Columbian Mesoamerica. Little attention has been given to the less spectacular remains of early hunters and gatherers or settled villagers who lived in the same area prior to 1,200 B.C. and provided the cultural foundation for later developments. Thus, in contrast to the well-publicized civilizations of the Maya or the Aztec, our knowledge of the rise of agriculture and settled village life remains in relative poverty...[5]

Twelve hundred B.C. is the date of the first known New World civilization, the Olmec civilization. In every area where a civilization arose in the New World there is a relative poverty of data and knowledge of the rise of agriculture. Nevertheless, archeologists do have some data concerning this subject, and theories of how this revolutionary change in lifestyle occurred in humans.

Before the late 1940s, it was assumed that agriculture had its beginnings in the lowland areas of the fertile crescent, a crescent-shaped area from Israel to southern Iraq and Iran, where the first known civilizations existed. But, starting in the late 1940's, archeologists discovered that the first evidence of agriculture was in the highlands of the fertile crescent. Most people know that farming is more difficult in the highland areas than in the more fertile, warmer lowlands. There is some evidence that food production had a few scant beginnings at about 10,000 B.C. in the highlands in and around the Middle East, those of Israel, Lebanon, Turkey, Iraq and Iran. By 8,000 B.C. several villages of this area had a simple but effective means of food production.

A similar phenomenon is observed in Mesoamerica, as the only evidence of any agricultural activities, or for that matter, any human activities, at around 8,000 B.C. to 7,000 B.C. and for the next several thousand years, is in the highlands of Mexico and Guatemala.[6] Much later in both the Near East and in Mesoamerica, agricultural peoples began to move to the lowlands.

Interestingly, there is little evidence of human activity of any kind, at about this time from the Near East. Perkins and Daly, while discussing animal domestication, state that "...there are few data from this period, roughly spanning 2,000 years, from 9,000 to 7,000 B.C."[7]

The "domestication" of grains involved the doubling of the number of chromosomes, a process that makes the plant extremely edible and useful to humans.[8] Wild cereals that still exist in the Near East, can hardly survive outside their natural habitats, even with human care.[9] The domestic varieties, however, can thrive in several habitats with intensive human care. All of the wild grains that were ancestors to "domesticated" grains, like einkorn wheat, emmer wheat, and barley are still quite widespread in the Near East. The wild forms exhibit an important difference from their "domesticated" offspring: the seed-bearing portion of the plant, or spike, of the wild form is brittle, and the spike of the "domesticated" forms is quite tough. The brittle spikes allow natural seed dispersion, while the tough-spiked "domestic" grain allows the seed to remain with the plant when it is harvested. The advantages of this are obvious. It is also obvious that the "domesticated" grains are completely dependent on humans for their survival.[10] Several other "domesticated" plants are similarly forever tied to humans because they have lost their independent power of seed dispersal and germination, including maize, bananas, and dates.[11]

Archeologists have provided several theories over the years to explain how humans "domesticated" plants and animals three times in three different locations in the past 10,000 to 12,000 years. These theories either involve conscious manipulation of animals and plants by humans, and/or environmental pressures, such as climatic desiccation, and/or human population pressure. There has been little data brought forth to demonstrate a major climatic desiccation in this period, so theories that depend on this factor have generally been discarded. Probably the most widely accepted theory today is proposed by Kent Flannery, whom I quoted above. Flannery's theory involves a crucial assumption: that hunting and gathering populations were increasing prior to food production. With increasing population pressure, some groups of hunters and gatherers were forced to move to the highlands and make a living in these marginal areas, Flannery contends. In order to "make ends meet" in these marginal areas, according to Flannery's theory, the human populations of the highlands would be forced to cultivate certain plants in order to make them more plentiful.[12] In fact, as we have seen, archeological data indicate that there was little human population and human activity of any kind in the lowlands of the Near East just prior to the first evidence of farming and herding.

Archeologist Robert Braidwood believes the beginnings of food production occurred when hunters and gatherers first began to consciously manipulate plants and animals. For example, Dr. Braidwood and other archeologists have speculated that some groups began to remove some plant species and replant them near their settlements where they could have easier access to them. Other groups began to selectively herd and hunt certain animal species, such as killing animals that were a certain species, and killing animals that were a certain age and/or a certain sex. This gradually led to "domestication".[13]

Archeologist Richard MacNeish, who has done extensive field work in the New World, including the Tehuacan Valley in Mexico, also has speculated about the beginnings of food production by humans. In the high Tehuacan Valley, MacNeish has distinguished several cultural phases, starting about 10,000 B.C., that demonstrate, he feels, how people gradually "domesticated" plants and settled in villages. MacNeish believes that the hunters and gatherers of the Tehuacan Valley became very knowledgeable of all the plants and animals they ate and adjusted their nomadic movement to take maximum advantage of the available wild plant food, as well as the wild animals they hunted. Year after year they returned to the same areas at about the same time of the year. As they consistently cleared the areas around their camps each year, some of the plants they exploited developed genetic mutations which made them more cultivable and manageable. Due to these accidental and "natural" mutations that occurred, people gradually developed agriculture. These gradual changes eventually led to the establishment of permanent villages in which the people lived year-round.

It is interesting to note that by the time the people had supposedly "domesticated" enough plants to provide about forty percent of their diets and allow them to live in permanent villages in the Tehuacan Valley (in what MacNeish calls the Ajalpan Phase [1,500BC - 850 BC]), the full-blown civilization of the Olmecs had been established in the Gulf coast lowlands of Mexico.[14]

The idea of accidental mutations, figures in Dr. Flannery's theory of the "domestication" of plants as well. He believes that accidental genetic mutations in a few plants started humans on the way to "domestication."[15]

Archeologist David Rindos believes that "domestication" is the result of a slow natural evolutionary process over time. He points out that hundreds of species of plants have been "domesticated" around the world, but only about twenty species make-up most of the plant food used by humankind today. He states that "remarkably, no major crop has been domesticated from the wild since the early days of agriculture."[16] Rindos says it is even difficult to improve existing crops, as often breeding crops increases their susceptibility to diseases and pests that were previously unknown or considered unimportant.

The Sumerian Story of the Beginnings of Agriculture

Let us make clear, once again, that we do not believe the ancient Sumerian and other Mesopotamian stories are "absolutely true" history. These stories that have come to us through thousands of years of oral tradition and writing are bound to be somewhat distorted - probably in some cases deliberately distorted by the Anunnaki. But, I feel these ancient stories are probably as close as we'll come to the truth today and Zecharia Sitchin's translations and interpretations of these ancient fragmentary texts are, in my opinion, the best available.

Sitchin presents the Sumerian story of the beginnings of agriculture by humankind in his book, *The Wars of Gods and Men*.[17] The story begins more than 400,000 years ago, when the Annunaki first came to Earth. Sitchin quotes the Sumerian text entitled by scholars, "The Myth of Cattle and Grain":

> When from the heights of Heaven to Earth
> Anu had caused the Anunnaki to come forth,
> Grains had not yet been brought forth,
> had not yet vegetated...
> There was no ewe,
> a lamb had not yet been dropped;

The Advent of Human Food Production

> There was no she-goat,
> a kid had not yet been dropped...[18]

So there were no domesticated grains and sheep and goats in existence at that time, but it wasn't long before the "gods" created something for themselves to eat:

> In those days,
> in the Creation Chamber of the gods,
> in the House of Fashioning, in the Pure Mound,
> Lahar (woolly cattle) and Anshan (grains) were
> beautifully fashioned.
> The abode was filled with food for the gods.
> Of the multiplying of Lahar and Anshan
> the Anunnaki, in their Holy Mound, eat -
> but were not satiated.
> The good milk from the sheepfold
> the Anunnaki, in their Holy Mound, drink -
> but are not satiated.[19]

Soon after the Primitive Workers (humans) had been "created" themselves, they were placed in E.DIN, the abode of the gods (the location of the later Sumerian civilization) to farm the grains and tend the "four-legged animals" in order that the gods might be satiated. The grains we know, however, did not yet exist. This same text lists the grains that had not yet been brought forth:

> That which by planting multiplies,
> had not yet been fashioned;
> Terraces had not yet been set up...
> The triple grain of thirty days did not exist;
> The triple grain of forty days did not exist;
> The small grain, the grain of the mountains,
> The grain of the pure A.DAM, did not exist...
> Tuber-vegetables of the field had not yet come
> forth.[20]

These grains and the tuber-vegetables were given to humankind after the Deluge, the Great Flood that destroyed most of humankind. Enlil, the chief administrator for the affairs of the Anunnaki, we recall, was at first furious that any humans at all survived the Deluge, as he had intended to allow all humans to be destroyed by the catastrophe. However, he soon saw the logic of Enki's argument that the

The Advent of Human Food Production

Anunnaki would have a much more difficult time living on Earth without humans to serve as a working resource. He then wished the human survivors of the Deluge to flourish and re-populate the Earth, at least in the Near East. It was then, after the Deluge, that he and the other Anunnaki gave humankind the grains referred to in the above passage. All the domesticated plants had been destroyed in the Deluge, but the Anunnaki had sent some seed to Nibiru, the "12th planet", before the Deluge. So these were now sent back to Enlil. So, in fact, agriculture was resumed on Earth after the Deluge, not initiated.

Enlil looked for a place suitable for the resumption of agriculture. Apparently, the lowlands were still covered with water and mud from the Deluge, so he picked the highlands in and around the Near East. A fragmented Sumerian tablet reported by Sumerian scholar, Samuel Noah Kramer in *Sumerische Literarische Texte aus Nippur* and quoted and translated from German by Sitchin, states that Enlil went to "the mountain of aromatic cedars."

> Enlil went up the peak and lifted his eyes;
> He looked down: there the waters filled as a sea.
> He looked up: there was the mountain of the aromatic cedars.
> He hauled up the barley, terraced it on the mountain.
> That which vegetates he hauled up,
> terraced the grain cereals on the mountain.[21]

Sitchin makes a solid case that the "Cedar Mountain" referred to is in Lebanon. He states that throughout the Near East there is only one unique Cedar Mountain of fame, and it is in Lebanon.

Sitchin then quotes a text that asserts Ninurta, the son of Enlil, drains the water of the lowlands and makes them habitable once again. Then the Anunnaki, "...from the mountain the cereal grain they brought down," and "the Land (Sumer) with wheat and barley did become acquainted."[22] The Sumerians revered Ninurta as the god that taught humankind farming; a Sumerian cylinder seal depicts him giving the plow to humankind. (Figure 7-1) The Akkadians whose empire succeeded the Sumerian civilization at about 2,300 B.C., called Ninurta, Urash, or "The One of the Plough."[23]

The grains that Enlil sowed on the Cedar Mountain of Lebanon (and also, archeological evidence indicates, in other highland areas in Turkey, Iran and Iraq) were not the grains that the historical people of Sumer used, but "the grain that multiplies", the grains with doubled, tripled and quadrupled chromosomes. These were created by Enki

The Advent of Human Food Production

with Enlil's consent:

> At that time Enki spoke to Enlil:
> "Father Enlil, flocks and grains
> have made joyful the Holy Mound,
> have greatly multiplied in the Holy Mound.
> "Let us, Enki and Enlil, command:
> The woolly-creature and grain-that-multiplies
> let us cause to come out of the Holy Mound."[24]

Enlil agreed and we read in a Sumerian text that abundance followed:

> The woolly-creature they placed in a sheepfold.
> The seeds that sprout they give to the mother,
> for the grains they establish a place.
> To the workmen they give the plough and the yoke...
> The shepherd makes abundance in the sheepfold;
> The young women sprouting abundance brings;
> She lifts her head in the field:
> Abundance had come from heaven.
> The woolly-creature and grains that are planted
> came forth in splendor.
> Abundance was given to the congregated people.[25]

In addition to the woolly-creature (sheep) and grains-that-multiply, Enki is credited with bringing into existence "the larger living creatures", domesticated cattle, which in addition to providing food for humans and the gods, served to pull the plow, which apparently was done by humans when the plow was first granted to them.[26]

Sitchin argues that Enki was responsible for draining and cleaning up the Nile Valley and preparing it for the great Egyptian civilization. Sitchin presents evidence that Enki and the Egyptian god Ptah were one and the same entity. The ancient Egyptians credit Ptah with reclaiming Egypt from the inundating waters.[27]

Some Correlations Between Archeological Data and the Sumerian Story of the Advent of Human Food Production

There is some archeological data that correlates with the various Sumerian texts that relate the story of the beginnings of human agriculture. These data by no means confirm that there was a Great Flood and that, afterwards, ETs provided humans with the means to practice agriculture, but the data are suggestive.

The Advent of Human Food Production

Let me make clear, first of all, that nowhere in the archeological literature will there be found any reference to the Great Flood, or to an advanced civilization from beyond this planet having influence on the rise of agriculture on Earth. Archeologists, as we have seen, and all academic scientists of the Western world (at least), believe that humans "pulled themselves up by their own sandal straps" and "domesticated" plants and animals all by themselves - three separate times in at least three different areas of the world, no less. In fact, archeologists, since Erich von Däniken's books became so popular in the late 1960s and 1970s, often go out of their way to state specifically that there was no outside influence in the rise of agriculture and civilization on Earth. The following, by English archeologist Colin Renfrew, is fairly typical. Renfrew is writing in reference to the huge stone monuments that appeared in the Old World from Japan to England prior to the advent of writing:

> In every part of the world where prehistoric monuments are found, these same questions were asked by the early antiquaries. Initially the answers given were often - to the Twentieth Century eye - exceedingly fanciful, the earlier counterpart of the modern notions of flying saucers and alien intelligence, with which the credulous are deceived and the unscrupulous enriched at the present day. So it was that until the middle of the last century, talk of Druids and of lost continents was taken very seriously.[28]

Furthermore, archeologists almost always dismiss anything written by the people of ancient civilizations about their own history as myths and fables which are not credible. This would include all Sumerian texts referring to the Deluge and the beginning of agriculture quoted above.

The first bit of archeological data concerning the rise of agriculture that correlates with the Sumerian story is the fact that agriculture began - at least after the Deluge - in the high country of the Near East. The Sumerian texts state that this was because the lowlands were uninhabitable because they were still wet and muddy. Archeologists, on the other hand, claim that this was because people living in these marginal areas had to do something so that they could survive, so they invented agriculture. The earliest date at which the lowlands in the Tigris - Euphrates Valley was occupied by farming people is 6,000 B.C.[29] But this was in the northern, higher portion of this valley

The Advent of Human Food Production

that was probably somewhat less affected by the Deluge than the southern portion of this valley where the Sumerian civilization was established in the Fourth Millennium B.C. The first culture known in the southern part of the Tigris - Euphrates lowlands, known as the Ubaid culture, was only established at about 4,300 B.C., a thousand years, or less, before the rise of the Sumerian city-states.[30] And, as we shall see in the next chapter, there is considerable doubt among many scholars that the Ubaid culture was the precursor to the Sumerian civilization.

Figure 7-1 Ninurta granting the plow to humans

Another, perhaps more important correlation of archeological data with Sumerian texts has to do with the paucity of archeological data showing any kind of human activity around the time of the Flood, especially in lowland areas. The reader will recall the date of the Deluge is far from concrete, not to mention the fact that a great number of people, including all academic scientists, don't even believe there was a world-wide Flood. Sitchin believes the Flood took place about 13,000 years ago or 11,000 B.C. We saw, however, that paleoclimatologists have data that indicate that the last ice-age, the Würm glacial period, ended at about 10,000 years ago, or 8,000 BC. The "Swiss Pleiadians" told Billy Meier in 1975 that the Great Flood was caused by a destructive comet (not Nibiru) 10,079 years ago, from 1975 or 8,104 B.C. This date is a true calendar date. The dates the archeologists give are radiocarbon dates. In order to calibrate radiocarbon dates with calendar dates, a brief discussion of radiocarbon dating is necessary.

Radiocarbon dates are based upon the proportion of carbon-14, a rare radioactive isotope of normal carbon, carbon-12, in the atmos-

The Advent of Human Food Production

phere. Carbon-14 atoms in the atmosphere combine with oxygen in the same way carbon-12 does to form carbon dioxide (CO2). Plants absorb CO2, including CO2 containing carbon-14 during the process of photosynthesis. Plants, plus animals that eat plants, or animals that eat other animals, end up with the same proportion of carbon-14 as is contained in the atmosphere. But, when a plant or animal dies, they no longer absorb any carbon-14. The amount of carbon-14 contained in the dead organism begins to decrease as the carbon-14 decays. Carbon-14 decays spontaneously, giving off an electron and becoming the stable element nitrogen. So, in order to date something using the carbon-14 method, it must first have been a living organism, meaning it absorbed carbon-14 during its lifetime. Then by carefully measuring the proportion of carbon-14 remaining and comparing it to the amount of the carbon-14 known to now be in the atmosphere, a radiocarbon date of its origin can be established.

Archeologists and other scientists began using the carbon-14 dating technique in the 1950s. It was assumed that the amount of carbon-14 in the atmosphere has remained constant over the years. However, in the 1960s, carbon-14 dates of some ancient trees of the western United States, such as the bristle cone pine, which might live as long as four to five thousand years, were compared to tree-ring dates, and found to be in error. Trees in temperate zones grow one tree-ring per year, so their exact age can be measured. Carbon-14 dates of tree-rings, whose age was known, showed carbon-14 dates to be significantly younger than the actual date after about 1,200 B.C., or a bit more than 3,000 years ago. For instance, a radiocarbon date of 3,000 B.C. is too young by as much as 800 years.[31] The actual date of something dated 3,000 B.C. through the carbon-14 method is actually closer to 3,800 B.C. It is not known for certain, but a date of 7,000 B.C. is most likely close to 8,000 or 8,500 B.C. in calendar years. Keep this in mind as we go back to the archeological record of the advent of food production.

While there is some evidence of "domestic" animals going back to 9,000 B.C. in the highlands of the Near East, archeologists admit that it is difficult to distinguish "domesticated" animals from their wild ancestors, such as sheep, goats and dogs. In fact, determining which animal remains are "domesticates" and which ones are wild animals that were hunted, often involves a lot of guesswork.[32] Unlike animals, domestic grains are easily distinguished from wild grains.[33] Domestic grains are larger and, as mentioned, have a tougher spike that allows the reaping of the grain without it being lost in the field. The earliest domestic grains are not found until around 7,000 B.C., and are not very common until a thousand years later. These data fit well with the evidence that the Great Flood occurred at around 8,000 BC, or close to the carbon-14 date of 7,000 B.C.

As we have seen, archeologists Lamberg-Karlovsky and Sabloff report that in the New World, the only evidence for human activities at around 8,000 BC or 7,000 B.C. and for several thousands of years thereafter, comes from the highlands of Mexico and Guatemala. In fact, there is no archeological evidence that the lowlands of Mesoamerica were occupied at all by humans until the Third Millennium B.C.[34] Perkins and Daly state that in the Near East, there are few archeological data from the period of 9,000 to 7,000 B.C.[35] Following the scenario presented here, the lack of signs of human activity in the lowlands was caused by the Great Flood: the younger carbon-14 dates of around 7,000 B.C. represent the time the Deluge actually occurred, while the period represented by older dates, from around 8,000 BC to 9,000 B.C., represents archeological data that has been covered by silt from the Flood, and not yet uncovered by erosion. If this scenario is correct, there would be a dearth of archeological data in all lowland areas around the world at these dates.

To those who have become convinced that humankind is a chance product of Darwinian evolution, and that humans are the only intelligent life form in the universe, the late 20th century myth of the beginnings of agriculture put forth by the archeologists is most likely adequate. But, for those of us who do not believe this theory (Darwinism) and do believe that the ancient Sumerian and other Mesopotamian texts represent historical texts, and not "pure fable," the archeological theories of the beginnings of agriculture can be considered no more than "pure fable." Let us move on to archeological speculations about the beginnings of human civilizations.

[1] Flannery, K. V. 1973. "The origins of agriculture." Annual Review of Anthropology, 2: pp. 271-310. Quoted in Lamberg-Karlovsky and Sabloff 1979. *op. cit.*, p. 228.

[2] Perkins Jr., D. and P. Daly 1974. "The beginning of food production in the Near East." In: *The Old World: Early Man to the Development of Agriculture*. R. Stigler (ed.), 1974. St. Martin's Press, New York, p. 74.

[3] *Ibid.*

[4] Lamberg-Karlovsky, C. C. and J. A. Sabloff 1979. *op. cit.*, p. 60

[5] *Ibid.*, p. 216.

[6] *Ibid.*, p. 217.

[7] Perkins Jr., D. and P. Daly 1974. *op. cit.*, p. 80.

[8] Stebbins, G. L. 1951. "Cataclysmic evolution." Scientific American, 184: pp. 54-59.

[9] Perkins and Daly 1974. p. 75.

The Advent of Human Food Production

[10] *Ibid.*, p.83
[11] Lamberg-Karlovsky and Sabloff 1979. p. 44
[12] Flannery's theory summarized in Lamberg-Karlovsky and Sabloff 1979, p. 50.
[13] Braidwood's theory summarized in Lamberg-Karlovsky and Sabloff 1979, p. 49.
[14] MacNeish's documentation and model of "domestication" in the Tehuacan valley summarized in Lamberg-Karlovsky and Sabloff 1979, pp. 220-225.
[15] Lamberg-Karlovsky and Sabloff 1979. *op. cit.*, p. 225.
[16] Rindos, D. 1984. *The Origins of Agriculture: An Evolutionary Perspective*. Academic Press, Orlando, Florida, p.284.
[17] Sitchin, Z. 1985. *op. cit.*, pp. 120-128.
[18] *Ibid.*, p. 120.
[19] *Ibid.*
[20] *Ibid.*, p. 121.
[21] *Ibid.*
[22] *Ibid.*, p. 124.
[23] *Ibid.*
[24] *Ibid.*
[25] *Ibid.*, pp.124-125.
[26] *Ibid.*, p. 125.
[27] *Ibid.*, p. 126.
[28] Renfrew, C. 1979. *Before Civilization: The Radiocarbon Revolution and Prehistoric Europe*. Cambridge University Press, Cambridge, p. 6.
[29] Stigler, R. 1974. "The later neolithic in the Near East and the rise of civilization." In: *The Old World: Early Man to the Development of Agriculture*, R. Stigler (ed.), 1974, St. Martins Press, New York, p.110.
[30] *Ibid.*, p. 112.
[31] Renfrew, C. 1979. *op. cit.*, p. 69.
[32] Lamberg-Karlovsky and Sabloff 1979. *op. cit.*, p. 45.
[33] Perkins and Daly 1974. op. cit., p. 83.
[34] See reference number 6, this chapter.
[35] See reference number 7, this chapter.

> #20 Kuan/Contemplation (View)
> "The wind blows over the Earth:
> The image of CONTEMPLATION.
> Thus the kings of old visited the regions
> of the world, Contemplated, And gave them instruction.
> I Ching

Chapter 8

The Rise of Ancient Human Civilizations

The fascinating information that follows will deal with the oldest human civilizations we know of, that of Sumer, Egypt and the Harappan civilization of the Indus River Valley in what is now Pakistan. Also, I will briefly consider the first known civilization of the New World, the Olmec civilization. I will give a brief archeological summary of each. In the books that deal with the possible ET influences in these first civilizations, there are usually statements that refer to the archeologists' surprise that these civilizations sprang up so suddenly, but there is little coverage of the actual archeological data. So again, as I did in the last chapter, I am going to try to give the archeologists their due in a brief summary of their views. We'll see that they are indeed surprised and somewhat puzzled at the quick rise of the first human civilizations, and I will provide several direct quotes of such puzzlement. But, as we'll see, all academic archeologists will say something to the effect, that despite the gaps in their knowledge, evidence points to a steady, if rapid, development of civilizations by humans. As noted in the last chapter, archeologists bristle at any suggestion of outside influence from advance beings from outer space or lost continents. After we cover the archeological myths of the rise of the first human civilizations, we will turn once again to Sitchin who provides the best historical account of the beginnings of human civilizations.

It must not be thought that the very first human civilizations were exceedingly crude and rudimentary compared to later civilizations. Logically, one would think that this would be the case. After all, these are the very first human civilizations. According to scientific lore, all of humankind had been only nomadic hunters and gatherers until a mere 4,000 or so years before the Sumerian and Egyptian civilizations sprang up. As Sitchin points out, the Sumerian civilization was characterized by,

> ...high-rise buildings, streets, market places, granaries, wharves, schools, temples, metallurgy, medicine, surgery, textile making, gourmet foods, agriculture, irrigation; the use of bricks, the invention of the kiln; the first-ever wheel, carts; ships and navigation; international trade; weights and measures; kingship, laws, courts, juries; writing and record keeping; music, musical notes, musical instruments, dance and acrobatics; domestic animals and zoos; warfare, artisanship, prostitution.[1]

The ancient Sumerian civilization had much more scientific knowledge of the heavens than many later civilizations. As most school children know, the Europeans of the Dark Ages and Middle Ages thought the world was flat! The ancient Sumerians not only knew that the world was round, but had intimate knowledge of parts of our solar system Middle Age Europeans didn't even know about. As pointed out earlier, the Sumerians knew about Neptune, which was re-discovered by a German astronomer in 1846 A.D. The Sumerians not only knew about Neptune, but they also knew that it was watery, something our astronomers didn't discover until 1989![2]

The Beginnings of the Sumerian Civilization

The Sumerian civilization was started in the southern area of the Tigris-Euphrates River Valley in what today is southern Iraq and Kuwait. This area is a strange place for the first human civilization. It is extremely hot in the summer and quite cold in the winter. Rainfall is both unreliable and unpredictable, and it is impossible to have agriculture there without irrigation, although the alluvium deposits make it quite fertile with water. Furthermore, there was no wood or stones for houses and buildings and no ores for metal production.[3] Some historians and archeologists suggest that harsh environments give rise to great cultures through the challenge of living in a harsh environment.[4] My understanding of this hypothesis is that the Sumerians had to develop and did develop a sophisticated civilization in order to live in the harsh conditions of Mesopotamia, the land between two rivers.

As we saw in the last chapter, there is no archeological evidence that people lived in southern Mesopotamia before remains of what is called the Ubaid culture are found, beginning about 4,300 B.C. Based on the study of pottery types and designs, there is no known precursor to the Ubaid culture, including the older cultures of around 6,000 B.C. of northern Mesopotamia. Archeologists often use pottery analyses to trace succeeding cultures. Pottery may change, but pottery of succeeding cultures is usually similar and can be traced to the pottery of older cultures. This is not so with the Ubaid culture. Archeologists have often attributed the origins of Ubaid culture to a new people coming into Mesopotamia, and they speculate that the new people may have come from the area that is today southwestern Iran[5]. Archeologist Robert Stigler reports that there "is a great need for more study in earlier levels of the Neolithic (farming cultures) in southern Mesopotamia, *for we can really say very little about the origins of settlement there.*" (my emphasis).[6] Archeologists Lamberg-Karlovsky and Sabloff say that archeologists have looked all over, but not found the ancestors to the Ubaid culture, and that the problem of the origin of the Ubaid culture remains.[7]

The Ubaid cultural period lasted from 4,300 B.C. to about 3,500 B.C., or just before the Sumerian civilization began to flourish (based on carbon-14 dates). The Ubaid culture consisted of small farming villages with little or no irrigation, no signs of centralized political control, and few characteristics of the later Sumerian sophistication, except temples, which dominated the towns. In fact, it is not believed by many archeologists that the Ubaid culture is a direct precursor of the Sumerian civilization.[8]

But the archeological remains indicate that something dramatic began to happen at about 3,500 B.C. At this time the resources that Mesopotamia lacked, such as wood, ores for metallurgy, and precious stones used for jewelry, were being imported into the land. So the age of internationalism had begun. A new culture, referred to as the Uruk period (3,500 to ~ 3,200 B.C.), suddenly appeared. The population of Sumer dramatically increased during the Uruk Period. Cities appeared, new forms of pottery appeared and writing began at about 3,200 B.C.[9] Shortly thereafter, in what the archeologists call the Proto-literate period (3,100 to ~ 2,900 B.C.), writing developed progressively, as did metallurgy, art and the bureaucracy of the government. Wheels on wagons appear, and the potter's wheel and all of the other sophisticated features of the Sumerian civilization listed above appear.

The sudden rise of the Sumerian civilization has been debated by archeologists and other scholars for years. This phenomenon was and still is, to a more limited extent, called the "Sumerian Problem." [10] E. A. Speiser tells us that early Sumerian scholars, in the face of the apparent cultural gap, often proposed that the first Sumerians to arrive in Sumer came from elsewhere, possibly by sea from a considerable distance.[11] Stigler, an archeologist, states that, "....the origins, linguistic and otherwise, of the Sumerians is a most complicated and debated question...alternative explanations involving invasions - migrations and in situ development have been put forth...".[12] Archeologists Lamberg-Karlovsky and Sabloff tell us that a, "minority of scholars have theorized that changes in Southern Mesopotamia, as in Egypt, were so rapid that the invisible hand of an invading culture must have been responsible..." [13]. I could give several more quotes from scholars that express questions and puzzlement (at least) of the sudden rise of the world's first known civilization. But modern day archeologists assure us that, in fact, the Sumerians established this astonishing civilization themselves, and that the scanty data for such still supports this contention. For example, Lamberg-Karlovsky and Sabloff state that a "...close examination reveals evidence for direct continuity in social organization: there was no real break in architectural or technological traditions from the earliest Ubaid times." [14]

The principal cities of Sumer were Ur, Nippur, Uruk, Kish, and Eridu (Figure 3-1). Each Sumerian city was devoted to one or two of the "great" Anunnaki gods. For example, Uruk was devoted to Anu, the chief celestial god of the Sumerians and the father of Enlil, and to E-anna (Isthar to the later Babylonians), the goddess of love. Eridu was the city of Enki.

In each city there were built enormous temples and places for the Anunnaki gods, the most famous types of which are structures known as ziggurats (see Figure 8-1). These temples were believed by the Sumerians to be the actual physical homes of the gods, at least part of the time.[15] It is difficult for archeologists and other scholars to understand what motivated Sumerian citizens to cooperate in the building of these monumental palaces and temples. Lamberg-Karlovsky and Sabloff state that it "is estimated that 1,500 laborers working a ten-hour day would have spent five years building one of the temple terraces at Uruk." [16]

The kings of the cities of Sumer derived their right to rule from the gods and goddesses of those cities. Sumerian "legends" say that kingship was given to humans as a gift of the gods. The kings served as intermediaries between the gods and humans. The cities themselves had populations of no more than 50,000.[17] Under the king were temple priests, and a further vast bureaucracy was under them. Excavations of cemeteries at Ur and other cities indicate that Sumer was hierarchically organized from king to slave, with tremendous differences between the haves and have-nots. There was much militarism, and data indicates that it increased throughout the Third Millennium BC. The people of Sumer did not dwell on philosophical questions, such as the meaning of life and the origins of themselves. They knew they were created by their gods in order to serve them.

The Beginnings of the Egyptian Civilization

If the archeological evidence for the rise of the Sumerian civilization seems sketchy and full of holes, the evidence for the rise of Egyptian civilization in the Nile Valley is even more so. Archeologists blame their lack of knowledge of the rise of Egypt on three things. First, they blame themselves for neglecting the prehistory of Egypt while concentrating on the magnificent remains of the civilization itself. Second, they blame the annual floods of the Nile Valley which have covered up prehistoric remains with layers of silt. Thirdly, they blame the centuries of human habitation in the Nile Valley that must have destroyed many prehistoric remains.[18]

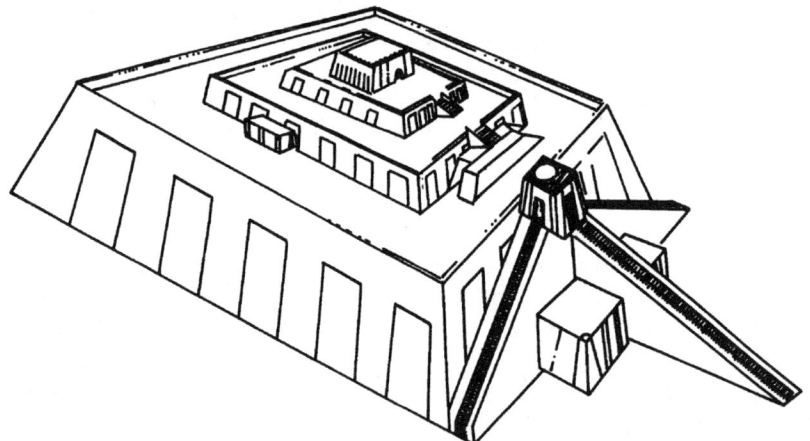

Figure 8-1 An archeologist's view of the enormous ziggurat of Ur. Much of this ziggurat still exists in the ruins of Ur. Drawn after Stigler, 1974.

There is virtually nothing in or near Egypt that indicates that there were people learning about and living with agricultural techniques before 4,000 to ~ 4,500 B.C., barely 1,000 to 1,500 years before the rise of the ancient civilization itself. By this time, as we have seen, people in the Near East had been practicing agriculture and living in villages for 4,000 years. Furthermore, the first farming, or Neolithic cultures, are quite simple until about 3,600 B.C. The first evidence of agricultural peoples in the Nile River Valley came from the Fayum Depression, some twenty miles from the Nile. In this 4,000 BC plus Fayum culture, people grew wheat and barley, herded sheep and goats and raised pigs. None of the "domesticated" plants and animals have wild forebears in Egypt or any place in Africa, so they are presumed to come from the Near East.

No evidence of houses or villages is found in this first farming culture of Egypt. The assumption is that only tents or other quite flimsy shelters were necessary in the mild climate.[19]

The first evidence that people were living on the Nile Valley floor itself is seen in what is called the Amratian culture (3,800 to ~ 3,600 B.C.). This culture, existing about 500 years before the magnificent Egyptian civilization, was still very simple. There is evidence that people lived in houses, but these were simple "wattle-and-daub" or "reed and mud" structures, and the settlements were only about 300 feet in diameter.[20] The simplicity and lateness of the first Egyptian farming (Neolithic) communities is "inexplicable" to archeologists.[21]

Things begin to pick up a bit in Egypt at about 3,600 B.C., as they do, we will recall, at about this same time in Mesopotamia. During this late Predynastic, or Gerzean culture, the entire valley floor was cultivated for the first time. Lapis lazuli from Afghanistan, used for jewelry, and lead and silver from Southwest Asia point to a rather extensive international trade. In addition to the stone tools characteristic of earlier Egyptian cultures, cast copper axes, daggers, and knives are found associated with Gerzean cultures. While no archeological evidence has been found for irrigation and flood control, archeologist Stigler assumes they almost certainly existed during the late Predynastic times.[22] Later dynastic writings indicate that there were two kingdoms in existence at this time, one in upper (southern) Egypt and one in lower (northern) Egypt, although there is no contemporary (Gerzean) archeological data to support this. As is customary, archeologists usually ignore historical records as too mythological to be reliable. Stigler does this and considers it more likely that instead of two kingdoms during this late Predynastic period, there were a series of small chieftainships from upper to lower Egypt. A chieftainship would have a chief in a small town who held political control over the town and surrounding villages.[23] This scheme fits better with anthropological theory for the rise of states or dynasties.

Nevertheless, Dynastic writings credit one man by the name of Menes, or Narmer, who became Pharaoh, for unifying upper and lower Egypt at around 3,200 B.C. or 3,100 B.C., founding the Dynasty of Egypt. The Dynasty is broken up into a number of dynasties by historians; the First, Second, and Third Dynasty are known as the "Archaic" period of Egyptian history, lasting from 3,200 to 2,600 B.C. Hieroglyphic writing appears at the beginning of the "Archaic" period which was to eventually provide historical accounts of this earlier, and later periods. As mentioned, archeologists (and other scholars) consider these historical accounts too mythologized to be reliable, and the "Archaic" period is seen as hazy and not well understood.[23] However, one thing all scholars agree on is that this "Archaic" period was a period of extraordinary technological advance. Stigler asserts "....that the 'Archaic' period witnessed an *explosion of technological advance.*"[24] (my emphasis) Lamberg-Karlovsky and Sabloff state that if

> our pictures were more complete, the transition from Predynastic to Pharaonic Egypt might appear more gradual than the sudden efflorescence which the distance of time and the absence of evidence suggest. As it is, the formation of Pharaonic Egypt under a single god-king, who assumes in his person the character of a god ruling over a unified state, his magical powers controlling the annual inundation of the harvest-giving Nile, *seems to arise suddenly, almost out of nowhere.*[25] (my emphasis)

The hieroglyphic system of writing appears fully developed from the beginning, unlike the Sumerian cuneiform writing. The Sumerian cuneiform writing, which appeared a century or two before the Egyptian hieroglyphic writing, underwent a rapid evolution from pictographs to the cuneiform style, which characterizes the writing of all Mesopotamian civilization (see Figure 3-2). There is a question among scholars as to whether or not the first writing of Egypt was influenced by the prior writing in Sumer. The Egyptian hieroglyphs are totally different from the Mesopotamian cuneiform style of writing. Lamberg-Karlovsky and Sabloff state that, if the invention of writing in Sumer and Egypt was autonomous, it is "...a remarkable coincidence that the independent invention of writing was so nearly simultaneous in both areas."[26].

Supposedly, the most remarkable and recognizable feature of Egyptian architecture, the pyramids, evolved from the compartmentalized trench tombs of the Late Predynastic period according to archeologists. They claim that the first pyramid was the huge Step Pyramid of Zoser, which was constructed about 2,700 B.C. "The archi-

The Rise of Ancient Human Civilizations

tectural progress from puddled-mud huts, within about five hundred years was enormous," Stigler understates.[27] The rapid developments and sudden rise of the Egyptian civilization were believed by many scholars to have been brought about by influences, and even invasions, by peoples of western Asia where the first human civilization, Sumer, originated. Yet, there is surprisingly little archeological data to support the idea that there was much contact between Egypt and Sumer. This fact is most puzzling to archeologists, for the two great civilizations coexisted for hundreds of years only a few hundred miles distant from each other. Lamberg-Karlovsky and Sabloff deny that there is any support in the archeological record for such an "Asiatic invasion." These archeologists call the Mesopotamian-Egyptian interaction "elusive":

> Though the two great civilizations were aware of each other's existence from at least 3,000 B.C., they charted their individual courses of development with seemingly far less contact than that between Mesopotamia and smaller communities on the Iranian Plateau. Just why this is so remains unclear. Perhaps the Egyptian concept of a god-king was such anathema to Mesopotamia that it acted as a taboo which reinforced conscious avoidance between both civilizations.[28]

Yes, the Egyptian pharaohs were considered actual divine beings, or gods, in the early Dynasties; whereas in Sumer, as we have learned, the kings were humans that were granted the right to rule by the gods. Stigler, commenting on the puzzling lack of evidence for extensive material object exchange and contact between Egypt and Sumer says, an "...Egyptian awareness and knowledge of Mesopotamia was probably much more important than the material objects themselves..."[29]

In addition to the contrast of Egyptian god-kings (or god-pharaohs) and Sumerian human-kings-through-grants-from-gods and the difference in writing, there are several other differences between the two civilizations. There was no urbanization in Egypt like there was in Sumer. The Nile Valley was covered with towns, villages and ceremonial centers, but no large cities like Ur or Lagash in Sumer. Also, there were no wheeled vehicles in Egypt until about 1,600 B.C., whereas wheeled vehicles were common in Sumer before 3,000 B.C. Finally, historian C.G. Starr claims that the Egyptians were more cheerful than the Sumerians, and "...brooded less darkly about the place of man before the gods."[30]

The Rise of the Indus Valley Civilization

The Indus Valley civilization, usually called the Harappan civilization, of what is now Pakistan, had its beginnings later than the Egyptian and Sumerian civilizations, around 2,400 B.C. to 2,300 B.C. The Indus civilization is considerably less well known to the general public than the first two human civilizations. Contributing to its relative obscurity is the fact that its writings have not been deciphered. Nevertheless, the Harappan civilization was a grand civilization that had fairly close relations with at least the Sumerian civilization, and it thrived for more than 600 years, until it mysteriously collapsed at about 1,800 B.C.

There were two large cities, Mohenjo-Daro and Harappa, located about 375 miles (~600 kilometers) from each other, and hundreds of smaller villages in the Harappan civilization. The villages and cities were laid out on a grid, apparently the first efforts of city planning. The cities and towns were constructed of baked brick, and had the best and most elaborate drainage and sewer systems in the ancient world. The domestic plants of this civilization included wheat, barley and rice, and the domestic animals included camels, pigs, cattle, horses, dogs, cats and probably elephants. The Harappan civilization carried out a tremendous amount of foreign trade, including a "staggering" amount of exports to Sumer.[31]

If there are gaps in their knowledge, lack of data and expressions of puzzlement among archeologists concerning the rise of Sumer and Egypt, there are even more of these concerning the rise (and fall) of the Harappan civilization. After briefly reviewing the scanty data relating to the beginnings of agriculture in and near the Indus region, Stigler concludes:

> As in the case of Egypt, the preceding Neolithic sequence has been viewed as too brief and too simple, at least in the material manifestations, to account satisfactorily at present for the complex cultural developments that followed virtually immediately.[32]

Elsewhere he writes:

> ...There is nothing in Pakistan like the elaborate Ubaid and Warka cultures of Mesopotamia in the

The Rise of Ancient Human Civilizations

Neolithic times. If such comparisons are valid at all, the Baluchistan sequence (and what is known of the pre-Indus sites in the Valley itself) appear more on the order and scale of the Neolithic in the Nile Valley, and again we are left with a presently rather inexplicable cultural explosion around 2,300 B.C., unsupported by what would appear to be inadequate previous development.This is not, of course, to suggest that some romantic archaeological mystery exists or that some unheralded burst of creative genius wrought civilization out of nothing. Rather, the problematical aspects of the emergence of civilization in both the Nile and Indus Valleys derive from the current state of archaeological knowledge...[33]

Lamberg-Karlovsky and Sabloff write:

The origin and formation of the Indus civilization have been the source of great speculation, but limited evidence. For decades, it was commonplace to maintain that the Indus civilization appeared suddenly, in a mature form, around 2,400 B.C., the result of diffusion from Mesopotamia. This view can no longer be maintained. Recent excavations on the Iranian Plateau at Tepe Yahya and Shahr-i-Sokhta confound any simple diffusionary mechanism for the rise of the Indus, for there is little doubt that both these non-Sumerian sites had contact with sites of the Indus civilization (These smaller cultures cannot be seen as the generators of the Indus, any more than can the Mesopotamian city- states.)
We still know very little about settlements in Pakistan and Northwestern India prior to the Indus civilization. Few excavations have examined such settlements - not because pre-Indus levels do not exist, but simply because so few have been dug. The limited evidence at hand, nevertheless, does not support the contention that the Indus civilization developed suddenly or was the result of diffusion from the West.[34]

The Rise of Ancient Human Civilizations

Later they add:

> There can be little doubt that when sufficient excavation is undertaken, we will comprehend more fully the independent genesis of the Harappan civilization - as independent a creation as that of Egypt and Mesopotamia. Though all three civilizations were contemporary, they were entirely distinctive in their form.[35]

The patterns of the archeological accounts of the rise of the first civilizations are clear. First there is a short period of time when very simple farming cultures existed, even when little or no evidence is available that they did exist, as in the case of the Indus civilization. These simple cultures do not satisfactorily account for the complex civilizations that were to follow. Then there follows a period of "sudden efflorescence" when the accouterments of civilization came into being, including writing and magnificent architectural structures. The archeologists assure us that this picture of the rise of civilization is not as it seems, and that a closer look at the scanty data available shows an overall picture of cultural progression toward civilization. Furthermore, they claim that much of the data is buried beneath flood deposits of these river civilizations and/or were otherwise destroyed and/or will be unearthed through further excavations. This same pattern is seen in archeological accounts of the rise of civilizations in the New World as we will see in an account of the first known American civilization, the Olmec civilization.

The Rise of the Olmec Civilization

The Olmec civilization was established by 1200 B.C. in the Gulf Coast lowlands of the modern country of Mexico. Like the Egyptians, the Olmecs did not build cities, but rather ceremonial centers with populations of only several thousand people. Most of the population were to be found living in the countryside surrounding these ceremonial centers, where they raised agricultural crops and provided labor for the priests of the ceremonial centers. The Olmec culture was a theocracy in which the peasants of the countryside provided food and labor, and the priests of the ceremonial centers, in turn, provided "security" through the rituals they carried out. "The Olmec priests evolved a religion and a complex iconography which centered on a number of gods, most notably a jaguar god who is depicted throughout Olmec art."[36]

The Rise of Ancient Human Civilizations

Although the ceremonial centers did not contain a large population, they feature monumental architecture and sculpture. For example, the ceremonial center of San Lorenzo sits upon an immense artificial platform, "an architectural feat that attests to the huge amount of human labor the Olmecs must have employed to construct and decorate their ceremonial centers."[37] The best known features of the Olmec civilization are the enormous heads found at San Lorenzo and other ceremonial centers. These heads weighed up to twenty tons and were carved from basalt that came from mountains that were located as far as 100 miles from the centers. Transporting huge basalt blocks to the ceremonial centers required "not only tremendous physical labor but also leadership that was equal to the task of orchestrating it."[38]

San Lorenzo itself was first occupied in 1500 B.C. "probably...as a farming community..."[39] Yet, in only 300 years the Olmec civilization with the colossal stone heads and imposing architecture appeared. "...This seemingly sudden growth of a complex society on the Gulf Coast has led to much speculation about the reasons for the rise of the Olmecs."[40]

And speculation is about all archeologists can do, given the lack of data and the premise they operate with, that humans in each area of the first civilizations are totally responsible for the rise of these civilizations without help from overseas or anywhere else. Lamberg-Karlovsky and Sabloff, whose archeological text *Ancient Civilizations* we are using to gain an archeological perspective on the origins of the Olmecs, comment on the proposal of influence from overseas:

> From an archeological point of view, the Olmec civilization suddenly appeared in a relatively well-developed form around 1,200 B.C. As we have seen, its antecedents were few. There seems to have been no long, slow sequence of local growth in the Gulf Coast lowlands prior to the rise of the Olmecs. If the Olmecs did not evolve in the same area where they later flourished, where did they arise? Were the elements of Olmec civilization developed elsewhere and imported to Mexico? *While we can dismiss the idea that the civilization originated in such places as Atlantis or outer space as totally without support or merit*, other hypotheses involving diffusion of ideas, art styles, and people from areas like the Far East cannot be rejected out of hand.

> There are some resemblances between Olmec and Chinese art. But there is nothing further to back the contentions of actual connections. Thus, for the moment at least, ideas of diffusion from overseas cultures must be viewed as conjecture.[41]

After considering and rejecting a hypothesis that the Olmecs may have been linked to a culture of the Peruvian highlands to the South, Lamberg-Karlovsky and Sabloff state:

> Thus, although there are suggestions that Olmec civilization might have originated outside the Gulf Coast lowlands of Mexico, the most reasonable position for the moment seems to be "wait and see." Without convincing evidence to the contrary, it seems best to assume that the Olmec civilization evolved locally, within Mexico. The assumption of local development, however, still does not answer the basic question regarding Olmec origins. Did civilization evolve in the Mexican lowlands or Mexican highlands? This question has split the ranks of Mesoamerican archeologists and reflects deep theoretical biases.[42]

After reviewing two hypotheses of the "highland school" and finding some, but little support for them, Lamberg-Karlovsky and Sabloff state:

> Although agriculture was practiced in the highlands for many thousands of years prior to the beginnings of Olmec civilization around 1200 B.C., and settled villages, too, had slowly emerged by the Third Millennium B.C., there seems to have been few major advances toward the growth of complex societies in the highlands between about 2,500 and 1,500 B.C.[43]

The authors then turn their attention to the "lowland school", which they favor. We recall from the last chapter that "...there is no archeological evidence for occupation of the lowlands of Mesoamerica until as late as the third millennium B.C."[44] Then there were people living in the Pacific coast lowlands by 2,300 B.C. Based on the crude pottery found, it is probable that these Pacific lowland people were living in settled villages. Lamberg-Karlovsky and Sabloff believe that maize was

> ...the trigger for the lowland advances which ultimately led by 1200 B.C. to the rise of Olmec civilization. Within the new environmental conditions of the lowlands, the highland maize *apparently underwent a series of genetic mutations* which led to a greatly increased size of cob. In turn, this important new food source offered sufficient food to allow for the growth of a more complex society in the lowlands.[45] (emphasis mine)

Maize had been grown in the highlands of Mesoamerica at least since 5,000 B.C., but the cobs were quite small, measuring less than 75 millimeters (~3 inches) in length. However, the maize grown by the Olmecs had apparently taken a sudden leap in size and approached the size of the corn-on-the-cob many of us eat and love today. As the above quote indicates, Lamberg-Karlovsky and Sabloff (and other archeologists) attribute this size increase to "a series of genetic mutations" that apparently occurred naturally.

Lamberg-Karlovsky and Sabloff continue their "lowland school" discussion:

> The lowland explanation seems plausible, but, unfortunately, the archeological picture is not sufficiently clear to link the rise of the Olmec civilization on the Pacific Coast directly with the rise of the Olmec civilization on the Gulf Coast. No Olmec remains lie directly above the early agricultural villages on the Pacific Coast, nor have deep early village deposits been found stratigraphically below Olmec remains on the Gulf Coast. Nevertheless, given our present state of knowledge, it is reasonable (if not totally satisfactory) to assume that the developments on the Pacific Coast provided the basis for the eventual rise of the Olmec civilization on the Gulf Coast.
> The 300-year period from 1,500 to 1,200 B.C. was sufficient time for the various aspects of Olmec civilization to evolve.[46]

The authors of the text *Ancient Civilizations* launch into a discussion of archeological hypotheses that attempt to explain why the humans of Mesoamerica developed civilization at all. They admit that if "hypotheses concerning the locale for the origins of the Olmecs are tentative and confusing, the reasons for the civilization's development are even murkier..." [47] After reviewing several hypotheses for the development of this first civilization of the New World, Lamberg-Karlovsky and Sabloff conclude:

> Whether trade, differential land productivity, a broad-based belief system, or other factors entirely were the ultimate causes for the rise of the Olmec civilization still remains unclear. Without a good deal more research, we are unlikely to uncover such causes. It does seem likely that some of the ideas noted above, particularly those concerning the role of long-distance exchange, may pave the way for future advances in this area.[48]

The Sumerian Historical Accounts of the Rise of Human Civilization

I recounted much of the Sumerian historical accounts in the last chapter and earlier chapters. We recall that the deluge destroyed almost all of humankind, plus the ancient Anunnaki cities in Mesopotamia, and just about everything on Earth that was not on high ground. The Anunnaki had expected all of humankind to perish in the Deluge. After the Great Flood, however, the Anunnaki, even Enlil, realized that they would need humans in order to continue to live on Earth. They then granted humans domesticated plants and animals that they had genetically manipulated, gave them the plow, and other agricultural implements, and taught them farming and animal herding. Humans practiced agriculture at first in the highlands in and around the Near East because the lowlands were uninhabitable for years after the Great Flood. Gradually the human population increased, and with the help of the Anunnaki, they were able to repopulate the lowlands.

We now, again, pick up the Sumerian and other Mesopotamian historical accounts from the translations and interpretations of Zecharia Sitchin. The Anunnaki deliberated after the Great Flood and decided to again establish a kingship in Mesopotamia, only this time they would allow humans to be kings and rule the cities that were to be established there. They would appoint human rulers who would act as intermediaries between them and the human population, and

The Rise of Ancient Human Civilizations

assure humankind's service to the Anunnaki.

In the Sumerian "Epic of Etana" we read that the "gods" decided to divide the area of the world they controlled into four regions. The first region was, of course, Sumer. Sumerian texts state: "When Kingship was lowered again from Heaven, the Kingship was in Kish.."[50] Kish became the first "human" Earth city. Soon the other great cities of Sumer were established. Eventually the Kingship passed from Kish to Uruk, and then to Ur and other cities of Sumer, then later, to Akkadian, Babylonian and Assyrian cities; as Sumer was succeeded by these later empires; Enlil and his descendants became overall suzerains, or feudal lords, of the Sumerian region, although other Anunnaki "gods" were granted Sumerian cities and lived, at least part of the time, in the palaces and temples of these cities.

The second region, established a few hundred years after Sumer, became the second human civilization, the Egyptian Dynasty. Sitchin presents evidence that Enki, called Ptah by the Egyptians, was the suzerain of Egypt. The third region and third human civilization, established about 100 years after Sumer, was the civilization of the Indus Valley.[51]

Sitchin argues and gives some evidence that Sumerian was the original human language, and that the language of ancient Egypt, that of the Indus Valley, and Chinese, as well, were all derived from Sumerian. This division of languages was a deliberate ploy by the Anunnaki, according to Sitchin, to keep humankind divided. Divide and rule was their policy. The story of the tower of Babel in the Bible is a reflection of this policy. The Anunnaki were apparently afraid that a united humankind would pose a real threat to their intended power and control over the world, or to the power and control of those four regions of the world they dominated.[52]

The fourth region established by the Anunnaki was not for humans. In fact, humans were generally forbidden to go there. The Sumerian name for this fourth region was TIL.MUN, which literally means "the place of the missiles." This was the post-Deluge spaceport of the Anunnaki, as their pre-Deluge spaceport in the pre-Deluge city of Sippar had been buried beneath silt deposits of the Great Flood. TIL.MUN, Sitchin believes, was located in the Sinai peninsula where a flat, dry lake bed provided an ideal location for a long runway for the landing of their space shuttle crafts, similar to the one in the Mojave Desert near Edwards Air Force Base in California at which place several American space shuttles have landed. Control of TIL.MUN was given to Sud, the half-sister of Enlil, who was later given the title and name Ninhursag. While Ninhursag was given nominal control over TIL.MUN, bickering for control of this spaceport by the sons of Enlil and Enki eventually led to a nuclear holocaust. This holocaust not only destroyed TIL.MUN, but the biblical cities of Sodom and Gomorrah, as well.[53]

In addition to bestowing kingship to the three regions of human civilization, Sumerian texts tell us that the Anunnaki gave humans laws to live by and knowledge and wisdom of the arts and sciences, as well. The texts tell of how "heavenly knowledge was at times given to humans through one man." For example, Sitchin provides part of a translated text, telling how the gods groomed a man, Enmeduranki, to be the first human priest. The gods,

> Showed him how to observe oil and water, secrets of Anu, Enlil and Enki. They gave him the Divine tablet, the engraved secrets of Heaven and Earth. They taught him how to make calculations with numbers.[54]

In addition to medicine and mathematics, Enmeduranki was taught cosmology, evolution, geology, geography and geometry.

Enki, whose main home base was in the Abzu, which Sitchin believes was in Africa, had in his possession a number of MEs, which Sitchin believes were something like computer chips containing much information of the arts and sciences. According to Sumerian texts, Inanna, the love goddess, paid a visit to Enki in the Abzu in order to get some of the MEs. She proceeded to get Enki drunk and convinced him to give her seven MEs, which she took back to her Sumerian city of Erech. The texts praise Inanna, "lady of the MEs," for distributing the knowledge of the MEs to the Sumerian people.[55] This may be a deliberate distortion. Possibly Enki actually gave humankind the knowledge of the arts and sciences contained in the MEs. After all, he is said to have been the chief engineer and chief scientist of the Anunnaki, and he had a long history of aiding humankind. Such a story might have been given to the Sumerians by the Anunnaki to diminish the reputation of Enki in human eyes. We will have more to say about Enki in later chapters. In any case, Sumerian texts make it clear that humans were given the knowledge of civilization by the gods.

Comparison of the Scientific and Ancient Historical Views of Humankind's Ascent to Civilization

In the previous two chapters we have seen a comparison of the materialist-scientific point of view and the ancient historical-mytho-

The Rise of Ancient Human Civilizations

logical writings, mostly Mesopotamian writings, of humankind's vault to civilization. There is no doubt which view I believe, of course, and I cannot honestly even claim to have given an unbiased view of both sides. It is clear that I have emphasized what the archeologist-historians do not know at the expense of what they do know. But I have done this for a purpose: I want to make abundantly clear to the readers what the archeologists do not know about our cultural leap from simple nomadic hunting and gathering people to the sophisticated people of the Sumerian, Egyptian, Indus River Valley, and Olmec civilizations. This astonishing cultural leap, remember, happened in only ten to twelve thousand years at least three separate times, according to the archeologists. Not only am I not convinced by the arguments of archeologists concerning the origins of human civilization, but I feel their case is extremely weak.

Over and over again, as we have seen in our brief review, archeologists and other scholars have commented on the rapid, sudden progress of humankind in the past 10,000 years, especially when the ancient civilizations, from Sumer to the New World, progressed very rapidly. The archeologists, at least those of the middle and late twentieth century, assure us that, although somewhat astonishing, the establishment of civilization by humankind these past few thousand years was all part of a natural process that has occurred and evolved here on Earth, beginning with the chemical "raw materials" of the ancient Earth more than 4,500 million years ago to the development of our modern civilization of the late twentieth century.

Again and again, archeologists assure us that the current lack of, or shortage of data they have to support their theories and explanations of how we humans developed will be made up in the future as more excavations are made giving them more data with which to work. This forthcoming data by future archeological digs will clarify the "natural," "random" forces that drove humankind from a nomadic hunting and gathering people to people living in quite sophisticated civilizations.

The archeological myth of the origins of civilization, which in many ways is the cornerstone of the materialist-scientific explanation of the current existence of humankind, might be the most inconsistent and weakest link of this fable. Three other major inconsistencies in the materialist-scientific view of human origins that might be as significant as those in the archeological stories we have examined, are the origin of life, the "Cambrian explosion", and the lack of a plausible explanation for the large biological changes (macroevolution) of the biological evolutionary record we examined in Chapter one.

Many will point out that my relying so heavily on the Mesopotamian records for the story of our origins is somewhat shaky. I admit that on the surface this charge is true. I repeat once again that I know very well the Sumerian texts and those of later Mesopotamian civilizations are not an "exact" account of what "actually happened", to the last detail, in ancient times. In the first place, I would argue that there exists no human historical account, modern or ancient, that is an exact account of what actually happened. Let us consider for a moment an historical incident with which we are at least somewhat familiar - the assassination of President John F. Kennedy in 1963. Was he murdered by a lone, unbalanced assassin, or by several professional assassins as part of a conspiracy? Many may think they know, but I would contend that there are so many unknown and questionable circumstances surrounding this recent historical event, that those of us who are part of the general public cannot know exactly what did happen. The same is true concerning historical events of ancient times, only even more so, especially if there was some sort of deliberate distortion or cover-up of the true origins and development of humankind by regressive ETs. We have already reviewed some evidence of such a cover-up, and much more will be presented in the next section of the book. But let us continue to examine ancient history to observe more human-ET interactions. After civilization was granted to humans by the Anunnaki, did relations between humans and their gods improve?

[1] Sitchin 1980. *op. cit.*, p.86.
[2] Sitchin 1990. *op. cit.*, pp. 4-21.
[3] Lamberg-Karlovsky and Sabloff 1979. *op. cit.*, p. 109.
[4] *Ibid.*
[5] Stigler 1974. *op. cit.*, p.114.
[6] *Ibid.*, p. 112.
[7] Lamberg-Karlovsky and Sabloff 1979. *op. cit.*, p. 108.
[8] *Ibid.*, pp. 112-114.
[9] *Ibid.*, p. 142.
[10] Speiser, E. A., 1969. "The Sumerian Problem Reviewed," in *The Sumerian Problem*, T. Jones (ed.). John Wiley and Sons, Inc., New York. pp. 93-102. Speiser's article was originally published in 1951.
[11] *Ibid.*
[12] Stigler 1974. *op. cit.*, p. 117.
[13] Lamberg-Karlovsky and Sabloff 1979. *op. cit.*, p. 142.
[14] *Ibid.*
[15] *Ibid.*, p. 172.
[16] *Ibid.*, p. 172.
[17] *Ibid.*, pp. 167-175.

[18] Stigler 1974. *op. cit.*, pp. 127-129.
[19] *Ibid.*, pp. 132-134.
[20] *Ibid.*, p. 137.
[21] *Ibid.*, p. 134.
[22] *Ibid.*, p. 140.
[23] *Ibid*, p. 141
[24] *Ibid.*
[25] Lamberg-Karlovsky and Sabloff 1979. *op. cit.*, p. 134.
[26] *Ibid*, p. 135.
[27] Stigler 1974. *op. cit.*, p. 142.
[28] Lamberg-Karlovsky and Sabloff 1979. *op. cit.*, p. 188.
[29] Stigler 1974. *op. cit.*, p. 143
[30] Starr, C. G., 1983. *The History of the Ancient World (3rd. ed.)*. Oxford University Press, Oxford.
[31] Lamberg-Karlovsky and Sabloff 1979. *op. cit.*, pp. 189-212.
[32] Stigler 1974. *op. cit.*, p. 151.
[33] *Ibid..*, p. 149.
[34] Lamberg-Karlovsky and Sabloff 1979. *op. cit.*, pp. 189-191.
[35] *Ibid.*, p. 192.
[36] *Ibid.*, p. 246.
[37] *Ibid.*, p. 244.
[38] *Ibid.*, p. 246.
[39] *Ibid.*, p. 247.
[40] *Ibid.*
[41] *Ibid.*, p. 248.
[42] *Ibid.*, p. 249.
[43] *Ibid.*, p. 251.
[44] *Ibid.*, p. 217.
[45] *Ibid.*, p. 252.
[46] *Ibid.*
[47] *Ibid.*
[48] *Ibid.*, p. 254.
[49] Excerpts from the "Epic of Etana" quoted by Sitchin 1976. *op. cit.*, p. 415.
[50] Sitchin 1976. *op. cit.*, p. 416.
[51] *Ibid.*, pp. 416-421.
[52] *Ibid.*
[53] Sitchin 1991. *op. cit.*, p. 206.
[54] Sitchin 1985. *op. cit.*, pp. 325-334.
[55] *Ibid.*, pp. 239-241.

"Thou shalt love thy neighbor as thyself."
Old Testament, Leviticus 19:18

Chapter 9

More "Heartless" Behavior of the "Gods"

We have seen that the Anunnaki of Pre-Deluge "Mesopotamia" treated humans with little evidence of respect, consideration, or compassion. According to Sitchin, who writes of the totally neglectful and horrendous manner in which the Anunnaki interacted with early human beings. These gods were jealous, conniving gods that were often fighting with each other. When these wars between the "gods" broke out, humans were often used by the "gods" to further their own ends. Both human soldiers and human civilians alike were often destroyed in the wars of the "gods," who apparently cared not the slightest for the fate of their human subjects. Human armies were taught how to use primitive weapons and use primitive tactics, but the "gods" themselves used much more advanced weapons, including nuclear weapons. The historical evidence indicates that humans at times suffered grievously from the use of nuclear weapons.

Much of the evidence of the extraterrestrial hypothesis has always come from the ancient religious writings of the people of the Indus River Valley civilization. As we saw, the ancient Indus or Harappan civilization was founded quite suddenly at around 2,400 B.C., and just as suddenly collapsed around 1,800 B.C. Archeologists have no idea of why, or even if, the Harappans collapsed suddenly. One of the ancient Indian writings, the Rig Veda describes the arrival of Vedic Aryans to the Indus Valley, who defeated in battle the "dark skinned" people that lived in walled cities there. The only city that offers evidence that something like this happened is the largest and most impressive of the Valley, Mohenjo-Daro. Archeological evidence at Mohenjo-Daro indicates that the city was destroyed and "...(b)roken skeletons of men, women, and children were discovered in several areas of the city, showing indisputable evidence of having been ruthlessly butchered...".[2]

More "Heartless" Behavior of the "Gods"

Hinduism was established in India at about 1000 B.C., but historians cannot help us much concerning the matter of the exact time of the origins of the Vedic writings and religious views. Historians assure us that the development of Vedic religious views into the historic Hindu outlook is obscure, and the 1,000 B.C date is approximate.[3]

Indian scholars informed Erich von Däniken that most of the Vedas were in their present form by 1,500 B.C., but that their history in the oral tradition, most likely goes back hundreds of years before 1,500 B.C.[4] So probably much, if not most, of the ancient Vedas and ancient religious epics are about events that took place during the Indus Valley civilization, dating from around 2,400 B.C. to 1,800 B.C., and possibly even before the Deluge.

These ancient Vedic texts and ancient epics clearly describe the flying vehicles, wars, and frightful weapons of the gods. They have been quoted and re-quoted many times in the UFO literature. These ancient writings recount in detail more than any other ancient writings, descriptions of the marvelous flying craft of the gods. Sitchin quotes from a translation of the epic tale of the *Mahabharata*, of gods arriving for a wedding feast in aerial cars:

> The gods, in cloud-borne chariots,
> came to view the scene so fair:
> Bright Adityas in their splendor,
> Maruts in the moving air;
> Winged Suparnas, scaly Nagas,
> Deva Rishis pure and high
> For their music famed, Gandharvas;
> (and) fair Apsaras of the sky...
> Bright celestial cars in concourse
> Sailed upon the cloudless sky.[5]

This sounds a bit like a description of Hollywood celebrities arriving at the annual Oscar Awards Ceremony, except that instead of arriving in limousines, the ETs arrived at this birthday celebration in "cloud-borne chariots." What is interesting about this quoted passage of the sacred epic, the *Mahabharata*, is that seven different groups of ETs arrive in their celestial cars representing at least two and, most likely, several species of ETs. Probably there were primate ETs of the planets of Lyra, Pleiades, and some other star systems, attending the wedding. The "scaly Nagas" would be a likely candidate for the lizzies, or reptoids, to which we've been referring. Other Hindu epics describe Sri Lanka (Ceylon) as the stronghold of the Nagas. Sri Lanka is said by ancient sources to be the home of strange reptilian-like creatures.[6] In the above quoted passage, all of the ETs seem to be getting along with each other. However, much of the ancient Indian Vedas and epics, like the ancient Mesopotamian texts and the ancient myths of the later Greeks, tell of wars of men and gods, where often terrible destructive weapons were used. The above quoted *Mahabharata* epic itself is about a bitter struggle and disastrous war between two branches of the same family, the Kurus. The war between the Kauravas and the Pandavas is started with conventional human weapons of the time, such as spears, bows and arrows, and swords. But the war escalates to much more powerful and sophisticated weaponry, provided by the gods of both sides. From the descriptions of these weapons by unsophisticated humans, it seems that in addition to flying vehicles, usually called Vimana, the gods of the *Mahabharata* used missiles, anti-missile missiles, and nuclear weapons that caused radiation sickness and sterility among the people of both sides of the terrible war.[7]

An Indian student told von Däniken in the early 1960s, after hearing one of von Däniken's lectures purporting our ET origins:

> "Do you really find anything new or shocking about what you have told us? Every half-educated Indian knows the main sections of the Vedas and so knows that the gods in ancient times moved about in flying machines and possessed terrible weapons. Really, every child in India knows that."[8]

Von Däniken quotes for us from an ancient Sanskrit text by one Maharshi Bharadvya, "...a seer of an early period..." The translation of this text from Sanskrit to English was provided by The International Academy for Sanskrit Research located in Mysore, India. This portion of the translation concerning flying vehicles and what they are capable of performing, was provided and checked by von Däniken:

> 6. An apparatus that moves by inner strength like a bird, whether on Earth, in the water or in the air, is called Vimana...
> 8. ...which can move in the sky from place to place...
> 9. ...country to country, world to world...
> 10. ...is called Vimana by the priests of the sciences...

More "Heartless" Behavior of the "Gods"

11. The secret of building flying machines
12. ...that do not break, cannot be divided, do not catch fire...
13. ...and cannot be destroyed...
14. The secret of making flying machines stand still.
15. The secret of making flying machines invisible.
16. The secret of overhearing noises and conversations in enemy flying machines.
17. The secret of taking pictures of the interiors of enemy flying machines.
18. The secret of ascertaining the course of enemy flying machines.
19. The secret of making beings in enemy flying machines unconscious and destroying enemy machines...[9]

Sitchin has pointed out that the names of the evil gods in ancient Hindu literature are often very similar to the gods of the ancient Babylonian, Assyrian, and Egyptian gods, and these Near East gods often became demons in Hindu translations.[10]

Von Däniken quotes a passage from the *Mahabharata* that almost certainly describes a nuclear holocaust:

> It was as if the elements had been unleashed. The sun spun round. Scorched by the incandescent heat of the weapon, the world reeled in fever. Elephants were set on fire by the heat and ran to and fro in a frenzy to seek protection from the terrible violence. The water boiled, the animals died, the enemy was mown down and the raging of the blaze made the trees collapse in rows as in a forest fire. The elephants made a fearful trumpeting and sank dead to the ground over a vast area. Horses and war chariots were burnt up and the scene looked like the aftermath of a conflagration. Thousands of chariots were destroyed, then deep silence descended on the sea. The winds began to blow and the Earth grew bright. It was a terrible sight to see. The corpses of the fallen were mutilated by the terrible heat so that they no longer looked like human beings. Never before have we seen such a ghastly weapon and never before

More "Heartless" Behavior of the "Gods"

have we heard of such a weapon.[11]

ET researcher Valdamar Valerian, in his book *Matrix II*, quotes British researcher David Davenport, whose years of study of the ancient Hindu texts and personal visits to ancient sites in India-Pakistan, has convinced him that Mohenjo-Daro, the largest city of the ancient Indus civilization, was destroyed by a nuclear blast.[12] Davenport found a similar pattern of destruction at Mohenjo-Daro that was observed in the destruction of Nagasaki, Japan after a nuclear blast had destroyed that city at the end of World War II. In both the modern and ancient city there was an epicenter fifty yards wide where everything was fused, melted and crystallized. Sixty yards from the epicenter, the bricks are melted on one side, the sides facing the blast. Davenport quotes part of the above quote from the *Mahabharata* and believes it may be a description of the end of Mohenjo-Daro. Davenport believes he has found evidence of collaboration between the Aryans and space aliens in the destruction of Mohenjo-Daro. The Aryans, according to Davenport, were then located in the mountains where space aliens were mining minerals. The aliens agreed to destroy Mohenjo-Daro, a city of the enemies of the Aryans, on behalf of the Aryans in order that they could continue to mine minerals in peace. According to Davenport, ancient Hindu texts tell us that the 30,000 plus residents of the city were warned to leave the city seven days before it was destroyed. Most people heeded this warning, but a few did not. Shortly after the ancient city was first discovered in 1927, 44 human skeletons were found. All of these skeletons were found face down, flattened to the ground.

This is pretty heady stuff, these clues - these not so subtle and very blunt records in the ancient literature of nuclear holocausts. Von Däniken himself traveled to India in order to check these ancient Hindu epics with Indian historians, Sanskrit experts, and every type of scholar who might know about the clear references to "aerial chariots" and weapons of terrible power. Von Däniken quotes the answer to an inquiry about references in the ancient Indian literature to advanced weapons and air travel of Sanskrit scholar Professor Esther Abraham Solomon of the University of Ahmedabad. Professor Solomon answered one of von Däniken's questions concerning ancient Indian literature:

> "They are just exaggerations of fanciful descriptions of an imaginary divine power. The ancients undoubtedly felt the need to endow their leaders and kings with a mystical, mysterious nimbus.

> They certainly invented the incredible and invulnerable attributes later - multiplying them with each generation."[13]

But later, Professor Solomon admitted that, "I don't know, I really don't know," when von Däniken asked her if the tales in ancient Indian literature could be "...descriptions of very remote real events..."[14]

Another graphic description of a possible ancient nuclear holocaust comes from the Bible. We westerners know about the destruction of the ancient cities called Sodom and Gomorrah, to some extent, through the historic account of the destruction of these cities in chapters 18 and 19 of the Book of Genesis. The story in Genesis revolves around Abraham, the grand patriarch of the Hebrews, and his son-in-law Lot, who actually lived in Sodom at the time of the destruction. Two to three "angels," as the Bible tells us, came to check on Abraham and save his son-in-law. Also, the "angels" were going to check to see if the inhabitants of two cities in the Valley of Siddim (called "the Plain" in the Bible) in what is today Israel, were as wicked as reported. After convincing Lot, his wife and two daughters to leave Sodom and flee to the hills, the "angels" allowed (almost certainly signaled) the Lord to rain

> ...down fire and brimstone from the skies on Sodom and Gomorrah. He overthrew those cities and destroyed all the Plain, with everyone living there and everything growing in the ground. But, Lot's wife, behind him looked back, and she turned into a pillar of salt. Genesis 19: 24-26 [15]

Meanwhile, Abraham, who that morning was located in the hills some distance, but within sight, of the Plain,

> ...rose early and went to the place where he had stood in the presence of the Lord. He looked towards Sodom and Gomorrah and all the wide extent of the Plain, and there he saw thick smoke rising high from the Earth like the smoke of a lime-kiln. Thus when God destroyed the cities of the Plain, he thought of Abraham and rescued Lot from the disaster, the overthrow of the cities where he had been living. Genesis 19: 27-29[16]

Are there any Mesopotamian accounts of this disaster? Surely an event of this magnitude would not go unnoticed in what was in ancient times the most literate society on Earth. The answer, of course, is yes. Sitchin compares the Biblical account of this disaster to the Mesopotamian account, which he pieces together from several fragmentary tablets from different time periods and written in different languages. Sitchin documents the incredible way by which he has pieced together this amazing story of the actual bombing of the cities of the Plain in *The Wars of Gods and Men*.[17] In Sitchin's detailed coverage of the Biblical historical rendition of the nuclear holocaust that destroyed Sodom and Gomorrah, he argues that the ancient Hebrew term *Netsiv melah* is better translated as "pillar of vapor" instead of "pillar of salt," as it has traditionally been translated in the Bible. He argues that Lot's wife turned into a "pillar of vapor" based on the evidence of this mistranslation, and his belief that the native tongue of both Abraham and Lot was Sumerian, not Hebrew. In other words Sitchin is arguing that the story of "angels" actively participating in the destruction of the cities of the Plain was originally written in the Sumerian language, and the Genesis version is, once again, a Hebraized version of a Sumerian and other Mesopotamian accounts, even though such Mesopotamian accounts have never been found.

Archeological-historical chronologies indicate that after the destruction of the Akkadian Dynasty at about 2,200 B.C., eventually a Sumerian "renaissance" period, known as the Third Dynasty of Ur (2,113 B.C. to 2,000 B.C.), came into existence. The Third Dynasty of Ur, according to academic scholars of today, was obliterated by an invasion of nomadic peoples.[18]

What do Sumerian and other Mesopotamian texts tell us about this event? Basically, they tell us that the destruction of Sodom and Gomorrah was a result of a severe conflict of interest between the firstborn son of Enlil, Ninurta, and the firstborn son of Enki, Marduk. The Mesopotamian texts tell us that the destruction of the cities of the Plain was a side-show. The primary nuclear blasts were blasts that destroyed the space port of TIL.MUN in the Sinai Desert. TIL.MUN, we remember, was one of the four regions of their world parceled out by the great Anunnaki. This was the region where humans were forbidden, and Sitchin makes a strong case that TIL.MUN was, in fact, the Anunnaki spaceport. The spaceport had come under control of Marduk, and the cities of the Plain had been persuaded to back Marduk and his ambitions, according to Sitchin. For this reason, plus the fact that the backers of Ninurta thought that Nabu, the son of Marduk, whom Ninurta and his backers wanted destroyed, was staying temporarily in one of the cities of the Plains.

As the conflict between the forces of Marduk and of Ninurta intensified, the council of the gods met to decide what action was to be taken. Again, it is Enki who attempts to step in on behalf of humankind. He argues that the use of the "Ultimate Weapons" which the gods were considering, "...the lands would make desolate, the people will make perish." [19] The council of the gods, the Great Anunnaki, who seemed to heavily favor the forces of Enlil, represented by his son Ninurta, approved the use of the "Ultimate Weapon" in destroying the ET base in the Sinai Peninsula, "Mount Most Supreme" and the rest of the spaceport. So Ninurta, known as Ishum to the Babylonians,

> ...to Mount Most Supreme set his course;
> The Awesome Seven, (weapons) without parallel,
> trailed behind him.
> At the Mount Most Supreme
> he then obliterated;
> in its forest not a tree-stem was left standing.[20]

Then Nergal (Erra in Babylon), a younger son of Enki, who had an intense grudge to settle with Marduk and Marduk's son, Nabu, destroyed the cities of the Plain:

> Then, emulating Ishum,
> Erra, the Kings Highway followed.
> The cities he finished off,
> to desolation he overturned them.
> In the mountains he caused starvation,
> their animals he made perish.[21]

Boulay, using his sources of basically little known religious literature, claims that actually five cities of the Plain, the Valley of Siddim, were destroyed. According to Boulay, the cities of Admah, Zebolym and Zoar were destroyed at the same time as Sodom and Gomorrah, in addition to the Spaceport and sanctuary of the gods.[22]

Where were Sodom and Gomorrah and the other cities located? Where are their remains today? Both Stichin and Boulay believe the remains of these cities have been covered by the waters of the Dead Sea, since the catastrophe occurred. Both cite evidence that the Dead Sea was either doubled in size or actually originated as a result of the catastrophe.[23]

In summary, the historical Mesopotamian evidence clearly indicates that:

1. The cities of the Valley of Siddim were destroyed as almost a side

show of the main event, the destruction of the spaceport located in the northern part of the Sinai Peninsula, by rival groups of ETs.

2. A primary reason for destruction of the cities is the off-chance of destroying Nabu, the son of Marduk and a longtime thorn in the side of the forces of Ninurta. (Nabu is reported to have escaped the catastrophe.)

3. The Anunnaki gods were willing to destroy the humans of the cities because some had become followers of a rival ET, Marduk.

Furthermore, both Biblical and Mesopotamian accounts of the holocaust make clear, as Sitchin points out, that the gods could have delayed, postponed and actually avoided this human disaster.[24] In the Biblical account, Abraham is able to bargain with the Lord, or the angels of the Lord, on behalf of the people that lived in the cities. The Lord and/or his angel/emissaries - there were some gaps left in the monotheizing process - had supposedly come to see if the people of Sodom and Gomorrah were as wicked as had been reported. Abraham asks if it is right for the Lord to punish the good with the bad:

> "...wilt thou really sweep away good and bad together? Suppose there are fifty good men in the city; wilt thou really sweep it away, and not pardon the place because of the fifty good men. Far be it from thee to do this - to kill good and bad together; for then the good would suffer with the bad. Far be it from thee. Shall not the judge of all the Earth do what is just?" The Lord said, "If I find in the city of Sodom fifty good men, I will pardon the whole place for their sake." Genesis 18: 23-26

Abraham eventually bargained the Lord down to ten good men - the Lord promised to spare Sodom if he found ten good men there. So, according to the ancient accounts of the Hebrews, the holocaust could have been canceled by the Lord. The possibility of avoiding this catastrophe is made evident in the Mesopotamian accounts as well. Nergal/Erra was not forced or ordered to destroy the cities of the Plain. According to Babylonian accounts, some of the gods tried to dissuade Erra (Nergal) from destroying the cities. One of the female Anunnaki asked Erra:

> Valiant Erra
> Will you the righteous destroy with

> the unrighteous?
> Will you destroy those who have
> against you sinned
> together with those who against you
> have not sinned?[25]

But Erra/Nergal was consumed with personal hatred: "I shall annihilate the son (Nabu), and let the father bury him; then I shall kill the father (Marduk), let no one bury him!"[26] So, it was basically the personal rivalry and personal hatred between two ET groups that resulted with the annihilation of at least two ancient cities toward the very end of the Third Millennium B.C.

There is much more than this to the tragedy, for according to Mesopotamian historical accounts, a nuclear cloud from these nuclear blasts was blown north and eastward and annihilated the people of the cities of Sumer through radiation poisoning. This, Sitchin argues, was the reason for the sudden demise of the Third Dynasty of Ur, not a sudden, overwhelming invasion of nomads and other neighbors from the highlands surrounding Sumer, as academic scholars speculate. Nomad remains have been found by the academic scholars in Sumer whose dates occur after the sudden fall of this dynasty. As we'll see, a better interpretation of all the known evidence indicates that these nomadic invaders came into Sumer after the Sumerian cities became virtually uninhabited, and after the radiation levels had subsided.

A close examination of the Sumerian texts of this period, leave little doubt that the Third Dynasty of Ur was destroyed by a radioactive wind or cloud:

> A storm, the Evil Wind
> went around in the skies.[27]

The radioactive wind formed over the Spaceport at Sinai and the "evil" cities of the Plain, and followed the prevailing winds to the west and the southern part of Sumer. Many well-preserved "lamentations" of southern Sumerian cities have been found. There are lamentations bewailing the fates of the Sumerian cities of Ur, Nippur, Uruk, Eridu, and all smack of desolation caused by nuclear fallout:

> Causing cities to be desolated,
> (causing) houses to become desolate;
> Causing stalls to be desolate,
> The sheepfolds to be emptied;

More "Heartless" Behavior of the "Gods"

> That Sumer's oxen no longer stand in their stalls,
> That its rivers flow with water that is bitter,
> that its cultivated fields grow weeds,
> that its steppes grow withering plants.[28]

Many of the people of Sumer died a horrible death:

> The people, terrified, could hardly breathe;
> the Evil Wind clutched them,
> does not grant them another day...
> Mouths were drenched in blood,
> heads wallowed in blood...
> The face was made pale by the Evil Wind.[29]

As Sitchin points out, before the nuclear holocausts of Hiroshima and Nagasaki in 1945, scholars could not understand the lamentations of Sumer or the destruction of Sodom and Gomorrah. Many Biblical scholars concluded the "the Lord" literally poured sulfur and brimstone on the two cities of the plain. But, now that we understand the destructive capabilities and the nature of nuclear weapons, there can be little doubt that Sodom and Gomorrah and the Third Dynasty of Ur (Sumer) were destroyed by a nuclear holocaust. Sitchin's analysis of the Sumerian "lamentation" texts leaves little doubt that Sumer was destroyed from the radiation fallout of the remains of the mushroom clouds spawned by several (seven!?) nuclear blasts to the west of Sumer. Sitchin writes:

> The source of the unseen death was a cloud that appeared in the skies of Sumer and "covered the land as a cloak, spread over it like a sheet." Brownish in color, during the daytime "the sun in the horizon it obliterated with darkness." At night, luminous at its edges ("Girt with dread brilliance it filleth the broad Earth") it blocked out the moon: "the moon at its rising it extinguished." Moving from west to east, "the deadly cloud" - "enveloped in terror, casting fear everywhere" - was carried to Sumer by a howling wind, "a great wind which speeds high above, an evil wind which overwhelms the land."[30]

Apparently, the "gods" underestimated the destructive power of these "awesome weapons without parallel" that Enlil had in his pos-

More "Heartless" Behavior of the "Gods"

session and allowed to be used only very rarely. The gods did not anticipate that the nuclear blasts in the Sinai Spaceport and Sodom and Gomorrah would destroy Sumer through nuclear fallout.

So, what did the gods do as the radioactive clouds and "evil wind" approached Sumer? They fled to safety leaving the human population to fend for themselves. The "lamentation" texts make it clear that the Anunnaki left very fast. Inanna, the chief Anunnaki "goddess" of the city of Uruk, fled Uruk complaining about the fact that she left so fast she had to leave her jewelry. From *The Eridu Lament* we learn that the head goddess of Eridu, Ninki, fled to Africa: "Ninki, its great lady, flying like a bird, left her city."[31] In fact, it is clear from the laments, some of them reputed to have been written by the gods themselves at a later date, that the Anunnaki gods and goddesses grieved over the loss of their beautiful cities and their beautiful temples. These same Anunnaki did not spend much time bewailing the terrible suffering and deaths of the humans of their cities and towns. We are writing about real people and real suffering here, not some fantasies dreamed up by the primitive, slightly insane people of the early civilizations, as most modern-day scholars have treated these historical texts, until Sitchin. Sitchin's detailed analysis shows little indication of Anunnaki compassion for the humans they left to suffer their horrible radiation sickness and death.

The fate of the city of Uruk, as reported in *The Uruk Lament* text, recounted by Sitchin, serves to demonstrate the fate of all the cities of southern Sumer.[32] The resident deities of Uruk raised an alarm in the middle of the night and warned the people to run away, but the gods themselves ran off right after giving a warning. *The Uruk Lament* text states:

> Thus all its gods evacuated Uruk
> They kept away from it;
> They hid in the mountains,
> They escaped to the distant plains

The people, left leaderless, began to panic and riot. Sitchin interpreting the ancient text, writes:

> "Mob panic was brought about in Uruk ...its good sense was distorted." The shrines were broken in and their contents were smashed as the people asked questions: "Why did the gods' benevolent eye look away? Who caused such worry and lamentation?" but their questions remained unan-

More "Heartless" Behavior of the "Gods"

swered; and when the Evil Storm passed over, "the people were piled up in heaps ... a hush settled over Uruk like a cloak."[33]

The people of Uruk might well ask who caused such worry and lamentation. Why had their Anunnaki gods caused so much fear and worry, death and disaster? Tragically, most of the people did not have too much time to concern themselves with questions about the behavior of their gods, as they "piled up in heaps." Later, the chief goddess of Uruk, Inanna, who had fled to Africa, bewailed the desolation of her city and her temple, not to mention her jewelry.

There was but one god who remained in the vicinity of the tragedy and tried to help the people, as once again Enki demonstrated more compassion than any of the other Anunnaki. Enki, whose city and part-time residence was Eridu, left the city only far enough to escape the danger of the Evil Wind, but close enough to see its fate.

> Its lord stayed outside his city...
> Father Enki stayed outside the city
> ...for the fate of his harmed city
> he wept with bitter tears.[34]

After the Evil Wind passed, Enki surveyed Eridu and found the city silent and most of the former residents "stacked up in heaps." Enki determined that the city was still too dangerous to re-inhabit, so he led the survivors to the desert where he used his engineering skills to make some sort of "foul tree" edible.

Sumerian historical accounts leave little doubt that the people of the Near East suffered from a nuclear holocaust around 2,000 B.C., 2,024 B.C., according to Sitchin's calculations. This Sumerian account and the one discussed earlier in this chapter from the Indus Valley are probably the best historical accounts of nuclear holocausts in ancient times, but there are numerous other possible references to nuclear blasts on Earth in ancient times.

Zeus, the chief god of the ancient Greek civilization, probably used nuclear weapons when he fought the Titans, according to Sitchin. Sitchin offers the following translation of *Theogony* (Divine Genealogy) written by the Greek, Hesiod, in the eighth century B.C. Hesiod wrote that Zeus, from his bastion at Mount Olympus in Greece, hurled a "thunder stone" toward his enemies, the Titans, at Mount Othyres:

> The hot vapor lapped around the Titans, of Gaea
> born; Flame unspeakable rose bright to the upper

air. The Flashing glare of the Thunder-Stone, its
lightning blinded their eyes - so strong it was. Astounding heat seized Chaos... It seemed as if Earth
and wide Heaven above had come together; A
mighty crash, as though Earth was hurled to
ruin.[39]

After this blinding flash and mighty crash of this apparent ancient description of a nuclear explosion, a wind storm formed:

Also were the winds brought rumbling,
Earthquake and duststorm,
thunder and lightning.[40]

So evidence indicates that the gods of the ancient Greeks, too, were ancient ETs who fought other ETs on Earth. Herodotus the famous Greek historian of the fifth century B.C., often called by modern historians, "The Father of History," was convinced that the Greeks had borrowed from the stories of the ancient Egyptian gods to obtain their own theology. During a visit to Egypt, Herodotus was apparently struck by the similarities of personal attributes and the tales about the gods of Egypt and the gods of Greece.[41] Whatever the case, apparently the Greek gods, like the gods of Sumer, India, the New World and all of the ancient ETs, were not adverse to using nuclear weapons in their quarrels with other ETs.

William Bramley believes one passage in *The Book of Mormon* describes a nuclear holocaust that occurred somewhere in Central America or Mexico in 34 A.D. Bramley argues in his book, *The Gods of Eden*, that *The Book of Mormon*, given to a New York State youngster, named Joseph Smith, early last century on metal plates by an "angel", is another of what he calls a custodial religion instigated by regressive ETs to further divide and mislead humankind concerning spiritual matters. Nevertheless, Bramley argues, that *The Book of Mormon* should be considered as "historical" as the Bible, as it is an American continuation of some of the "custodial" events of the Old Testament. After Bramley quotes a lengthy passage from *The Book of Mormon* (3 Nephi 8: 5-23) which he believes describes a nuclear holocaust, he writes,

The rumblings, flashes of lightning, rapid incineration of cities, all within three hours, followed by three days of thick heavy darkness combine to accurately depict a nuclear strike followed by the inevitable thick lingering cloud of soot and debris....[35]

As we've stated, scholars have interpreted these amazing tales of gods and their shenanigans as myths that ancient people somehow dreamed up. These myths of cunning, flying gods are not considered by most modern scholars to have any basis in actual reality. This position was perhaps understandable in the nineteenth and the first half of the twentieth centuries. However, this position can no longer be supported now that human science has given us such marvels as space travel, genetic engineering, and, unfortunately, thousands of nuclear weapons. It is now clear that the tales, and the myths of ancient civilizations do have much substance and validity.

If some readers still need convincing that ETs were here and in a dominant position throughout ancient times, the historical records referenced herein, especially Sitchin's books, are packed with much more information than I have covered here. Sitchin also covers the regions of the first civilizations of the Americas. I feel that new discoveries and new scientific investigations will more and more support the view that ETs are and have always been involved in worldly affairs.

In this regard, one of the most fascinating areas of ancient history that I have not touched upon is Sitchin's well-presented argument that the pyramids of Giza and the world-famous Sphinx, were not built by the early Pharaohs of the Old Kingdom of Egypt, but by ETs thousands of years before the historical dynasty of Egypt began just before 3,000 B.C. The twenty or thirty other pyramids of Egypt, Sitchin claims, are imitations of the immaculate pyramids of Giza, and were constructed by the Pharaohs of the Old Kingdom. Sitchin's argument is quite detailed and convincing.[42] Sitchin wrote this in 1980. Interestingly, Sitchin has recently received some support from modern science.

Geologist Robert Schoch presented evidence to the 1992 annual meeting of the American Association for the Advancement of Science in Chicago that the Sphinx was carved between 5,000 and 7,000 B.C. Archeologists believe the Sphinx was carved about 2,500 B.C. by pharaoh Khafre. Today, Egyptian guides tell tourists that the face on the Sphinx is a depiction of Pharaoh Khafre. Schoch and two colleagues found that the weathering patterns on the Sphinx are much older than those of nearby monuments of the historical era of Egypt. Schoch has received harsh criticism from archeologists, whose chronology of ancient history does not match his claims. Professor Mark Lehner of the University of Chicago is quoted as asking:

"If there was another civilization that built the Sphinx so much earlier, where is the evidence? Where is a single potsherd (bit of broken pottery); where is any sign of such a people?" Schoch, he said, has tried to turn the chronology of Giza upside down on "one piece of ambiguous evidence, his interpretation of erosion. That is not how good science works."[43]

Speaking of "good science," the Chairman of Stanford University's geophysics department, Amos Nur, has concluded that Sodom and Gomorrah were destroyed by an Earthquake. Nur presents no data to support this view, except that the whole area is in an Earthquake zone where many Earthquakes have occurred in the past. The newspaper article in which Dr. Nur's hypothesis is reported even quotes the passage from Genesis, quoted above, where Abraham observes "...the smoke of the country went up as smoke from a furnace," and the cities of the Plain were obliterated. I will repeat again, in defense of Dr. Nur and Dr. Lehner and all scientists, that there is much evidence of the truth of the ET influence on this planet Earth that has been cleverly covered up by the very ETs who wish to continue to control humans. Nevertheless, it must be said that scientists' explanations of the happenings of the ancient world are sounding more and more absurd. I believe that more observations and discoveries in the near future will tend to support the hypothesis that ETs have had a major part in forming humans and this world, thereby further undermining the materialist-scientific view of reality. Watch for them!

[1] Stevens, W. 1988. *op. cit.*, pp. 129-130.
[2] Lamberg-Karlovsky, C. C. and J. A. Sabloff 1979. *op. cit.*, p. 205.
[3] Starr, C. G. 1983. *op. cit.*, p. 166.
[4] Von Däniken, E. 1989. *In Search of the Gods: Chariots of the Gods?; Gods from Outer Space; Pathways to the Gods.* (Three volumes in one). Avenel Books, New York, pp. 320-321.
[5] *Mahabharata, The Epic of Ancient India*, translated by R. Dutt, quoted by Sitchin, Z. 1985, *op. cit.*, p. 62.
[6] Boulay, R. A. 1990. *op. cit.*, p. 109.
[7] *Ibid.*, p. 111.
[8] Von Däniken, E. 1989. *op. cit.*, p. 319.
[9] *Ibid.*, pp. 318-319.
[10] Sitchin, Z. 1985. *op. cit.*, p. 64.
[11] Translation of the *Mahabharata* by C. Roy (1889), quoted by von Däniken, E. 1989. *op. cit.*, pp. 79-80.

More "Heartless" Behavior of the "Gods"

[12] The research of David Davenport is summarized in Valerian, V. 1991. *Matrix II: The Abduction and Manipulation of Humans Using Advanced Technology*, 3rd Ed., Leading Edge Research Group, Yelm, Washington, p. 370D.

[13] Professor Solomon is quoted by von Däniken 1989, *op. cit.*, p. 324.

[14] *Ibid.*

[15] *The New English Bible* 1971. Oxford University Press, New York, Genesis 19: 24-26.

[16] *Ibid.*, Genesis 19: 27-29.

[17] Sitchin, Z. 1985. *op. cit.*, pp. 310-344.

[18] Lamberg-Karlovsky, C. C. and J. A. Sabloff 1979. *op. cit.*, pp. 165-166.

[19] Sitchin, Z. 1885. *op. cit.*, p. 327.

[20] *Ibid.*, pp. 328-329.

[21] *Ibid.*, p. 329.

[22] Boulay, R. A. 1990. *op. cit.*, pp. 261-289.

[23] Sitchin, 1985, pp. 316-317, and Boulay, 1990, pp. 271-272.

[24] Sitchin 1985. *op. cit.*, p. 311.

[25] *Ibid.*, p. 327.

[26] *Ibid.*, p. 328.

[27] *Ibid.*, p. 334.

[28] *Ibid.*, p. 335.

[29] *Ibid.*, p. 337.

[30] *Ibid.*

[31] *Ibid.*, p. 340.

[32] *Ibid.*, pp. 339-340.

[33] *Ibid.*, 340.

[34] *Ibid.*

[35] Bramley, W. 1990. *op. cit.*, p. 343.

[36] Stevens, W. C. 1989. *op. cit.*, pp. 44-47.

[37] Stevens, W. C. 1988. *op. cit.*, pp. 135-136.

[38] Valerian, V. 1991. *op. cit.*, pp. 375-376.

[39] Sitchin, Z. 1985. *op. cit.*, p. 55.

[40] *Ibid.*, p. 56.

[41] *Ibid.*, p. 49.

[42] Sitchin, Z. 1980. *op. cit.*, pp. 253-282.

[43] Petit, C. 1992. "Controversy over age of Egypt's Sphinx". *San Francisco Chronicle*, Feb. 8, p. A-4.

[44] Stevens, W. C. 1989. *op. cit.*, p. 24.

"Is there any question that the watchword "E pluribus unum" (one of many) adopted during the successive stages of evolution in those stirring times is not a direct outpicturing of the Spirit of Truth? It certainly did not emanate mechanically from man's mortal mind. Then the emblematic phrase - In God We Trust - does not that show the most sanguine faith or trust in God, the creator of all? Then the choice of the eagle, the bird both male and female, complete in one, as the emblem. It shows that these men were deeply spiritual or they built better than they knew. Can you doubt for a moment that all were guided by the whole of the God spirit in creative action? Does it not bespeak that America is destined to be the guide to the whole world?"
Emil speaking from the
Life and Teachings of the Masters of the Far East.

Chapter 10

Esoteric Correlations with Historical Data

Our investigations of the beginnings of human food production and the rise of human civilization have again demonstrated how woefully inadequate are the explanations of materialistic science, especially when they are compared to the few historical accounts we have of these enormous cultural leaps. In the course of our investigations, however, we have uncovered evidence that the historical accounts we do have may have been deliberately distorted. Therefore, as we did in Chapters 4 and 5, we will investigate a few esoteric sources in order to try to gain a more complete and accurate picture of our ancient history. We will find that the esoteric sources used tend to correlate with - and clarify the historical data we have reviewed.

We have seen that the entity Enki (or Ea) is portrayed in the Mesopotamian historical epics as having played a crucial role in the genetic engineering of modern humans, and he consistently championed the cause of humans, usually in the face of considerable opposition from other "gods," the Anunnaki. Enki, who is identified as the chief genetic engineer of the Anunnaki and half-brother of Enlil in the Mes-

Esoteric Correlations with Historical Data

opotamian creation epics, is identified as a Sirian, "who protects humanity," in *The Prism of Lyra* by Royal and Priest.[1] This channeled information certainly fits in well with the historical information we have reviewed. The reader will recall that we have evidence of entities from the vicinity of the double star Sirius being involved in the ancient affairs of humans from the oral traditions of the Dogon tribe of west Africa.[2] This, plus historical data indicating that ancient Egypt was involved with entities from near Sirius leaves little doubt of a Sirian connection with ancient humankind.[3]

Enki's frequent compassionate behavior that he displayed towards the human slaves of the Anunnaki bespeaks of an entity that was much more spiritually evolved than the Anunnaki. The Anunnaki would not want their slaves to know of an entity who had their best interests in mind and was actually trying to help them. On the other hand, the Anunnaki apparently needed the skills of Enki, especially in the genetic engineering of their slave species. It seems probable, therefore, that the Anunnaki would try to distort accounts of Enki, the Sirian, even to the point of making Enki one of their own.

In *The Prism of Lyra*, the Sirians are depicted as interfering with the plans of a Lyran group that was trying to create a species that was to have no knowledge of good and evil. How "the Lyran group" was related to the Anunnaki of the Sumerian texts is not made clear. Perhaps, the Lyrans of *The Prism of Lyra* and the Anunnaki of the Sumerians are one. Or, perhaps we are looking at different levels, where the Anunnaki of the "12th planet" in our solar system are actually carrying out the wishes of a "Lyran group" unknowingly.

In any case, *The Prism of Lyra* states that the Lyrans and Sirians, who had worked together in the creation of humans, the primitive workers, disagreed philosophically concerning their creation. While the Lyrans wanted to create a species devoid of the knowledge of polarity - or of "good" and "evil" - the Sirians saw that humans could not evolve spiritually without this knowledge. Royal and Priest point out that Enki in the Sumerian texts is sometimes portrayed as a serpent - an evil serpent - and that perhaps this was a ploy by the Lyrans to keep humanity from following the instructions of the Sirians who were attempting to help humankind. Royal and Priest imply that it was the Sirian group that encouraged Adam and Eve to eat of the Tree of Knowledge in the story of the Garden of Eden, as contained in the Bible.[4]

The Sirians may have been at least temporarily thwarted by the Lyrans (the Anunnaki?) in the Garden of Eden, but it seems that the Sirians had the last laugh. According to *The Prism of Lyra*, the Sirian group inserted a latent DNA code in humans:

> The code is triggered by an accelerating vibration that occurs when a civilization begins to evolve spiritually. As Earth accelerates toward self-awareness and fourth density (which is occurring presently), the code is activated. Once activated, the human race unwinds its limited vision like a coil until the expanse of All That Is becomes visible. It was their way of allowing humanity to eat from the tree of life after all.[5]

Perhaps your latent DNA code implanted by Enki or other members of the "Sirian group" is becoming active even as you read this book. In any case, it is safe to say that entities from a planet near Sirius were very much involved in our biological, cultural, and spiritual evolution/creation.

Another esoteric source whose information concerning the creation of modern humans that correlates with the historical information we have reviewed, is known as The Awareness. The Awareness describes Itself as a non-physical entity and has been communicating to a human group known as Cosmic Awareness Consciousness since 1963 through carefully trained channels. Cosmic Awareness Consciousness has been putting out a newsletter, published about every three weeks, with messages of The Awareness and Its answers to questions submitted by people from all over the world for almost three decades now. See the introduction for a discussion of The Awareness and channeling in general.

The Awareness informs us that Enlil, the leader of the regressive reptilian gods -the Anunnaki of the Sumerian texts and the "lizzies" of the Pleiadian Plus group - created humans as prototype slaves. These first prototypes - these first Adams - were incapable of reproducing themselves and were not good slaves. They had no self-motivation and were not easily programmed to follow instructions.[6]

A second prototype was created with the help of Enki of Sirius, who infused other genes and genetic substances into the new Adams (and Eves) which allowed them to reproduce. This pleased Enlil, but the infusion of genes by Enki also allowed the Adams and Eves the ability to make choices and have what was called knowledge. This did not please Enlil, who The Awareness states was the god depicted in the early Chapters of Genesis. And, Enki was represented as the serpent in the garden of Eden and was generally the advocate of the creatures, humans, he helped create. For example, Enki helped Noah and others survive the flood, which was caused by the melting of ice in Antarctica.[7]

Esoteric Correlations with Historical Data

The Awareness states that Enki's infusion of Sirian and other genes allowed humans to become less and less like Enlil, their reptilian creator, with the passing of each generation.[8] Thus, early humans, because they had more of the genes of Enlil, lived long lifetimes, such as Noah's, nearly 1,000 years. But, with the passing of each generation, Enlil's (the reptilian) genes influenced humans less and less and the human life span became more like that of living humans, less than 100 years. So, humans were denied a form of physical near-immortality (from the Tree of Life), but they were granted the ability to evolve mentally and spiritually from The Tree of Knowledge. Thus, while human life spans became shorter, the ability to reason, the powers to make choices, and other powers of the mind were evolving and developing. These powers have evolved to the point where humans are capable of challenging their own creators with compassion, understanding and intelligence, although these attitudes and behavior are still weak in humans in many ways. The Awareness states that we are spiritual beings and that It, Enki, and many others of what It calls, The Family of Light, are, and have been trying to help us evolve spiritually for thousands of years.[9]

The information provided by The Awareness about the "lizzie" nature of the Anunnaki and their extremely long life-spans correlates to the historical[10] and the esoteric[11] information we reviewed in Chapter 5. The transmissions of The Awareness summarized here also correlates with the Mesopotamian historical information we reviewed of the Great Flood and Enki's role in saving a few humans from that catastrophe.

We recall from Chapter 8 that the Mesopotamian historical texts state that the world was divided into four regions at the inception of the first human civilizations; three of them under nominal human control, but all of them controlled directly by one or more members of the high hierarchy of the Anunnaki. Enki was the "Anunnaki god" named to be in charge of the Egyptian civilization. This correlates with the probability that Enki was a Sirian. As we've seen, the ancient Egyptians revered the planet Sirius, not Nibiru, the 12th planet revered by the ancient Sumerians. The likelihood that Enki was a Sirian, however, calls into question whether the Anunnaki lizzie "gods" of Sumer had any control and influence in ancient Egypt at all. We observed that archeological-historical evidence does not support a close relationship between these two ancient civilizations. Some of the contrasts we noted between the ancient Egyptian and Sumerian societies are differences in the degree of urbanization, the difference in the godly-human status of the kings or pharaohs of the two areas, and the archeological evidence of hardly any material exchange between the two ancient Dynasties, even though they were located only a few hundred miles from each other. Perhaps the most significant difference, noted by a historian, was the greater humor, apparently happier, less foreboding attitude that the ancient Egyptians had with regard to the "gods" above them. The "gods" of Egypt may have actually shown some compassion toward their human subjects, in sharp contrast to the Anunnaki "gods" of Sumer. The fact that the Anunnaki of the Sumerian texts claim sovereignty over Egypt is probably a deliberate distortion of these texts perpetrated by these lizzie ETs. This, plus the claim that Enki was an Anunnaki, was possibly an attempt by these lizzie ETs to assure their human slaves that, in addition to controlling them, they controlled everybody and all known civilizations of the world. And certainly the Anunnaki did not want their slaves to know that there were civilizations whose gods had at least a somewhat loving relationship with their human subjects.

This is not to suggest that the Anunnaki did not, at some point, gain partial or complete control of ancient Egypt. Unfortunately, historical epic texts, such as those found in the remains of ancient Mesopotamian cities, either never existed in Egypt, or have never been found, or were destroyed. Sitchin, however, makes it clear in *The Wars of Gods and Men* that the gods of these two ancient civilizations were often at odds with each other, often to the point of armed conflict, such as the destruction of a spaceport in the Sinai peninsula and the cities, Sodom and Gomorrah, by nuclear bombs we recounted in the last Chapter.

Concerning possible nuclear blasts in the ancient world depicted by ancient Mesopotamian texts, we again have correlating evidence from esoteric sources. Billy Meier was told by the "Swiss Pleiadian" group and their allies, that there have been many nuclear wars here on Earth. The "Swiss Pleiadians" tell of many times in the past when their ancient ancestors and other ETs had to flee the Earth in order to escape nuclear war. Apparently every time a civilization would arise here on Earth - and, again, they are usually talking of civilizations that are much more ancient than Atlantis and Lemuria - the Earthmen, ETs, or whatever, quarreled and eventually destroyed their civilizations. For example, the "Swiss Pleiadians" told Meier of a nuclear war that completely depopulated North America some fifty thousand years ago. They told him that Sodom and Gomorrah were destroyed by nuclear blasts. The "Swiss Pleiadians" talked to Meier about ancient nuclear blasts in India, in Lebanon, Australia, and Ecuador.[12]

Wendelle Stevens, who has been responsible for publishing much of what the "Swiss Pleiadians" told Meier, states in a footnote of his book, *Message from the Pleiades*, that:

> ...the Takauti Documents of Japan, which pre-date all other records there and all records in the western world, going back over 24,000 years, support these claims of earlier man-made atomic devastation of this world. These Takauti Documents, on which the ancient Shinto Teachings and the old Kojiki history are based, describe a great worldwide atomic war of that ancient time, and even contain maps showing the locations of each of the atomic blasts, and the cities destroyed. The symbols on the maps are a mushroom-shaped cloud...[13]

Valdamar Valerian in *Matrix II*, using esoteric sources, presents a chronology of Earth history going back seventy-five million years in which several nuclear wars have been fought.[14]

Conclusion

What ancient historical information we have, plus esoteric sources that correlate with ancient historical data, indicate that the Earth has been inhabited and fought over by various ET groups for millions of years. Modern *Homo sapiens*, as we call ourselves, was perhaps, ultimately created by the Divine Creator of this Universe, but some of the ET groups who have been encroaching on this planet for millions of years, have genetically manipulated and engineered the physical aspects of the human species. ET genetic manipulation of the human species is astonishing news, but what is more shocking, is that some of these ETs, such as the Mayan and Mesopotamian "gods", engineered humans so that human spiritual knowledge would be limited. Thus humankind could be more easily manipulated and controlled.

Evidence indicates that the Anunnaki and other ET species were directly and overtly involved with the early Dynasties and Kingships of Mesopotamian and other early human civilizations until around 1,000 B.C., when the ETs left Earth, or literally went underground. For many ETs their own consciousness-spiritual evolution took them to a state of consciousness where it was considered unethical to interfere with the affairs of Earth humans, and they longed to return to the planet(s) of their origins where a high spiritual state had been reached that emphasized love.[15] For more regressive ETs, however, this stage of Earth history at around 1,000 B.C. represented an opportunity to move underground where they would secretly continue to control and manipulate the course of human history.

Let us look for information that human history and knowledge has been secretly managed and controlled by ET species, especially with an understanding of the human spiritual nature.

[1] Royal, L. and K. Priest 1989. *op. cit.*, p. 77.
[2] Temple, R. K. G. 1987. *op. cit.*
[3] *Ibid.*
[4] Royal, L. and K. Priest 1989. *op. cit.*, p. 77.
[5] *Ibid.*, p. 75.
[6] *Revelations of Awareness* 1991. Issue 91-16. Cosmic Awareness Communications, P.O. Box 115, Olympia, Washington, 98507, pp. 3-5.
[7] *Ibid.*
[8] *Ibid.*
[9] *Ibid.*
[10] Boulay, R. A. 1990. *op. cit.*
[11] Marciniak, B. 1992a. *op. cit.*
[12] Stevens, W. C. 1989. *op. cit.*, pp. 44-47.
[13] Stevens, W. C. 1988. *op. cit.*, pp. 135-136.
[14] Valerian, V. 1991. *op. cit.*, pp. 375-376.
[15] Stevens, W. C. 1988. *op. cit.*, pp. 129-130.

"I claim not to have controlled events, but confirm plainly that events have controlled me."
Abraham Lincoln, *Speech 1864*

Chapter 11

Hoodwinked and Controlled

If our "creator gods" genetically designed us to be functional, but ignorant slaves, especially ignorant of spiritual knowledge, what means might they have used to keep humans unenlightened for thousands of years? And what historical data do we have of such a diabolical conspiracy?

We continue our historical investigation of this conspiracy from the book of Genesis of the Bible, where we read the famous story of the Tower of Babel:

> Once upon a time all the world spoke a single language and used the same words. As men journeyed in the east, they came upon a plain in the land of Shinar (Sumer) and settled there. They said to one another, 'Come, let us make bricks and bake them hard;' they used bricks for stone and bitumen for mortar. 'Come,' they said, 'let us build ourselves a city and a tower with its top in the heavens, and make a name for ourselves; or we shall be dispersed all over the Earth.' Then the Lord came down to see the city and the tower which mortal men had built, and he said. 'Here they are one people with a single language, henceforward nothing they have a mind to do will be beyond their reach. Come let us go down there and confuse their speech, so that they will not understand what they say to one another.' So the Lord dispersed them from there, all over the Earth, and they left off building the city. That is

why it is called Babel, because the Lord there made a babble of the language of all the world; from that place the Lord scattered men all over the face of the Earth.[1]

Several comments about this passage are in order. First, the "us" is clearly yet another oversight by the Hebrew editors who were probably "monotheising" ancient Mesopotamian accounts of the early history of humans. The "us" most likely refers to Enlil and his Anunnaki comrades. Secondly, what is wrong with "making a name for ourselves?" The Lord (Enlil?) answers this rhetorical question: "henceforward nothing they have a mind to do will be beyond their reach." The "Lord," Enlil and his comrades, would not be able to control and manipulate their slaves. So, he confused their speech and scattered them.

Sitchin not only believes that the plural "Lord" represents Enlil and the other high ranking Anunnaki, but that the humans were building a "launching tower" for a rocket ship. Mr. Sitchin bases his belief on Mesopotamian stories that have a similar theme as the Biblical story, and what he believes is a mis-translation of the ancient word shem. Sitchin argues that shem, which is translated as "name" in the Bible (let us... make a name for ourselves), should be translated as "skyborne vehicle", or rocket ship. Sitchin points out that throughout the Mesopotamian texts it is made clear that flying machines were meant for the gods and not for humankind.[2]

The babbling-of-the-languages is also seen in the New World. The *Popul Vuh*, the book that contains the only version we have of the knowledge of the ancient Mayan mythical-historical accounts, also speaks of human trouble with different languages. In this case, however, the Mayan "gods" spoke different languages, and the people had to learn a new language every time their tribe fell under the rule of a new "god."[3]

William Bramley, in his recently published book, (1989) *The Gods of Eden: A New Look at Human History*, has found strong evidence of a conspiracy by UFO associated beings to control humans by keeping human knowledge of spiritual consciousness and spiritual behavior at low levels.[4] Mr. Bramley, writing in a rigorous, scholarly manner and using only historical references, has presented a novel and radical way of looking at human history, as is indicated in the subtitle of his book. Among other things, Mr. Bramley argues that other intelligences, ETs or aliens, have tampered with all of the major and minor religions of Earth in order to prevent humankind from gaining spiritual knowledge and development.

Mr. Bramley began his research for *The Gods of Eden* with the intention of investigating the causes of war in human civilizations. Through his research, Bramley first found strong evidence that human wars are caused, and then financed by a small group of men in order to control and profit from the wars of human societies. This small group of men, of course, operated in secrecy and often employed secret societies to help carry out their deceit and manipulation of the human population. As Bramley continued his investigations he started coming across the ET factor again and again, until he concluded that ETs have been behind almost all events throughout our history to keep humankind ignorant, divided and in more or less constant conflict with each other. In the beginning, Bramley had no intention of dwelling on the UFO-ET-alien phenomenon at all, as he had not come to this study of history as one interested in UFOs, as was the case when I set out to investigate the truth of our origins. He was a rational man, he says, and "...(r)ational minds tend to seek rational causes to explain human problems."[5] Certainly rational minds do not go looking for UFOs, (at least until now). But, the more he uncovered evidence of conspiracies to control and dupe humankind, the more the UFO-ET-alien phenomenon appeared.

Bramley began his investigation with a single idea that was not even a novel idea: *"War can be its own valuable commodity"*(his emphasis).[6] Of course, he was originally thinking of those who profit monetarily from wars, such as those who sell military hardware to warring nations or lending institutions who make loans to warring governments. But, he could also see, through his studies, that war can also be a very effective "....tool for *maintaining social and political control over a large population*" (his emphasis).[7] Bramley discusses the theories of the 16th century Italian author and statesman, Niccolo Machiavelli, whose name has become synonymous with less-than-moral politics - some might call it ruthless politics. Machiavelli was simply reflecting the turbulent politics of his times when the 15th and 16th century city-states of Italy were seemingly constantly deceiving and warring with each other. After reviewing some of Machiavelli's writings, Bramley summarizes the political writer's observations concerning how a third-party instigator can secretly divide and control large groups of people by instigating conflicts between the people:

1. Erect conflicts and "issues" which will cause people to fight among themselves rather than against the perpetrators.

2. Remain hidden from view as the true instigator of the conflicts.

3. Lend support to the warring parties.

4. Be viewed as the benevolent source which can solve the conflicts.[8]

Every reader has heard some version of the old adage, 'divide and rule', but how about, 'remain secreted, cause conflicts among the people, support all sides of any conflict, and appear to be what you are not, benevolent.' There is no evidence that Machiavelli knew anything at all about UFOs and aliens. He appears to have been merely a keen observer and communicator of the manner in which the politics of his day were carried on in his city-state and neighboring city-states. However, Bramley believes that Machiavelli has expressed in human terms precisely how ETs have operated throughout human history.

Bramley uses some of the references we have used, such as Sitchin's views of the Sumerians and other 'ancients,' as well as the Bible, to demonstrate that the human race was created as slaves and treated very harshly by the early 'gods' of ancient history. Bramley does not like the terms 'gods' or 'extraterrestrials' as both words, he reasons, are used in many ways, so he refers to the ancient 'gods' of the Sumerians and the Bible as 'Custodians.' The Custodian 'gods' influenced human affairs directly for some time, argues Bramley, and then went 'underground' to manipulate and control humans by working within certain secret societies that have flourished in many human civilizations since those of ancient Sumer and Egypt. Bramley points out that the secret societies of civilizations through the centuries had several levels, or degrees, of initiation and secrecy. As the members advance to successively higher levels, they are exposed to more secrets of the society to which they belong. The highest and most secret level of some of these secret societies are where the 'Custodians' - the ET or alien entities - have exacted their power and influence. Following this scenario, a few humans who have ascended to the highest level or degree of these secret societies, would then teach some sort of discord, such as a form of racism and ethnocentrism to members of the secret society who were at lower levels of understanding. The secret society would then secretly agitate for a cause that would eventually lead to armed conflicts. Other secret societies, or splinter groups of the same secret society, would secretly support the "other side" of whatever cause was being contested. In addition to power access to the secrets, the 'custodians' provide the humans of the highest level of these secret societies with power and control of their own. These humans would often be the recipients of enormous monetary gains through "helping" the financing of any wars that might be fought, as a result of the agitation of the secret societies. Bramley cites several examples of human conflicts that he believes were almost certainly greased by the actions of one or more secret societies, including the "religious" wars of Europe that occurred after the Protestant reformation, and the numerous revolutions that flared-up around the world in the 18th, 19th and 20th centuries, including the American Revolution.

Bramley recognizes that the greatest harm the 'Custodians' have done to the human species, is to hinder their spiritual development. Bramley documents how the custodians spread misinformation in order to obfuscate any true spiritual truths that might have existed in any of the major world religions.

Bramley contrasts "Custodial" religions with what he calls "maverick" religions. Maverick religions, according to Bramley, are basically religions inspired by the teachings of a highly evolved spiritual teacher that quicken the spiritual truths, in contrast to custodial religions, in which spiritual truths have been twisted and distorted until they are all but unrecognizable.[9] For example, Bramley identifies Gautama Siddharta - the Buddha - and Jesus as representatives of spiritually high teachers whose teachings established "maverick" religions that were soon "Custodialized." Through studies of these and other "maverick" religions and their "Custodialized" offspring, Bramley has drawn-up a comparison of "Custodial" and "maverick" religions.[10]

Basically, maverick religions emphasize the individual human as the basis of spiritual evolution, not some supernatural force, some supreme being, or some 'god' as the center and basis of spiritual knowledge, as is emphasized by custodial religions. The differences between maverick and custodial religions is nowhere seen more clearly than when comparing the reasons an individual should follow a particular doctrine or religion. In a maverick religion, "...observation and reason are held to be the proper foundation for adhering to a doctrine."[11] On the other hand, most of what Bramley would label custodial religions, teach that adherence to a particular doctrine is based upon "...faith or obedience alone."[12] We will have occasion to refer to Bramley's interesting comparison in our future investigations, but let us look at these comparisons concerning the important question as to who or what is responsible for one's own spiritual evolution and salvation. To the maverick religions:

> Spiritual recovery and salvation are entirely up to the individual to achieve through his or her own self-motivated effort.[13]

But to the Custodial religions:

> Spiritual recovery and salvation depend entirely upon the grace of "God" or other supernatural entity.[14]

Bramley has uncovered evidence indicating that custodians (ETs) have caused plagues and diseases that have punctuated human history. One such plague was a little known plague called "Justinian's Plague," which struck what was known as the Eastern Roman Empire centered in Istanbul in the middle of the 6th century A.D., then spread to Europe. Bramley quotes historians and other sources from that time that indicate that there were frequent appearances of unusual aerial phenomena that sound much like modern-day UFO phenomena, located in regions where outbreaks of the Plague occurred. These observations by humans "....chillingly suggest the unthinkable: that Justinian's Plague was caused by biological warfare agents spread by Custodial (ET) aircraft."[15]

Evidence is also strong, according to Bramley, that ETs caused The Black Death of Europe. The Black Death, so named because of the dark color of its victims after death, is said to have begun in Asia and had spread to southern Europe by 1347 A.D. By the time the Plague had spread through northern Europe four years later, at least a quarter to as much as one-half of the population of Europe had died of the Plague. Whole villages were wiped-out by the Plague. The hardships, pain and suffering the Plague caused is unimaginable.

The Plague struck Europe every 10 to 20 years after this initial outbreak, although never as severely as the first outbreak, for at least four hundred years. Although exact figures are impossible to calculate, Bramley estimates that well over 25 million people were killed by the initial outbreak of the Plague and a total of more than 100 million Europeans died of it over a 400 year period.[16]

The Plague is caused by a rod-shaped bacterium (*Yersinia pestis*), which often infest rats and squirrels and other rodents. There were a few reports of outbreaks of rodent infestation in Europe before the Plague struck, so it has generally been concluded that The Black Death was and is spread by rodents infested with this type of bacterium.

The Plague germ is known to infect the body in three ways, two of which were known to occur during the plagues of Europe. In the first type, the germ enters the lymph glands which causes swelling (called buboes) in the armpits and groin. Other symptoms are fever and vomiting, and death occurs within a few days, if not treated. This type, known as the Bubonic Plague, is not communicable between humans. However, the second type of infection which, according to Bramley, some authorities believe was responsible for most of the deaths of The Black Death, infects primarily the lungs, is known as the pneumonic plague, and is highly contagious.[17] The pneumonic plague causes high temperature, the coughing of blood and almost always death within a few days.

As with the Justinian Plague 700 years earlier, lights, usually called comets by the populace, were often associated with outbreaks of the Plague. The people of Medieval Europe were much more superstitious than today's population, and meteors were seen as portents of later disasters, like a plague. However, Bramley makes a good case that many of the so-called comets were actually what we would today call UFOs.[18] In fact, Bramley reproduces a drawing of a "comet" published in 1557, drawn from eyewitness reports, that clearly is not a comet, and is what we would today call a "cigar-shaped" UFO, or a rocket ship.[19] (see Figure 11-1)

Bramley notes several eyewitness accounts that tie "strange lights" of the sky with the plague. From a book written by Johannes Nohle, who in 1926 assembled a number of reports of strange aerial phenomena which were associated with outbreaks of the plague, Bramley quotes this 1568 account from Vienna, Austria that begins a lot like a modern UFO report:

> When in sun and moonlight a beautiful rainbow and a fiery beam were seen hovering above the church of St. Stephanie, which was followed by a violent epidemic, in Austria, Swabia, Augsburg, Wuerttemberg, Nuremberg, other places carrying off human beings and cattle.[20]

Bramley points out that "vile-smelling" mists were often associated with an outbreak of the plague. A modern historian and student of The Black Death writes that,

> ...people were convinced that they could contract the disease from the stench, or even, as is sometimes described, actually see the plague coming through the streets as a pale fog.[21]

Several of these "evil smelling mists" were seen and associated with an outbreak of the plague, and some "mists" were directly associated with bright lights in the sky. Sometimes strange men dressed in black and other terrifying figures, usually considered "demons," were seen spreading "mists" and were associated with the spread of the plague.

Hoodwinked and Controlled

These "demons" were often reported carrying "brooms," or "swords," or "scythes," and Bramley believes the "image of death" carrying deadly scythes that are used to harvest people, may be associated with these "demons." Europeans drawing paintings of this time had their origin, Bramley believes, with these strange, unknown figures that were apparently spreading the plague and death.

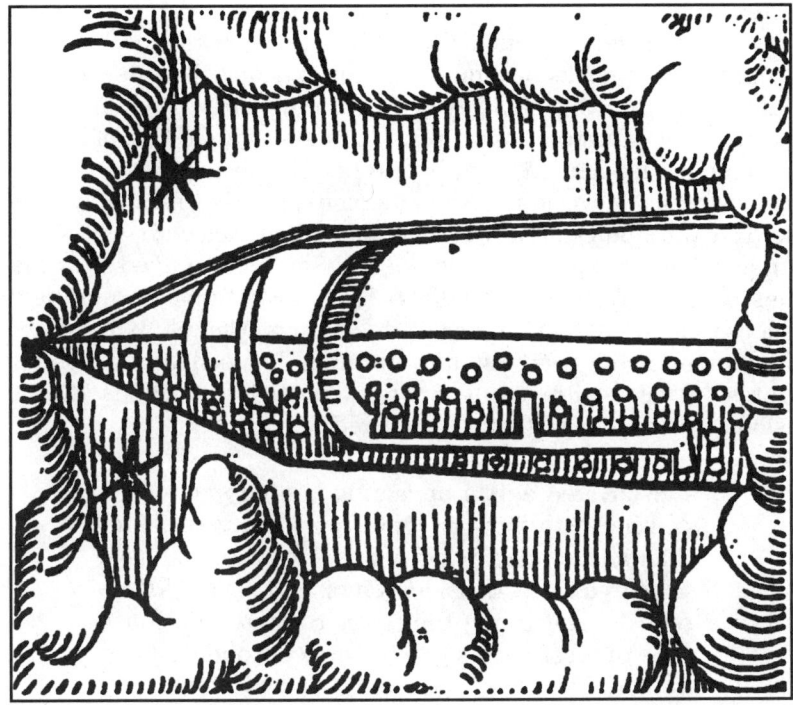

Figure 11-1 In European Medieval times, any unusual flying object was called a comet. This illustration, first published in 1557, was based on eyewitness testimony of a "comet" observed in 1479 in Arabia. Reproduced from "A Chronicle of Prodigies and Portents..." by Conrad Lycosthenes. From Bramley, 1990, courtesy of Dahlin Family Press.

Bramley believes that what he calls the "Custodians," were carrying out germ warfare on the humans of Europe and elsewhere during The Black Death, and during other outbreaks of epidemic diseases throughout the world. Plagues and other epidemic diseases have been prevalent, on and off, throughout all periods of human history, sparing no creed, color nor people of any human dogmas. He points out that the current governments of America and Russia and other parts of what was the Soviet Union, have developed and stockpiled biological weapons containing the germs of the Plague and those of several other epidemic diseases, which when used, appear as visible mists. When these modern biological weapons are released from their canisters, where the germs or other biological agents are kept alive, they form mists which when simply inhaled by the "enemy," or any human being, will most likely cause infection, disease, and in the case of the Plague, at least, death in a few days after the disease is contracted, unless treated.

Bramley presents evidence that ETs use sophisticated holograms to deceive and mislead humans. As an example, Bramley cites the founding of the Church of Jesus Christ of Latter Day Saints, also known as the Mormon Church. The Mormon ChurchBook of Mormon is today headquartered in the state of Utah in the western United States, and has a world-wide membership of close to six million people. The Mormon Church was founded in the nineteenth century by a young religious fanatic named Joseph Smith, who lived near Manchester, New York. In the 1820s, Joseph Smith saw a number of visions that led to the founding of the Mormon Church. While praying in his room, Smith was visited by an "angel" named Moroni. Bramley quotes Joseph Smith words written almost two decades after the visions were supposed to have occurred:

> I discovered a light appearing in my room, which continued to increase until the room was lighter than at noonday, when immediately a personage appeared at my bedside, standing in the air, for his feet did not touch the floor. He had on a loose robe of most exquisite whiteness beyond anything Earthly I had ever seen; nor do I believe that any Earthly thing could be made to appear so exceedingly white and brilliant. (*Joseph Smith* 2:30-31)

Moroni told Smith where to find some ancient metal plates which contained a history of a group of ancient Hebrews that were transported from the Near East to North America at about 500 BC., and the subsequent history of these people in North America. Smith was instructed by Moroni to find these tablets and present them to the world. After the message, Moroni disappeared in a peculiar way:

> ...I saw the light begin to gather immediately around the person of him who had been speaking to me, and it continued to do so until the room was again left dark, except just around him; when,

> instantly I saw, as it were, a conduit open right up into heaven, and he ascended till he entirely disappeared... (*Joseph Smith* 2: 43)

Only a few minutes after he had disappeared, Moroni came into his room again. Smith relates:

> He (the angel) commenced, and again related the very same things which he had done at his first visit, without the least variation; which having done, he informed me of great judgments which were coming upon the Earth, with great desolations by famine, sword, and pestilence; and that these grievous judgments would come on the Earth in this (Joseph Smith's) generation. Having related these things, he again ascended as he had done before. (*Joseph Smith* 2: 45)

This apparition apparently came and went several times that night and even the next day when Smith was returning home after a day working in the fields. Bramley believes that the fact that the young Smith saw and heard an identical message delivered in an identical way, that Smith was seeing a hologram with a recorded message, albeit a message to which additions could be made. Bramley believes the Mormon Church, as well as all human religions and most of the their varieties, were "inspired" by regressive ETs - what he calls "Custodians" - in order to divide humankind and keep humankind in a deep state of spiritual ignorance.[22] Bramley argues that many of the policies of the Mormon ChurchBook of Mormon are examples of "spiritual misinformation," such as the racial policy of the church and their belief that one's immortal spiritual self can and should be united to a physical form. Until the last decade, black people were not allowed into the priesthood of the Mormon Church because blackness of skin color was believed to be the result of punishment by God. This belief is expressed in the *Book of Mormon*, the Bible of the Mormons, in II Nephi 5:21-23. To the credit of the Mormons, as of the last decade, they have officially dropped this racial belief and now admit black people to their priesthood.

After pointing out other Mormon beliefs that he feels are "spiritual misinformation" spread by regressive "Custodians," Bramley points out some of the good humanitarian work done by the Mormon Church, and states that no human organization is all bad or all good. He states that, "...(t)he real trick in judging a person or group is to determine whether more good is being done than bad, and how the bad may be corrected without destroying whatever good there might be..."[23]

On a related subject, Bramley presents evidence that the prophetic writings, usually called apocalyptic, or end-of-the-world teachings, are the work of "Custodians," or what we are referring to as regressive ETs.[24] Apocalyptic predictions have been a regular aspect of human history since at least 500 B.C. when monotheistic religions were well established in Persia (Zoroastrianism) and Israel (Judaism). Bramley summarizes apocalyptic prophesies:

> Mankind will suffer upheaval during a future global cataclysm. The cataclysm will be followed by a Day of Judgment in which God or a representative of God (of a monotheistic religion) will decide the fate of every person on Earth. Only those people who are obedient to the religion preaching the apocalypse will be granted mercy on the Day of Judgment. Everyone else will be doomed to death or eternal spiritual damnation...[25]

Apocalyptic prophesies, Bramley argues, provide one more distortion of spiritual truths, and channel people into obeying a particular religion or "god". End of the World predictions create much fear in people, and discourage them from exploring other religious systems.

Bramley has also found considerable evidence that the world monetary system was also established with the help of human secret societies by the "Custodians," or regressive ETs, for purposes of manipulation and control.[26] Bramley demonstrates that the international "funny money system", as he calls it, was organized from the middle of the 19th century into the 20th century so that financial control of a few private individuals and their banks are able to dominate the political system of each nation and the economy of the world as a whole. The economy of the world as a whole in the 20th century is characterized by inflation, massive debts, and fluctuations; fluctuations from recession to better times to recession etc.

Esoteric Correlations with Historical Sources

The Hidden Hand

We have quoted William Bramley at length because his version of history, emphasizing secret spiritual distortions and other clandestine manipulations of human affairs by regressive ETs, or custodians, is the tract followed here. Needless to say Bramley's version of history is not taught in any school of this or any other country, as far as I know. If this version of history is at least partially correct, it has been covered up and hidden from humanity by the very regressive ETs that have been manipulating human affairs behind the scenes. At present, we can only receive partial conformation of radical historical works used here, like those of Bramley and Sitchin, from esoteric sources. We will observe in this section that Pleiadian esoteric sources to which we have been referring that have correlated with the historical evidence presented in previous chapters, support and corroborate the essential thesis presented by Bramley in *The Gods of Eden*.

The Pleiadian Plus group, channeled through Barbara Marciniak, tells us that extraterrestrials - off-planetary beings - have negatively influenced almost all human institutions, especially our religions, in order to keep us ignorant and controllable.[27]

The "Swiss Pleiadians" and their friends through Billy Meier, also talk of control of humans by ETs through "Earth religions and secret circles."[28] They talk of ancient times, many thousands of years ago, when many humans were in regular contact with ETs and shared much of their knowledge. But, gradually, these contacts became less frequent until they ceased altogether due to the "...interference...of evil inclined elements, of Earthly and extraterrestrial origin."[29] So, eventually the knowledge of the ETs was lost. Secretly, some of the regressive ETs built stations beneath the Earth and on nearby planets, fostered cults and, in general, secretly manipulated humans.[30]

The "Swiss Pleiadians," as well as the Pleiadian Plus group freely talk of the involvement of their own ancient ancestors in nearly all of the ancient civilizations of humankind. At times, they were allies to the creator gods of the ancient Sumerians and at times they were their enemies.[31] We noted in a previous chapter that the ancient Pleiadians were at least some of the "sons of god" that mated with some of the "daughters of men," mentioned in Genesis and other ancient human records, and thereby contributed to human biological evolution. The "Swiss Pleiadians" tell of some of their ancient ancestors who carved their own dominions, enslaved races of humans, demanded to be treated as "gods," and were very cruel to humankind. Eventually, the spiritual evolution of these ancient Pleiadians created cravings within them for home, and they left the Earth and thereby left the development of Earth humans to their "natural" course. Evidently, it is unethical to interfere with the Earth affairs of humans or any other primitive species on any planet without being invited by the primitive species, because it hinders further development and spiritual evolution of the primitive species and the more advanced species as well.

The Pleiadians that did return to their home planets, according to the "Swiss Pleiadians", found that their own people had evolved spiritually to more advanced, loving states, as they are today.[32] By the time the Pleiadians were driven from the Earth, mostly by about 1,000 BC., most of the Earth humans were controlled by regressive ETs, who were not concerned with human or their own spiritual evolution, and by this time were operating in secrecy from bases located underground or on nearby planets. The direct contacts with ET "sons of heaven" by humans were mostly over by this time.

The Pleiadians have special interests in the human condition here on Earth, because of the involvement of their ancient ancestors in the affairs of humans. The "Swiss Pleiadians" summarized their mission here on Earth at this time to Billy Meier as follows:

1. To break the bondage of Earth humans by the self-proclaimed "Gods" and "Sons of God"...in all mythologies and religions of the last 70,000 years of our ancient past...

2. To expose politics as an outgrowth of early religions, and to show that it should be recognized for the tool of enslavement that it really is...

3. To show religions as the father of politics, and fundamentalism as the greater tool of enslavement by the ancient visitors to this planet. To show religions as the institution for preserving this enslavement, primarily through fear, ignorance and intimidation...

4. To proclaim The Creation (Prime Creator), impersonal and infinite, as the true source of all life and being, with which we all are in most intimate contact every moment of existence, and which is the sum of all its infinite parts. It is in All, and All is It.[33]

Bramley's claims of the ETs (Custodians) using holographic inserts to hoodwink we humans receives support from the Pleiadian Plus group. They tell of recent events in our history that were caused by holographic inserts. They claim that the vision of the children of Fatima was a holographic insert.[34]

We late 20th century humans have hologram technology, although it is apparently not nearly as sophisticated as that of the ETs. A hologram is a three-dimensional picture of something formed by using a beam of light that is set up by using a split, but coherent beam of light. The split light beam is arranged in such a way as to cause a pat-

Hoodwinked and Controlled

tern of interference so that the image formed appears to be an actual, solid three-dimensional object. The Pleiadian Plus group admits that the ancient Pleiadians used holograms to influence the civilizations they wished to control.

Fatima is a small town in central Portugal. In 1917, three young peasant children saw an apparition of what was thought to be the Virgin Mary on at least six different occasions. On October 13, 1917 a large crowd of around 70,000 people gathered at Fatima to see what they could see, as the children prayed to the Virgin Mary. The huge crowd saw a "miraculous solar phenomenon" immediately after the lady had appeared to the children.[35] In 1930 the Catholic bishop of the nearby city of Leiria declared the children's visions as true appearances of the apparition of the Virgin Mary. A Portuguese national shrine has since been built at Fatima and is visited annually by thousands of people. Most agnostics and non-Catholics haven't spent much time considering what happened in Fatima in 1917, but there are millions of Catholics that have, especially in Europe, and do believe that an apparition of the Virgin Mother of Jesus did appear to the three children of Fatima in 1917. This apparition left a prophecy with the children. This prophecy has been kept secret by the Church. Apparently, the high church hierarchy feels that the prophecy given is too sensitive for Catholics, as well as the rest of humankind. In any event, the prophecy has not been publicized.

For those of you who think this whole Fatima phenomenon is nonsense, I urge you to think again. What was the prophecy? When will it be revealed? If it was a hologram designed by ETs who wish to control us, who is responsible for it? When Pleiadian Plus was asked this, they said it was not their task to save us (humankind) and implied that we would have to figure this out ourselves.[36] We will discuss the Fatima prophecy in more detail in the next chapter, but the evidence indicates that it was one of the apocalyptic prophecies instigated by regressive ETs, Bramley discusses.

As Bramley's research suggests, the Pleiadian Plus group states that ETs are responsible for spreading diseases among humankind:

> ...We have said from the beginning that extraterrestrials - off-planetary beings - have been behind every major shift in consciousness. They have been behind diseases on your planet - your black Plague was instigated by extraterrestrials - your AIDS. They were behind the installation of certain religions. They have been behind many things.[37]

179

Hoodwinked and Controlled

The Pleiadians say they are here to help us save ourselves, not to save us directly. While they may have evolved to a much higher spiritual level than we Earth humans, they warn us that there are several Pleiadians in this universe who are renegades, some of whom might work with ET groups who wish to continue our enslavement. There are also those who have little or no Pleiadian genes and cultural training, who say they are Pleiadians just to deceive or "get on the good side" of Earth humans. The Pleiadians warn us not to necessarily believe ETs or any consciousness just because they seem more intelligent or more technologically advanced, and give the impression that they are more spiritually advanced. Pleiadian Plus, in answer to a question about humans being hoodwinked by intelligences greater than ours, said:

> "You have been led to believe on this planet - and this is a 'hood-winked' - that if intelligence is greater than you then it must be 'gooder'..."[38]

We've seen evidence that not only are such intelligences often not 'gooder,' they may be downright deceitful in order to keep us under their control.

The Pleiadians claim to have sent prophets to Earth in order to try to help humans evolve toward The Light. The Pleiadian Plus group claims that many members of The Family of Light, who might have less than a "prophet status," have been born in human form in order to help humans evolve. One of the main stated goals of the "Swiss Pleiadians" is an attempt to restore the true teachings of Jesus, whom they call Jmmanuel (pronounced "Immanuel"). According to the "Swiss Pleiadians," Jmmanuel, the entity we know as Jesus, was a very high teacher of The Truth (The Light) and not God almighty, or the son of God almighty. They claim that the true lessons of Jmmanuel (Jesus) became distorted and changed in time until they are practically indiscernible.[39] Many religious historians, including Bramley, make the same assertion.[40] We will discuss these claims, as well as the early development of the Christian religion in later chapters.

Let us end this grim chapter on a positive note. The good news is that the Pleiadians and many other members of the Family of Light are here now to help us evolve spiritually and physically past the Family of Darkness into a much higher level of existence, characterized by unconditional love and bliss. Despite the fact that level of humankind's spiritual consciousness on the Earth is still quite undeveloped, the Pleiadian Plus group assures us that the time period in which we are living is a time of extremely rapid spiritual consciousness growth potential for every human and ET alike, and will be more so as we approach the turn of the century and beyond. Throughout their transmissions, the Pleiadian Plus group emphasizes that we humans need only to live impeccably in the knowledge of wisdom and light during every second of every day, and thus we will be able to move onto a much more elevated level of consciousness - to a higher level of frequency, and move right past those devious ETs. We will have much more to say about this in the succeeding chapters, but a large part of this rapid spiritual evolution is learning greater discernment - learning to discern, for example, which extraterrestrial intelligences have the best interests of the human species in mind, and which do not.

Let us look for some more evidence that humankind has been- and is being duped. If humankind has been duped and manipulated by unseen forces for thousands of years as the evidence indicates, there should be some evidence that this is happening today!

[1] *The New English Bible* 1971. Oxford University Press, New York, Genesis 11: 1-9.
[2] Sitchin, Z. 1976. *op. cit.*, pp. 147-153.
[3] Goetz, D. and S. G. Morley 1950. *op. cit.*
[4] Bramley, W. 1990. *op. cit.*
[5] *Ibid.*, p. 3.
[6] *Ibid.*, p. 2.
[7] *Ibid.*, p. 3.
[8] *Ibid.*, pp. 94-95.
[9] *Ibid.*, pp. 113-120.
[10] *Ibid.*, pp. 117-118.
[11] *Ibid.*, p. 118.
[12] *Ibid.*
[13] *Ibid.*
[14] *Ibid.*
[15] *Ibid.*, pp, 162-163.
[16] *Ibid.*, pp. 197-198.
[17] *Ibid.*, p. 198.
[18] *Ibid.*, pp. 200-205.
[19] *Ibid.*, p. 201.
[20] Nohl, J. 1926. *The Black Death, a Chronicle of the Plague*. George Allen and Unwin Ltd., London, pp. 56-57. Quoted in: Bramley, W. 1990. *op. cit.*, p. 204.
[21] Deaux, G. 1969. *The Black Death*, 1347. Weybright & Talley, Inc. New York, pp. 2-4. Quoted in:Bramley, W. 1990. *op. cit.*, p. 207.
[22] *Ibid.*

[23] *Ibid.*
[24] *Ibid.*, pp. 121-131.
[25] *Ibid.*, p. 121.
[26] *Ibid.*, pp. 359-364.
[27] Marciniak, B. 1990 "The Pleiadians." Manuscript of a tape of material channeled on November 15 and 16 at Stanford University, J. Horneker, (ed.), Bold Connections, Raleigh, North Carolina.
[28] Stevens, W. C. 1990. *op. cit.*, p. 94.
[29] _____ 1988. *op. cit.*, pp. 384-385.
[30] Marciniak, B. 1990. *op. cit.*, p. 7.
[31] *Ibid.*, p. 14.
[32] Stevens, W. C. 1988. *op. cit.*, pp. 129-130.
[33] Stevens, W. C. 1989. *op. cit.*, pp. 49-50.
[34] Marciniak, B. 1990. *op. cit.*, p. 14.
[35] *Encylopedia Britanica,* 15th ed., p. 697.
[36] Marciniak, B. 1990. *op. cit.*, p. 14.
[37] *Ibid.*, p. 38.
[38] *Ibid.*, p. 33
[39] Stevens, W. C. 1990. *op. cit.*, p. 163.
[40] Bramley, W. 1989. *op. cit.*, p. 133.

> *"Suddenly the inventor (Nikola Tesla) became aware of strange rhythmic sounds on the receiver. He could think of no possible explanation for such a regular pattern, unless it were an effort being made to communicate with Earth by living creatures on another planet. Mr. Nikola Tesla has announced that he is confident that certain disturbances of his apparatus are electrical signals received from a source beyond the Earth," Professor Holden, the former Director of the University of California Lick Observatory.*
> Margaret Cheney, Tesla: Man Out of Time

Chapter 12

The Late Twentieth Century UFO-ET Phenomena

When we first began to conceive this book, we had no intention of covering the UFO phenomenon of the late 20th century, because we didn't think such a review was necessary for an understanding of how the ancient ETs created the final stages of biologically modern humans and granted humans civilization. Besides, this modern-day UFO-alien phenomenon is so filled with unsavory elements, such as cover-ups, hoaxes, disinformation, scandal, assassinations, degradation, disgrace, duplicity, shame and a host of other types of dishonorable behavior. However, we hadn't yet discovered evidence of the conspiracy of regressive ETs secretly trying to manipulate and control humankind throughout human history.

Since the modern UFO era began in 1947, there have been millions of sightings of UFOs by millions of people, millions of alleged abductions of human beings by ETs, thousands of books and articles written about the UFO phenomena, with hundreds of different opinions and analyses. And throughout these latter decades of the 20th century, the American and most other governments of highly technological countries, have established a policy of cover-up, which includes ridicule of their citizens, and even death to its citizens, at least in the U.S.A. No wonder I was not anxious to tackle the modern-day UFO phenomenon. But, as I studied how the ETs of old created humans as slaves, then later help to found the earliest human civilizations, I began to see a pattern that characterizes human history - a pattern of domination and control, of secrecy and cover-up and disinformation - telling it like it isn't! This stimulated my interest in the present-day human situation vis-à-vis ETs, and I began to see, through reading the likes of Bramley and several others, that what was true in the past, is still true in the present. We human beings of the surface of the Earth are being manipulated and largely controlled by entities that are not originally of this world, with the help of a small group of humans. In the ancient days the custodian-gods first applied their domination and control of humans directly, then through a human or half-human, half-god middle-man, called king or pharaoh. We have uncovered evidence that the ET "gods" began to leave this planet in a period ranging from at least 2,000 B.C. to 1,000 B.C., or literally went underground, in order to secretly be able to exert their negative influence on human affairs. So was there evidence of ET manipulation and control in the late 20th century?

Fortunately, at about the time I began thinking about writing this book, a flood of information about the UFO phenomenon had become available which has cleared up many of its puzzling aspects. So, I've decided to write a brief, personal review of the modern UFO phenomenon, with the emphasis on trying to find evidence that humankind is being manipulated and controlled by regressive ETs and their human allies, using some of the flood of information that has become available in the last few years. In doing so, I am going to have to leave out the work of many UFO researchers, who have done a tremendous amount of research against almost impossible odds, to try to bring to light what was covered up and hidden in darkness and, therefore, usually very confusing. I am not interested in trying to convince anybody of the veracity of UFOs and ETs because, as far as I am concerned, readers that do not believe in UFOs are not ready for this book! Besides, opinion polls have shown that up to 80% of Americans believe in UFOs,[1] so I'm simply not interested in convincing anybody of the veracity of UFOs.

The modern UFO phenomenon began in 1947, when Kenneth Arnold, flying his private plane near the spectacular, dormant volcanic mountain, Mount Rainier in Washington state, witnessed the flight of nine "flying saucers." Mr. Arnold was flying on a business trip and had some extra time, so he decided to look for a crashed plane that was missing in the Mount Rainier area. Mr. Arnold was a member of a search and rescue group, the Idaho Search and Rescue Mercy Flyers. Arnold first saw a couple of extremely bright flashes, then nine disks that appeared each individually "like a saucer would, if you skipped it across the water."[2] Arnold, using his skills gained in over 4,000 hours of flying experience, estimated that the "saucers" were traveling well over 1,000 miles per hour, faster than any airplane could fly in 1947.

While several of the news reporters that interviewed Arnold are reported to have been skeptical, they were so impressed with the man and his credentials for being a "solid citizen," this story was widely reported and the term "flying saucer" was coined.[3]

The United States Air Force (USAF) said in their official explanation of the incident that Arnold had seen a mirage. An academic astronomer, J. Allen Hynek, who was hired by the USAF to investigate UFO reports, used his own logic to conclude that Arnold probably saw a fleet of airplanes that were traveling at a much lower speed than Arnold had estimated.[4]

Here we see the first example of a pattern of UFO sightings and alien contacts in the U.S.A. by reputable citizens, whose reports are put into question and usually denied outright by some agency of the U.S. government, usually the USAF. Hynek, the official scientist-astronomer of the USAF, in this case, admitted later in his life that he had been hired by the Air Force as an "official debunker" of UFO sightings and was always supposed to find a possible "natural explanation" for them.[5] At least Arnold was not heavily ridiculed, as ridicule became the official government policy toward those who claimed to have had a UFO experience.

Of course, it has not been lost to UFO researchers, at least since von Däniken began to publish his books, that the UFO phenomenon dated back well before 1947, back to the beginning of humankind, but the 1947 report marks the beginning of widespread sightings and interest in this country and throughout much of the world.

I had just turned five and was living in a small town in eastern Kansas, when Arnold saw the "flying saucers," and I don't remember a thing about Arnold's sighting from 1947. But I do remember some talk about "flying saucers" and possible visitation to Earth by people from outer space in my childhood. It was all vaguely scary, especially when I was closer to the age of ten and began to see movies with frightening outer space themes. Since I never experienced a UFO sighting myself, and didn't know anybody who talked about them, I didn't spend much time worrying about UFOs. However, I was somewhat fascinated by movies that depicted crafts that flew at speeds far beyond the capabilities of the fastest jets in the USAF, which, I was sure, had the fastest jets in the world.

The UFO phenomenon was extremely bizarre. Not only could UFOs move very fast, but they also reportedly could make right-angle turns while traveling at breathtaking speeds. They could stop suddenly and hover, and sightings were often accompanied by electrical disturbances and extremely bright lights. Occasionally people would see or have contact with creatures associated with UFOs. Usually such creatures seemed strange and bizarre to humans, whereas sometimes such entities appeared quite human-like. The UFOs themselves were sometimes small "saucers," and sometimes very large saucers, and sphere-shaped and "cigar-shaped" crafts were occasionally seen. Sometimes UFOs would simply disappear before astonished, and usually very frightened human observers. There were reports of telepathic communication between humans and the "space beings" of UFOs. There were reports of space beings and humans floating through the air. Some people claimed to have had much contact with these "space beings," and some people even claimed to have traveled in some of these UFOs to Venus or to other planets. Some even claimed to have been given the "true religion" by these ETs. Were these people crazy, nut-cases, or what, much of the population of the country wondered?! The UFO phenomenon was indeed bizarre. No wonder uninformed scientists and other educated people considered the UFO "arena" the exclusive haunt of the "lunatic fringe". Through all of this, the American government maintained that the UFO phenomenon was all nonsense - that virtually all supposed UFO sightings could be explained through "natural phenomena."

As I was growing up in Kansas, I would not have believed that the government of the United States of America could tell a lie to the American people. I believed that the American government was the best government that had ever graced this Earth and part of this "greatness" I was led to believe in through church and state and the mass media, was that "Uncle Sam" would not tell a lie. My blissful naiveté began to crumble when the U-2 incident occurred in 1960, when I was a Senior in high school. As I remember, word came over the news that an American U-2 spyplane had been shot down by the Russians - the Soviets. The first statement from President Eisenhower indicated that the American pilot, Gary Powers, had gotten just a little off course and had barely crossed over the Soviet border. For that, the Russians had shot down our U-2 spy plane (which I had never heard of, until this incident). I had about a day to rent my rage at those "dirty rotten commies," when it was announced that, in fact, Powers had piloted the U-2 spyplane half-way across the Soviet Union in a south to north direction, before the Soviets had shot it down. I was crushed. Ike, our President, and fellow Kansan to me, had lied! After that incident, I had the misfortune of seeing several instances where the American government possibly, or probably, was not telling the truth and/or was putting out misleading information.

Certainly, by the late 1960's, I knew beyond any doubt that the American government was lying "through its teeth" when it said that UFOs did not exist. I was confused about UFOs and I didn't know what was happening, but I knew something was happening - that UFO reports were not all the figment of the imaginations of crazed people, or simply misidentifications of "natural phenomena." I wondered what the government was hiding - what they were covering up?

J. Allen Hynek, the USAF scientist-astronomer that investigated the Kenneth Arnold UFO sighting of 1947, claimed to have become a "turncoat" and openly admitted that he was hired by the government to debunk UFO sightings and spread misinformation, but that gradually he came to believe in UFOs and deplore the unscientific manner with which the government approached the UFO question. Hynek was able to give us several insights into the manner in which the government approached the UFO question.[6]

The United States Air Force (USAF) was put in charge of "investigating" early UFO reports. The USAF created a project called Sign in 1947, supposedly for the purpose of investigating UFOs. Project Sign was replaced by Project Grudge in 1948, which was in turn replaced by Project Blue Book in 1952. Throughout these projects, according to Hynek, who was hired onto Project Sign by the USAF and remained an official debunker through Project Grudge and most of Project Blue Book, the unofficial policy of the Air Force was basically an "anti-UFO policy."[7] The Air Force simply wanted to get rid of UFO reports as soon as possible by almost any means, according to Hynek. The Air Force officially finally did get rid of any duty of investigating UFOs when it abandoned Project Blue Book in 1969, after a "scientific" study they sponsored, the Condon investigation. The final, infamous Condon Report concluded that the so-called UFOs were not a subject worthy of scientific study.

Hynek observes that the Air Force, throughout the existence of Project Blue Book, did not conduct investigations using anything close to good scientific methods. Their methods were scientifically "sloppy," at best, and often their conclusions insulted the character of the witnesses.[8] Hynek observes that there was usually no detailed analysis of reported UFO sightings in Blue Book reports and the explanations presented in these reports "appear to have been straws grasped simply to close the case - quickly and quietly."[9]

In 1966, the Air Force, reacting to some pressure from Congress, engaged a team of scientists who were to investigate UFOs. The team of scientists would assemble at the University of Colorado in Boulder, under the leadership of Dr. Edward Condon, a well-known and respected physicist. The final Condon Report, released in 1969, has become notorious in UFO literature as a prime example of the Air Force and government cover-up.

Hynek is very critical of the Condon Report from a scientist's point of view. Hynek points out that beyond Dr. Condon's very negative conclusions in the summary of the report, which Condon wrote, the Condon Report could be viewed as "...a powerful document in favor of the reality of the UFO phenomenon." [10] Hynek further points out that Dr. Condon showed a blatant disregard of the contents of his own report in his summary, and Hynek even wonders whether Dr. Condon had ever read it.[11] Indeed, the report without Condon's summary contains many cases where the committee tentatively concludes that the probability of the existence of UFOs is quite high. Hynek criticizes Condon for dismissing out of hand thousands of UFO reports from credible witnesses from all over the world and thinks Condon was predisposed to reject any UFO hypothesis. In short, Hynek thinks the report of the Condon committee, especially Dr. Condon's own summary, to be an example of poor, shoddy science. Nevertheless, the National Academy of Science (NAS), a powerful body of "elite" scientists, were quick to endorse the Condon Report, and Hynek wonders whether the scientists of the Academy who had the responsibility of reviewing the Condon Report actually read it.[12] Indeed, statements from a majority of scientists about the UFO phenomenon indicate that we (as a group) are the least informed and most ignorant of this phenomenon among all the segments of our society. Scientists, buried in their own specialties of our materialistic science, as was I for many years, exhibit a kind of intellectual laziness, as well as laziness in their appreciation of their logical abilities. They exhibit a basic philosophy of materialists, which might be summarized as follows:

> From a materialist point of view, any phenomenon that appears to behave in a manner that one cannot explain through standard materialistic science, supports the belief system that simply cannot be. Therefore, it is not.

Hynek quotes what has become a cliché, attributed to Dr. Carl Sagan, whose efforts to popularize materialistic science in books, TV programs, and articles in popular magazines has made him sort of a spokesman for earth-bound science. Concerning UFOs, Dr. Sagan said, "...there are no reliable reports that are interesting and no interesting reports that are reliable."[13] Science had spoken as of 1969: there was nothing worth investigating in this whole phenomenon.

But, apparently those responsible for the UFO phenomena didn't read the Condon report or any of the statements of the "lofty" scientists, because the decades of the 1970s, 80s and now the 90s have seen as much and even more UFO related activity, as the 1950s and 60s.

Hynek, while deploring the lack of objectivity and lack of scientific methods exhibited by the Air Force in its investigation of the UFO phenomena, comments on the strangeness and weirdness of reported UFO sightings. He believed, in 1977, that the USAF had somehow determined that the UFO phenomenon did not threaten national security and was quite bewildered by the whole phenomenon. Hynek states:

> The United States Air Force, as well as the military in other countries, does not appear to be guilty of some sinister cover-up; rather they appear to be honestly baffled. Since the UFO phenomenon cannot be solved easily...the military, in their bewilderment, does its very best to wave it away.[14]

I wanted to believe something like this; the Air Force and the government didn't want to confess their lack of understanding of this phenomenon. Or, perhaps, the government was trying to protect the public from the shock of knowing that extraterrestrial intelligence existed, and the panic liable to result from such knowledge. Yet, the explanations for possible UFO sightings and abductions given by government agencies continued to be, it seemed to me, "insultingly stupid." Did the government really want us to believe that the thousands upon millions of reported UFO sightings and contacts by mostly credible witnesses all over the world could be explained by "natural phenomena," or the "temporarily distorted mind" of the witnesses? I knew that there was a cover-up, but I could not understand why the government would go to the extreme measures it was apparently taking - even to treat their honest, good citizens as if they were "obviously" mistaken, even possibly deranged, just because they had felt obligated to report strange and bizarre happenings to their government. I wanted to believe that the government had good reasons that would largely justify its shoddy treatment of so many of its own citizens, even into the early 1980's, but I never received much encouragement from what I read about the UFO phenomenon.

The abduction of Betty and Barney Hill on a lonely highway in the White Mountains of New Hampshire on the night of September 19, 1961, became the first widely publicized abduction case in the

U.S.A.[15] This abduction followed patterns that were to be reported again and again, to the present-day. While driving along a deserted, rural highway at night, Mr. and Mrs. Hill first saw a flying disk, then experienced "missing time." Ten days later, Betty began to have vivid dreams that she and Barney had been taken aboard a flying saucer and medically examined. More than two years after the abduction occurred, Betty and Barney underwent a series of hypnotic regression sessions in which they both recalled being taken aboard a saucer and being given "medical examinations" by small four and a half foot tall aliens with large heads, large, black and strangely shaped eyes, and greyish skin, ETs that have become known as "the greys." These particular greys, and many of the greys "encountered" by humans since then, have become associated with two stars named Zeta 1 and Zeta 2 in the constellation Reticulum, visible on Earth only from the southern hemisphere. Betty and Barney Hill had been terrified by the ordeal. The ETs had somehow blotted the two-hour incident from their conscious minds, at least for a couple of years, but when Barney began to recall the abduction under hypnosis, he re-experienced the terror and hysteria he had experienced during the abduction. The ETs, he said, reminded him of "German Nazis."

In addition to the uninvited, often terrifying human abductions, there were reports of thousands - even millions - of mysterious animal mutilations, that usually occurred in association with UFO sightings. Probably the most publicized of the millions of domestic animal (mostly cattle) mutilations that have occurred, happened in Colorado - the mutilation of Snippy, a three-year old Appaloosa saddle-horse. The body of Snippy was found in Alamosa County of south-central Colorado on September 15, 1967. No blood remained in the carcass of Snippy and his vital organs had been removed. The incisions made on Snippy were so smooth that they appeared to be done with an advanced surgical technique. There had been a rash of UFO sightings in the area, and there were subsequent mutilations elsewhere in the county, and Snippy's plight was publicized nationwide. Soon, Alamosa County was flooded with investigators and sightseers. Snippy's skeleton is on display today for tourists in a pottery shop in Alamosa, Colorado.[16]

Certainly, there was enough information floating about by this time to indicate that the UFO phenomenon, whatever it was, had many negative aspects. Jacques Vallée, a French astrophysicist who has become one of the most widely respected UFO investigators, makes a strong case in his book *Messengers of Deception*, that humankind is being manipulated by something or somebody through the UFO phenomenon, using unknown techniques of mind control and mind manipulation.[17] Vallée notes that some of the contactees are given "spiritual messages," which lead to irrational expectations of ET intervention. For example, Vallée discusses the case of the French contactee, Claude Vorilhon, who began to "preach" worldwide after being instructed to do so by "Elohim" from another planet.[18] Vorilhon was taught that we humans are now living in "...the last days of an age..." and that we must make drastic political changes if we are to survive. The changes the "Elohim" instructed Vorilhon to advocate included a one world government, with a single language, the elimination of democracy, a new monetary system, and the elimination of military service here on Earth. If we humans successfully carry out these "reforms," we will be rewarded with the benefit of their superior science when the "Elohim" return; if not, they will wipe out our scientific centers, "...as they did once with Sodom and Gomorrah."[19] Vallée points out that this late 20th century apocalyptic prophet and others are usually not successful in forming a large following of people with such messages, especially among the establishment. Vallée points out, however, that there are sects and cults forming around contactee prophets, beyond the scrutiny of scientists, scholars, and other establishment personnel, that are expecting alien intervention and are thus, slowly undermining world order. Vallé valle argues that it is not necessarily an alien race that is trying to manipulate humans, but that other humans might be using and creating part of the UFO phenomenon in order to undermine the establishment authority systems of the world.

Dr. Vallée points out in his books that the UFO phenomenon is not new and that, as we have seen, there are several reports from all over the world hundreds and thousands of years old that might describe, and almost certainly do describe UFO phenomena.[20] Also, Vallée is the first (I read) to have pointed out that there are too many UFO reports to support the "extraterrestrial hypothesis," the idea that every one of the millions of UFO sightings and contacts represents a visit from outer space.[21]

In the 1960s and the early 1970s, I was one that believed that if an extraterrestrial intelligence existed, it had to be "gooder" than we primitive humans, and would have the best interests in mind of a technologically inferior species, such as humankind, and would act accordingly. Gradually, however, through reading Vallée and others, the negative aspects of the UFO phenomenon reared its ugly head and could not be ignored. I was confused and disturbed by this, but I was not obsessed with the UFO phenomenon, and didn't spend a lot of time indulging in my UFO hobby. I was busy, even obsessed, with my fascinating profession - biological anthropology. This was especially true because almost everybody I knew, especially my friends within my profession, did not profess any interest or belief in the UFO phenomenon. Today, in academic circles, at least since the Condon report of 1969, interest in UFOs is taboo and considered the exclusive range of total whackos and other "fringe people" of our culture, so what interest I had in UFOs I kept carefully in my "closet." In the late 1980s, Timothy Good's book, *Above Top Secret: The Worldwide UFO Cover-Up*, established unprecedented documentation of the cover-up of the UFO phenomenon.[22] Almost all books and manuscripts written about the current UFO phenomenon talk about the government cover-up, but Good's book is the best written on the subject. Good, as the title of his book suggests, has documented a worldwide cover-up of the UFO phenomenon involving (at least) the governments of the former USSR, China, Canada, Australia, France, Spain, Portugal, Italy, Great Britain and, of course, the United States. Good provides information that indicates that the "intelligence" organizations of the governments, especially those of the United States, are the organizations that are the most involved with the UFO phenomenon, including the cover-up of this phenomenon. As part of this cover-up, these intelligence organizations - especially those of the USA - will first deny that they have any interest in the UFO phenomenon, then heap ridicule on those citizens that say they have had some direct UFO experience. For example, The Federal Bureau of Investigation (FBI), the agency responsible for enforcing federal laws among the population of the USA, had always denied that they had any interest at all in UFOs, until 1976. In 1973, for example, the director of the FBI, Clarence M. Kelley, sent a letter to a citizen who had inquired about UFOs, and "informed" him that "...the investigation of Unidentified Flying Objects is not and never has been a matter that is within the investigative jurisdiction of the FBI...".[23] Only three years after Mr. Kelley wrote this in a letter, Dr. Bruce Maccabee, a physicist who worked for the US Navy, filed a Freedom of Information Act request for information the FBI might have about UFOs. The Freedom of Information Act was passed in the early 1970s to allow the public of the USA to gain access to information our secret services, such as the FBI, had gathered and had once classified as "top secret," which had been declasssified later. For his efforts Dr. Maccabee received about 1,100 pages of documentation on the UFO phenomenon.[24]

After the passing of the Freedom of Information Act, the only information secret services could legally withhold from American citizens were those matters that concerned an individual's constitutional rights, or the national security of the country. Nevertheless, we

Americans cannot be sure to receive all the information from our secret services that we are legally allowed to receive. Some suspect that the secret services of our country might sometimes, or often, stretch the definitions of individual constitutional rights, and especially national security, in order to keep some secrets from the public. After all, one must ask oneself, what agencies might be keeping from the public if they are capable of telling bold-face lies, at least with regard to UFOs, as former Director Kelley was caught doing. Figure 12-1 shows pages thirteen and fourteen of a twenty-one page "above top secret" affidavit released by the National Security Agency (NSA) after censoring it. This twenty-one page affidavit was outlining the reasons why NSA withholds some of its documents on UFOs from the public. We can have no idea what was written on pages thirteen and fourteen of this affidavit as, obviously, the NSA didn't want the American public to know.

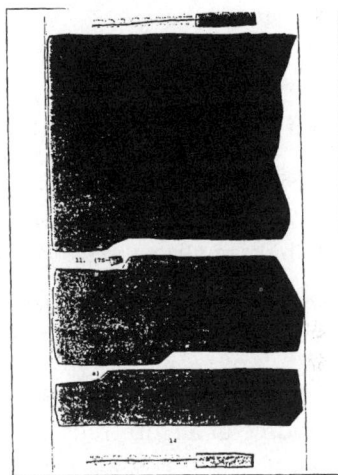

Fig. 12.1   *Two pages of a twenty-one page "above top secret" affidavit released by the National Security Agency (NSA) after censoring it.*

There is, unfortunately, evidence that the various intelligence organizations might hold back or "lose" documents they may not want to be seen by the general public. Dr. Hynek relates in a footnote that all of the files of the seventeen years of the existence of the Air Force's Project Blue Book are available, but that some of the cases were either lost or misplaced, and much of the supplementary material, such as photographs and teletype messages, are likewise missing. Dr. Hynek believes that most of the "omissions" from the USAF Project Blue Book files are probably lost through carelessness.[25]

The Late Twentieth Century UFO-ET Phenomena

Many Americans, however, have serious doubts about American intelligence agencies, and would be surprised if missing files and other missing evidence were a result of "carelessness." American security agencies have been implicated in the series of the "lone-assassin" political assassinations, such as the Central Intelligence Agency (CIA) in President John F. Kennedy's assassination in 1963, and the FBI in the assassination of Martin Luther King in 1968. Let me just say that American intelligence agencies have not inspired a lot of trust and confidence over the decades of their existence, especially with regard to UFO phenomena.

Mr. Good provides some interesting documentation on the importance the United States authorities attributed to the UFO phenomenon in 1950. The Canadian government released a memorandum from Wilbert B. Smith, a highly respected engineer and scientist. Mr. Smith stated in his memorandum dated November 21, 1950, that through discrete questions he made at the Canadian Embassy in Washington, D.C., he learned that:

a. The (UFO) matter is the most highly classified subject in the United States government, rating higher even than the H-bomb.

b. Flying saucers exist.

c. Their modus operandi is unknown, but concentrated effort is being made by a small group headed by Dr. Vannevar Bush.

d. The entire matter is considered by the United States authorities to be of tremendous significance.[26]

Mr. Smith hinted at the possibility that this flying saucer matter might be the source of an introduction of new technology.

Let us move to a source that might help explain the secrecy and duplicity of the American Government. William (Bill) Cooper is a former member of the United States Naval Intelligence Service. Between 1970 and 1973, Mr. Cooper, having a very high security clearance, had access to much top secret material, which revealed the U.S. Government's secret involvement with an ET species. Cooper has carried out much research on this subject since and first began to divulge the information about the "secret government" of the United States and its involvement with ETs in 1989. In 1991, Cooper published what he knew about the secret government and a host of other conspiratorial information in his book *Behold a Pale Horse*.[27] Since some of the information Cooper relates is derived from sources he cannot divulge, and from published sources which he cannot vouch for, he presents the chapter in his book concerning the secret government as hypothetical. He believes that the hypothesis he presents is true, because it is supported by most of the available historical information. While it would seem impossible for anybody to get all of the details of this subject absolutely correct, given the "above top secret" nature of this material and the volumes of disinformation put out by government agencies, we are following Cooper's hypothesis here because most of the puzzling aspects of the UFO phenomenon are explained through it. The extremely secretive behavior of the United States government, especially, "falls into place" - makes sense, if Cooper's hypothesis is followed. As Cooper states, most of the available historical information supports his hypothesis.

Cooper claims that the American secret government had its beginning in the late 1940s when alien craft - UFOs - either crashed or were downed in the southwest part of the country, mostly in the state of New Mexico. Between 1947 and 1952 the government recovered at least sixteen such craft along with sixty-five alien bodies and at least six live aliens, according to Cooper. On some of the crashed disks a large number of human body parts were stored. The "beast" had appeared and paranoia took hold of the government officials in charge of the recovery of these disks. The government immediately put the clamp of secrecy on these UFO crashes and these events have "...become the most closely guarded secrets in the history of the world."[28] Certainly, there has been much written about possible UFO crashes in the southwest, especially the first crash that was alleged to have occurred near Roswell, New Mexico in 1947.[29]

Cooper states that the Central Intelligence Agency (CIA) was established for the purpose of dealing with the alien presence. Also, top scientists were organized in 1947 under project Sign to study the alien phenomena and to debunk the (by then) numerous sightings of UFOs. Project Sign evolved to Project Grudge in 1948, which evolved into Project Blue Book in 1952, a progression of projects we have already seen from Hynek's work. In 1952, according to Cooper, the National Security Agency (NSA) was formed by a secret Executive order in order to establish a dialogue with the ETs. The NSA has now become the premier agency in the American intelligence network and is still primarily concerned with alien projects, according to Cooper and other sources.

In 1953, face to face contact was made with aliens and the government in a desert of the U.S.A. Diplomatic relations were established with this group of ETs and they promised that they would return and formalize a treaty.[30]

While this was going on a group of "humanoid" ETs landed at

Homestead Air Force Base in Florida and communicated with the U.S. Government. This group of ETs warned against the other group of ETs, the greys, that was contacting the government. This group of ETs also offered spiritual help; they offered to assist us in our spiritual development and awakening, if we would dismantle our nuclear weapons. This "humanoid" group declined to exchange technology with the Government, whereas the other group, the greys, promised to do so. The "humanoid" group said that we were spiritually unable to handle the technology we already possessed. The U.S. Government rejected these overtures from this "humanoid" group because the high officials "in the know" thought it would be foolish to disarm in face of an uncertain future.[31] Furthermore, the government was very much involved with technological development and not much involved with spiritual development and awakening. In 1953, remember, the Second World War was a vivid memory and the cold war with Communist countries was at its height. Cooper doesn't say much about this "humanoid" group of ETs who wanted to help us with our spiritual progress, presumably because he doesn't know much about them. This may have been the chance for the human species to escape the wars and suffering that has characterized our history and begin to make giant strides in our spiritual development, but it was not to be. What we got instead, unbeknownst to all but a handful of "elite" American officials, was a treaty with the greys.

According to Cooper, President Eisenhower signed a formal treaty with the greys on February 20, 1954 at Muroc, now Edwards Air Force Base, California. Eisenhower had arranged to be on vacation in Palm Springs, California during this time and then was rushed to the nearby air force base on the day of the signing. The press was told that Ike had to go to the dentist on that day.

The treaty with the aliens stated that their presence on Earth would be kept a secret, and that underground bases would be constructed for their use. Two of these bases would be constructed for joint use by the aliens and the U.S. Government. They agreed to furnish the Government with advance technology, and they would be allowed to abduct humans for "medical examinations." The treaty stipulated that the humans would not be harmed and would be returned to the place of abduction with no memory of the event. Also, the treaty stipulated that the aliens were to make no other treaty with other Earth nations.[32]

By 1955, it had become apparent to the secret government that the aliens had broken the treaty. Mutilated animals had been found across the country, as well as mutilated humans! Mutilated animals, as far as is known, were not in the treaty, and certainly mutilated humans were not part of the treaty. Also, it was suspected that not all abductees had been returned at all, let alone to the point of abduction. A bit later, it was found that the aliens were using hormonal secretions, glandular secretions, blood plasma and enzymes of animals and humans, probably in genetic experiments. Furthermore, it was suspected that the aliens had been in contact with the Soviet Union, in violation of the treaty.

When confronted, Cooper contends, the aliens admitted all of their treaty violations. With regard to the biological secretions and blood plasma, they explained that their own biological beings had deteriorated to the point where they could no longer reproduce, and that if they did not improve their own genetic structure, they would become extinct.

The government officials were very suspicious of this and other explanations given by the ETs, and didn't appreciate the treaty being so blatantly broken. But, the weapons of the U.S. had been shown to be totally useless against the aliens, so there was nothing to do but go along with them. Meanwhile, the government ridiculed and debunked U.S. citizens who tried to point out that there might be an ET presence.[33]

Interestingly, the government suspected that the events in Fatima, referred to in the last chapter, might be an ET manipulation. Using its Vatican moles, the U.S. Government soon obtained everything the Vatican knew about Fatima, including the prophecy.[34] The readers will recall that three Portuguese children saw what they thought was a vision of the Virgin Mary several times, and finally, about sixty or seventy thousand people saw bright lights in the sky shortly after the children received a prophecy from the "Virgin."

The prophecy berates humans for their evil ways and demands that humankind place themselves at the feet of Christ, or the events described in the Book of Revelation would occur. The prophecy predicts that World War III would begin in the Middle East, and that before it was over, most life on Earth would perish in a nuclear holocaust.[35]

The aliens were confronted by the government with this finding, and the greys admitted that it was true that they had manipulated humankind throughout its history. *They had manipulated humankind through religion, witchcraft, magic, Satanism, the occult and holograms* (emphasis mine). They said that they were capable of time travel and that the events described in the prophecy would come true if the conditions were not met.[36]

Cooper, as did Bramley, writes of how the secret societies of the world have been - and are - currently being used by the ETs in order to control humankind. Like Bramley, Cooper traces secret societies back to the ancient times of the first civilizations, to the Brotherhood of the Snake, also called the Brotherhood of the Dragon, which still exists under many different names. Most secret societies, according to Cooper, claim to be guarding the "secrets of the ages," and most claim communication with a higher, divine source. Basically, members of a secret society believe that there are only a few people in the world that possess minds mature enough to understand the "secret of the ages," and that they are the ones with such minds. Most secret societies claim that their special knowledge is for the benefit of all humankind, but Cooper claims that, in fact, most secret societies ultimately, are tools of the dark forces, and that they worship such forces as Lucifer, or Satan. Most members of secret societies never advance to the "higher," most secret levels of the society, and never learn the real purpose of the group. Cooper maintains that the real purpose of the secret societies from ancient times is to be but a vehicle for the ETs to control the masses, with a handful of "privileged" humans.[37]

In the government of the United States today secrecy is "built-in" through the numerous "intelligence" organizations. In addition to all of the known intelligence organizations where secrecy is the rule, such as the NSA, the CIA, the FBI, the Army, Navy and Air Force intelligence services, Cooper claims that there are secret organizations in the government that a vast majority of the American people have never heard of. For example, according to Cooper's hypothesis, President Eisenhower in the mid-1950s, through a secret Executive memorandum, formed a secret committee known as Majestic Twelve (MJ-12) to oversee and conduct all covert operations concerned with the ET question. Cooper presents evidence that Majestic Twelve, which is the most secret and most powerful group in the government, is controlled by the Illuminati, an international secret society dedicated to greater world economic and political power for its members and the ETs. According to Cooper, Eisenhower was the last President to know fully about the interaction of the secret government and the aliens, or ETs. All Presidents since Eisenhower have been told only "what they need to know" about the aliens by the secret government, and Congress has been told next to nothing. Other organizations of the U.S. Government where many secret decisions are made and much secret power is held that Cooper names, are: the Executive Committee of the Council on Foreign Relations, the Jason Group, and the Trilateral Commission. Many of the members of these groups are recruited from the secret societies of Harvard and Yale, such as the Scroll & Key and Skull & Bones, which are themselves branches of the Illuminati. Other members of the Council on Foreign Relations, the Trilateral Commission, and the Jason Group are recruited from members of the U.S. Senate and Congress, the U.S. Military, large U.S. corporate executives, presidents and professors of the most prestigious American colleges and universities, members of the leading American television, newspaper and magazine news media, and executives of the largest American banks.[37]

The secret Government of the United States has many ties to a secret world Government, according to Cooper. In addition to the Illuminati, Cooper mentions a few other secret societies of the secret world government, the Club of Rome, the Knights of Malta and the most powerful of the secret organizations of the world, the Bilderberg Group.

Mr. Cooper's hypothesis is much more elaborate than this very brief summary presented here. As we have seen, Cooper's hypothesis seems to fit the known historical facts, especially with regard to the ET presence and the government cover-up of their presence. Also, Cooper's hypothesis fits the historical pattern of alien/ET control of humankind that can be derived from the earliest writings of humankind, which we have reviewed here. Furthermore, Mr. Cooper's hypothesis is supported by the work of many other UFO/alien researchers, many of whom were once working directly or indirectly on projects of the secret government and/or ETs before they became disillusioned and disgusted with the Government, and others who simply knew of such projects.

One of the latter is Michael Lindemann, author of the manuscript *UFOs and the Alien Presence: Time for the Truth*.[38] For years, Mr. Lindemann worked as a researcher and educator on subjects like U.S.-Soviet relations, the arms race and the like. During the late 1980s Lindemann learned through the course of his research that the United States was "engaged in topsecret development of new weapons that would seem utterly useless in an era of diminishing superpower confrontation."[39] This was especially true with regard to a "burrowing bomb," supposedly designed to burrow beneath the earth's surface, before exploding in order to destroy deep bunkers under the Kremlin in the Soviet Union. But why was the U.S. so intent on building this weapon in the post-Cold War era, Lindemann asked himself, especially since the government was supposedly trying to cut the defense budget? Through further research he learned that the government might be in a hurry to develop a "burrowing bomb" for use against ET-aliens who occupy underground colonies constructed under the United States.

Mr. Lindemann describes himself as having been a false skeptic in the UFO/alien controversy until his research led him to the "burrowing bomb" in 1989. Mr. Lindemann states that a true skeptic must be informed on the subject of his or her skepticism in order to draw rational conclusions on the subject.[40] Since 1989, Lindemann has been a UFO/alien researcher and writer. He recognized that because of the extreme secrecy, duplicity and the disinformation released about the subject, it is next to impossible at the present time for anyone to know exactly what is happening in this area. However, he has concluded that many of the claims of Bill Cooper are likely to be true. Much of his own research seems to corroborate that of Cooper, including the possible existence of Majestic Twelve (also called Majic, Majority or simply MJ-12) as head of the secret government of the United States.[41]

Unfortunately, Cooper's hypothesis does tend to confirm the quote of the Pleiadian Plus group with which we ended the last chapter concerning the spread of AIDS by the aliens. According to Cooper, the Club of Rome secretly recommended to the Bilderberg Group that some sort of plague be introduced to the human population for the purpose of controlling and reducing the world's population. Remember that Cooper identified the Club of Rome and the Bilderberg Group as part of the secret world government, most likely in direct contact with the ETs at the highest levels. The order was given by the Policy Committee of the Bilderberg Group of Switzerland to develop a microbe that could be used for population control, according to Cooper. Ultimately, the Department of Defense of the United States was given the task of developing such a microbe. Funding for the development of this microbe was obtained from the U.S. Congress in 1969, as it provided $10 million dollars to the Department of Defense's 1970 budget for its development. According to Cooper, representatives of the Department of Defense testified before a committee of the U.S. Senate that they intended to produce "...a synthetic biological agent, an agent that does not naturally exist and for which no natural immunity could have been acquired."[42] The ruling "elite" decided to target what they considered undesirable elements of society for this new plague. Specifically, they targeted blacks, Hispanics and homosexuals.

The synthetic biological agent which many believe causes AIDS (the HIV virus?) was developed by the late 1970s by U.S. Government scientists. The African continent was given a smallpox vaccine that contained whatever causes AIDS in 1977. The vaccine was administered by the World Health Organization. The U.S. population was infected by an experimental hepatitis B vaccine administered by Centers for Disease Control in San Francisco, New York and four other cities in 1978, 1979, 1980 and 1981. The experimental hepatitis B vaccine was manufactured and bottled in Phoenix, Arizona.[43]

Cooper doesn't state that the aliens themselves were directly involved with the development and spread of the AIDS epidemic. However, we recall that it is highly probable that some secret societies and secret organizations of the world are in direct contact with ETs at the highest levels, that wish to control us, and, if Cooper's hypothesis is correct, it is possible, if not probable, that ETs, as well as their human allies are behind this tragic epidemic. Here we have an Earthman source (Cooper) and an esoteric source (Pleiadian Plus) that suggest the regressive ETs are behind the spread of AIDS throughout the world in the late 20th century.

Evidence of a Fascist Element in the U. S. Government

Is it possible that the United States Government could be such an integral part of such a diabolical plot and have worked with such negatively oriented ETs for more than 40 years without the knowledge of a vast majority of its citizens? If so, why hasn't some news organization dug up the details of such a plot and exposed it? A majority of Americans would not believe that their Government could act and behave with such duplicity as Cooper, Lindemann and others have suggested. The United States was supposedly founded on the principles of the freedom and well-being of its citizens and, theoretically, all people. Most Americans believe that their Government is set up to serve the American people, and all people of good-will. However, the evidence is that there are more sinister aspects to the U.S. Government that do not have the best interests of Americans and the rest of humankind in mind. Let us step outside of the ET phenomenon for a moment, and see if we can find evidence that a sinister, menacing element exists in the U. S. Government.

A look at the history of the United States since 1945, especially the wars in which the nation has been involved, the political insurrections "intelligence" agencies of the government have been reportedly involved in throughout the world, and the political assassinations in this country where "intelligence" agencies were implicated, all point to an increasing fascist element in the United States Government during the past 45 years. Perhaps part of why most Americans don't see, or can deny such fascist tendencies, is that most of the information, or "news" presented to them through the mass media of America is controlled.

The highly respected linguist, Noam Chomsky of the Massachusetts Institute of Technology, has argued for years that the mass media in the United States is controlled. In recent reviews of a film documentary produced to air his political views, Chomsky argues "....that the mass media shape the information fed to the public in ways that basically reinforce the status quo..." [44] Instead of a free press, Chomsky believes, "...we have a press that provides unquestioning support for U. S. policy even when its contradictions and moral failures are obvious..." [45] Chomsky gives the recent Gulf War as an example. He states that 99.9% of the discussions in the U. S. media of the Gulf War "...excluded the possibility of a peaceful settlement, while the newspapers kept writing articles about (President) Bush going the last mile for peace. We went to war in the manner of a totalitarian state, thanks to media subservience." [46] One of the main reasons for the failure of the mass media in the U.S. to report more accurate accounts of the events of the world, Chomsky believes, is the fact that twenty-three corporations in the U.S. own 50% of the T.V. stations, radio stations, newspapers, book publishers and film companies. Chomsky believes that the goal of these twenty-three corporations is to stay in power, and they do this by using the media to manipulate public opinion and, "marginalize dissenting voices." [47] Certainly the ET presence has been "marginalized."

On the evening of the day in November, 1993, I had reviewed the opinions of Dr. Chomsky, CBS TV, one of the three major American television networks, broadcast a news documentary about the assassination of President Kennedy in 1963. CBS purported to give equal coverage of the position that there was a conspiracy to kill JFK on the one hand, and, on the other hand, the position of that only one, lone, assassin, Lee Harvey Oswald, killed Kennedy. Dan Rather, the CBS spokesman, concluded that despite numerous claims of conspiracy, none has been demonstrated, and the evidence pointed to the lone assassin, Oswald, as having killed Kennedy. He stated that the evidence was quite conclusive that only three shots had been fired, all from behind the President's motorcade where Oswald had apparently been located at the time of the shooting.

My wife, Lynette, and I were appalled at the one-sided (lone-assassin) view presented by CBS. There was an enormous amount of evidence of a conspiracy that CBS TV failed to report. For example, CBS failed to mention the results of a unique United States Congressional inquiry of the late 1970s, under the auspices of the Select Committee on Assassinations. A distinguished acoustics expert was appointed by the Committee to analyze a sound recording made on the day of the assassination. A motorcycle policeman escorting the President's motorcade had left his radio on, and the sounds of the shots that had killed President Kennedy were recorded at Dallas Police Headquarters. The acoustics expert concluded that four shots had been

The Late Twentieth Century UFO-ET Phenomena

fired, and at least one of the shots came from in front of the President.[48] In 1979, the Committee reported that President Kennedy,

> "was probably assassinated as a result of a conspiracy.." The scientific evidence, said the report, compels acceptance of the fact and, "demands a re-examination of all that was thought to be true in the past."[49]

Is the mass media of the United States simply orchestrating the news in order to maintain the status quo? This TV program presented by CBS on the assassination of JFK, that didn't even mention the conclusions of this important Congressional Committee, is but one small example of such orchestrating. Certainly, the status quo of America would be upset if it were demonstrated conclusively that there was a conspiracy to murder one of the most popular United States Presidents that ever lived, and that the conspiracy may have involved elements of the American Government.

This is not the place to discuss the JFK assassination, but the CBS documentary provided some information that was new to me: roughly 90% of Americans do believe there was a conspiracy. This figure demonstrates once again, as in the UFO cover-up, that although American people are ignorant, largely because of the efforts of their own government and their mass media lackeys, they are not stupid.

A large part of the reason American people do not receive objective news, especially on matters concerning their government, is because the American news media has joined hands with the government. A partial list of television news broadcasters who are familiar names to American TV news watchers, who are or have been members of the Council on Foreign Relations (CFR) include Dan Rather, Harry Reasoner, and Charles Collingwood of CBS; Ted Koppel and Barbara Walters of ABC; John Chancellor, David Brinkley and Irving Levine of NBC; Daniel Schorr of CNN; and Robert McNeil and Jim Leher of the Public Broadcast Service.[50] Most likely these individuals join the CFR for prestige or to further their careers and are not privy to the most secret machinations of the secret government, but they clearly have collaborated with the U.S. Government on some level. Most Americans are not even aware of the CFR, or the Jason Group, or the Trilateral Commission, let alone of what they actually do, because they are secret societies.

If the mass media is mostly controlled, how about scientific progress? I've already pointed out how the official U.S. Government policy, with regard to the origin and existence of humankind, is the anthropological-Darwinian point of view of matter randomly becoming conscious. This is a huge part of the cover-up. If virtually all very intelligent scientists could be convinced that humans and all else in the universe is matter randomly arranged according to some natural laws, how much easier it is to claim that there is no one in the universe except us, and that we humans have no spiritual existence. If anything, information is more controlled in the scientific world than in the mass media. For example, if, in the days when I was a dues-paying anthropologist in good standing, I had applied to one of the grant-giving institutions of the government, say the anthropological division of the National Science Foundation, for a grant to investigate the possibility of intervention by ETs in the development of humans, this grant application would not only have been rejected, but I am sure much ridicule would have been heaped upon me. This is one of the ways information is controlled by the government. Essentially, all of the scientists dependent upon government money - a vast majority of all scientists in the United States - are lackeys of the government. This is amazing, as very few scientists (to my knowledge), except probably a few that have been hired by the Government, are aware that they are part of a cover-up or know that they are part of a system to propagate misinformation and maintain that status quo.

In fact, it is quite astounding that the American society has been controlled without walls. One must hand it to the members of the secret world government and their ET masters; they are very clever, indeed.

The dark side of the United States Government can be gleaned through casual readings of newspapers. Even though nearly all of the newspapers in the United States are mostly controlled, the activities of the fascist element of our government are so numerous and so dark, that articles occasionally appear that clearly reveal the dark side. Usually there is little or no elaboration, or follow-up to these "tales of darkness," by the media reporters, but they are there if one looks for them, even on a casual basis. For example, recently USA Today stated that Newsweek magazine reported that a probe of the 1963 files of the assassination of President Kennedy released for the first time in the summer of 1993, "...found the FBI and CIA withheld and destroyed evidence, as they focused more on protecting their agendas than the truth..."[51]

In another story, David Hulen of the Los Angeles Times reported that the U.S. Atomic Energy Commission in 1962 secretly buried, in the Arctic tundra of Alaska, 15,000 pounds of soil contaminated with radioactive fallout from the Nevada nuclear explosions. This burial of radioactive soil was not discovered until 1992, when a researcher from the University of Alaska (Fairbanks) read federal archives describing the burial at Cape Thompson in northern Alaska. Government officials rushed up there and assured the small local Eskimo population that there was no danger. However, quite understandably, the Eskimos were upset at the news of the radioactive soil on land they used to hunt and gather food. The mayor of a nearby small Eskimo village, Mr. Ray Koonak Sr., is reported as stating:

> They're (the government) stating that there's no immediate danger to health, but how can we trust them after what they buried there thirty years ago? ... people are very upset...They feel they've been lied to.[52]

The American people witnessed a pack of lies from high government officials during the investigations of the failed arms-for-hostages deal secretly conducted by the Reagan Administration in 1986. The administration first defied the Constitution of the United States, breaking a law by secretly negotiating the sale of weapons to Iran in exchange (they hoped) for American hostages being held by Iran, and they were to use the funds generated by the arms sale to break another law that forbade the government to aid rebel forces in Nicaragua. Then government officials lied to and tried to cover up their illegal dealings with Iran and Nicaragua to Congressional investigating committees. Finally, a special prosecutor, Lawrence Walsh, was appointed to investigate the matter. Walsh's six-year investigation ended in December of 1992, when President George Bush pardoned six former government officials for any role they may have played in the scandal, including former Defense Secretary, Casper Weinberger. Bush, who was Vice-President at the time of the arms-for-hostages deal, is believed by many to have been culpable himself. In prosecutor Walsh's final report, he concludes that "...former Attorney General Edwin Meese concocted a 'false account' of an Iranian arms-for-hostages deal to cover up President Ronald Reagan's role..." Furthermore, Walsh's final report concluded, there was a broader cover-up carried out by senior cabinet officers that tried to make 'scapegoats' out of lower government operatives, such as Oliver North and John Poindexter. A newspaper report stated that on the day of the release of prosecutor Walsh's final report, in

> ...separate statements, Reagan, Bush and several others named in the report called Walsh's conclusions inaccurate and unfair. Reagan termed the

report "an expensive, self-administered pat on the back and a vehicle for baseless accusations that he (Walsh) could never have proven in court."[53]

We will never know whether Walsh's accusations could have been proven in court, because Bush's presidential pardon of all involved in the scandal barely a month before he was to leave office short-circuited the normal course of judicial proceedings.

Phillip Hilts, of the New York Times, reports that the U. S. Department of Defense lied to American veterans of the Persian Gulf War concerning the possible use of chemical warfare, until investigative teams from another country (Czechoslovakia) reported finding evidence of chemical warfare. Even after more than 10,000 veterans reported strange symptoms that could have been related to toxic chemicals, the U. S. Government denied any evidence of chemical warfare and dismissed the veterans reports. The newspaper report states that:

> For two years, the Pentagon had denied having any evidence that Iraq had used chemical or biological agents, but last Tuesday it acknowledged the validity of work by Czech teams of chemical weapons experts in the week after the air war started in January 1991. They found that there were traces of nerve gas and a blister agent in the Gulf region.[54]

Thomas Lippman, of the Los Angeles Times-Washington Post Service, reports on the outrage living survivors of strong doses of radiation received from aboveground nuclear tests in southwestern Utah feel towards the United States Government. The residents of St. George and other small communities in southwestern Utah were subject to considerable radioactive fallout from aboveground nuclear bomb tests conducted in neighboring Nevada between 1951 and 1962. Not only did the United States Government not inform the residents of St. George of the danger at the time of the blasts, but it then denied culpability after cancer, mental retardation, miscarriages, and birth defects suddenly began to appear among residents at alarming rates. Finally, in 1989, as a result of legislation introduced by a Utah Senator (Hatch), the government is paying the survivors money as compensation. The compensation program is generally viewed by the "downwinders" as "too little, too late and too grudgingly given." One of the "downwinders" who survived is quoted:

The Late Twentieth Century UFO-ET Phenomena

"...they (the government) knew, they absolutely knew at that time, although they claimed they did not," said Elmer Pickett, 72 year old, proprietor of Elmer's Home Center, whose wife died of cancer at 39, leaving six children. "Knowing what they were doing to us, they were in a sense murdering us. They are just as guilty of murder in my estimation as the people who were running the death camps in Nazi Germany."[55]

These are strong words by Mr. Pickett, but clearly they fit into the scenario presented here. Perhaps the most prevalent theme revealed in this small sample of newspaper articles of 1992 and 1993, is that the United States Government routinely lies and behaves in a deceitful, non-loving way toward its citizens - loyal, honest, hard working citizens. Then it uses a controlled media to brainwash the people into believing that the federal government loves them and is serving them. And, of course, the biggest lie and deception of all is the one featured in this book: the past and ongoing involvement of ETs in the affairs of humans.

More Modern Day ET Activity

The human abductions and animal mutilations continue at an astonishing rate in the United States. John Lear, a former airline pilot and UFO researcher, stated in a 1988 paper that there have been more than 14,000 cattle mutilations since 1973.[56] Linda Moulton Howe, an award-winning professional documentary-maker and author, has probably done more research on the thousands of animal mutilations that have occurred in the United States than anyone else. She reports that one of the strongest and strangest hallmarks of the animal mutilation mystery is the complete lack of blood in the carcass of the cow, horse, sheep, or whatever animal found mutilated.[57] Precision incisions, the removal of internal organs from the carcass, including the sexual organs, are also characteristics of animal mutilations. No footprints are found near the carcass, including the animal's own footprints. In addition to flying disks that are sometimes associated with animal mutilations, dark, noiseless helicopters are sometimes seen where animal mutilations occur. There have been many sightings of creatures known as "Bigfoot," large six to eight foot tall hairy creatures, sighted on ranches and other areas where animal mutilations were found.[58] Howe reports that this is a worldwide phenomenon, as animal mutilations have been reported in Canada, Australia, Europe,

South America, Central America and other places.⁵⁹ She states that she and others have received reports of human mutilations, but has been unable to verify these reports, as she has with animal mutilations.⁶⁰

The human abduction cases are at least as strange and bizarre as the animal mutilation cases. Budd Hopkins, a leading human abductee investigator and author, estimates that at least one million Americans have been abducted.⁶¹ Donald Ware, a retired Air Force pilot and UFO researcher, states that a U.S. Government intelligence operative told Timothy Good, author of *Above Top Secret*, that six million Americans have been on board a UFO, whether they know it or not.⁶² That is one in 40 Americans. Valdamar Valerian, a UFO researcher and author, states that the greys and other alien species regularly and systematically abduct 10% of the entire human population, and that this percentage number is steadily rising.⁶³ 10%! What is going on?

To answer this question has been very difficult even though human abductions by aliens have apparently been occurring at a high rate for almost 40 years, because of the aliens' ability to erase the memory of the abduction. In many cases, such as that of Barney and Betty Hill, many details of the abduction can be remembered through regressive hypnosis, but most abductees do not consciously remember being abducted by aliens. Abductees usually experience some "missing time", often have nightmares and feel anxiety, fear and self-doubt for unknown reasons, but they do not remember the abduction. However, the memory-erasing techniques of the aliens, whatever they are, are far from perfect, as many abductees remember their abduction consciously, even without hypnotic regression. Therefore, by the late 1980s and 1990s there were hundreds of abduction cases on the records of abductee investigators like Budd Hopkins. Hopkins was the first to claim and document in his 1988 book, *Intruders: The Incredible Visitations at Copley Woods*, that the main purpose of alien abductions was to create alien human hybrids on Earth.⁶⁴ By now, abductees have observed many developing fetuses in the underground bases of the aliens and a few drawings of what they have seen exist (see Figure 12-2). Also, Hopkins was the first to document cases of women who become pregnant, often unexpectedly, and then mysteriously lose their fetuses about three months later.⁶⁵ These women reportedly are impregnated by the aliens during an abduction, which the women do not remember. It is not clear whether the aliens use artificial insemination or in vitro fertilization and implantation of the fertilized egg in the women's womb, or both methods, but alien and human genes are somehow combined. After roughly three months the aliens remove the hybrid fetus to their underground colonies where they continue to develop in some sort of fluid encased in a see-through container (Figure 12-2). After the fetus is removed there is no indication that the women were ever pregnant as they have no conscious memory, and no blood or fetal tissue is ever found, although the women often "feel" acutely or "dream" that they have lost a baby.[66]

In addition to the hybrid breeding program, the aliens tend to dazzle their abductees with their technical and paraphysical abilities. For example, the aliens have the capability of materializing and dematerializing at will. Abductees are often impressed with the aliens in that they use telepathy in order to communicate directly with them. They can walk through walls and other "solid" objects. They can do this, they explain, by simply changing the rate at which they vibrate.[67] Evidently all life-forms vibrate at a certain rate; we three-dimensional beings vibrate at one rate, while higher dimensional beings vibrate at a higher rate, which makes them invisible to us. The Pleiadian Plus group also talks about vibrational frequency and how this Earth is being kept at a certain vibrational frequency by the ETs in order to control us.[68] The aliens use a technical device small enough to be attached to a belt to change their frequencies in order to walk through walls and move to higher dimensions.[69] Aliens use this small technical device to change the frequencies of humans and move them from their beds and houses or wherever, into their flying disks (UFOs). The abductees can sometimes see their bodies they leave behind as they are floating away toward the alien crafts. These experiences are equivalent to what have been called out-of-body experiences (OBE's). OBE's have been reported by more than two million Americans, according to a recent national survey conducted by the National Opinion Research Council (NORC).[70] It is interesting to note in passing that there is evidence that humans might have a latent capability of OBE travel without the clever little technical devices of the aliens. But, most humans either haven't had an OBE experience, or they don't remember them, so they are amazed at the routine OBE capabilities of the aliens. What I'm suggesting here is that we humans might have natural capabilities, even though they are mostly latent at the moment, to do things that the aliens need their marvelous technology to accomplish. We will be examining this possibility in greater detail as we discuss our own spiritual evolution.

Bob Lazar, a physicist who claims to have worked on a flying disk project for the secret government near Las Vegas, Nevada, believes that alien telepathic capability is a technological development on their part, as well.[71] While telepathic abilities of humans are not well known, here again there is evidence that humans actually have more mental telepathy capabilities than the greys. There is at least one human population that is reported to communicate using mental telepathy, a small group of Australian aborigines. Marlo Morgan, an American woman who spent more than three months walking across the Australian desert with this group of 62 aborigines, reports that the adults of this small group of aborigines train their children from infancy to communicate mentally, and all of the adults communicate with each other in this manner.[72]

Nevertheless, most of we humans at our present stage of spiritual and technological evolution, are awed by the technical and paranormal abilities of the aliens. We humans, like children, are in awe of anybody who can do things that humans cannot do. Many abductees are in awe of the aliens and would revere them like "gods." For example, Raymond Fowler, a well-respected UFO/abductee researcher, who has written extensively about the ongoing abductions of Betty Andreasson, now Betty Luca,[73] gives us "the message" of the aliens.[74] Many abductees, Fowler reports, awed by the technical and paranormal abilities of the grey ETs, might "swallow this message whole." This message berates we humans for the increasing destruction of the life-supporting environment of the Earth and states that humans are going to become sterile and extinct. This is why, the aliens suggest, the UFO activity has increased dramatically since the 1940s, as they are caretakers of this planet and are carrying out their genetic activity in order to preserve life on this planet after the human species becomes extinct. Also, the message "confirms" that humans are much more than flesh and bones, and that we do have a spiritual existence.

The "confirmation" that we humans have a spiritual existence should come as no surprise to most of the readers, as this has been proclaimed by literate and non-literate cultures for thousands of years. And here we again have a message of doom, such as we have seen in the Fatima prophecy, and the prophecy given to French citizen, Claude Vorilhon, noted above, and many others.

For one more perspective on this subject, let us turn to Valdamar Valerian, UFO/alien researcher and author. Few have presented more information about the aliens than has Valerian in his book, *Matrix II: The Abduction and Manipulation of Humans Using Advanced Technology*.[75] Valerian, also, presents evidence that stems from the "Meier aliens" (whom I have been calling the "Swiss Pleiadians") and other esoteric sources, that indicates that the malevolent ETs (the greys and their allies) are indeed spreading end-times predictions as part of their psychological-control-of-humans strategy.[76] Valerian also presents evidence from Jason Bishop III and others, that the greys are interfacing "...with humans in 'secret societies' and within the military/governmental complex..." in order to manipulate surface Earth cultures.[77] The information from the Meier aliens ("Swiss Pleiadians,") some of which has been withheld from the public by its editor, Wendelle Stevens, indicates that the greys and their allies have been controlling and manipulating humans for thousands of years.[78]

Figure 12-2 Depiction by an artist of an underground laboratory of a grey species, based on several eyewitness accounts of abductees. From Matrix II: The Abduction and Manipulation of Human Beings Using Advanced Technology, 1989/90 by Valdamar Valerian, Leading Edge Research Group, P.O. Box 481-MU58, Yelm, Washington 98597. With permission.

Valerian states that "innumerable witnesses" have confirmed the existence of the underground bases that the greys use for their genetic manipulations, mostly abductees. The existence of these underground bases is also confirmed by a few U.S. military and civilian personnel who have worked with the secret government and the greys, and have become disgusted with their programs and come forward with information. One such individual, Thomas C., was working in what is probably the most visited of all the underground bases by abductees, a joint U.S. secret government-grey alien base under Dulce, New Mexico. Mr. C. took out a number of black and white photos, a videotape, and a set of papers about the facility. He made five copies of this material, which became known as the Dulce papers, and gave them to a few close friends. Apparently, Thomas C. lost his wife and son to experiments of the greys and secret government for his "indiscretion". Among the horrors that Mr. C. witnessed at Dulce were row after row of humans and human remains in cold storage.[79] The facility at Dulce is connected to other facilities by tunnels and underground "trains," which travel at tremendous speeds.[80]

Valerian also presents evidence of mind control efforts of the greys and the secret government.[81] He also gives some biological information on the greys and the "lizzies" (reptilians), and some information on the historical interaction of the malevolent aliens and humans.

Clearly, if Valerian's information is mostly correct, and the hypotheses of Cooper, Lindemann and others are mostly correct, the fear factor of the current UFO/ET question would be astronomical to most humans. But, this is exactly what the ETs and their allies in the secret government want to create - FEAR!! How might we primitive humans avoid fear when we face such dastardly ETs that possess amazing (to us) technology and mean us no good? The key to avoiding fear and panic in the face of these seemingly overwhelming odds, as I have alluded to, is our individual and collective conscious and spiritual development, and this will ultimately be the most important message this book is meant to convey. On a higher spiritual level, these regressive ETs may be here, in fact, to HASTEN our own evolution toward the light and away from the darkness that these aliens and our secret government represent!!

Valdamar Valerian is one of the few UFO/alien researchers that tries to present spiritual information, which, as we've suggested, might be the key to understanding the historical ET phenomenon on Earth. Valerian states that as we humans develop the ability to contact our higher spiritual Self, we will have the tool we need to switch from being a victim of the aliens, the secret government, or anything or anybody, to being an awakened creator.[82] We have seen evidence that indicates that true spiritual knowledge is precisely what the ETs have distorted the most throughout our history. Valerian presents evidence that indicates that one of the methods the aliens use to deceive us is to "...(d)eny there exists choice that can be spiritually exercised by humanity that can obviate the entire alien games and reality set."[83] As stated, we will suggest some choices we can make that will enable us to move spiritually right past and beyond the ETs in future chapters.

Here I have only brushed the surface of the current UFO/ET phenomenon, looking specifically for evidence that the ETs have been trying to manipulate and control humankind with considerably less than honorable intentions. I am satisfied that I have found such evidence, and that such evidence corroborates the historical and esoteric evidence I have presented that we humans have been manipulated and controlled by ETs since our "creation" by the reptoids, and others, within the last three hundred thousand years.

In order to better be able to understand both the reptoids and the greys, and better understand how they have endeavored to control humankind, let us take a break from the ET hypothesis and review some basic Earth human behavioral biology.

[1] Gallup poll (1984) of the percentage of Americans believing in UFOs quoted by Schellhorn, G. C. 1989. *Extraterrestrials in Biblical Prophecy.* Horus House Press, Inc., Madison, Wisconsin, p. 251.
[2] Sachs, M 1980. *The UFO Encyclopedia.* Perigee Books of G. P. Putnam's Sons, New York, p. 207.
[3] *Ibid.*
[4] *Ibid.*
[5] *Ibid.*
[6] Hynek, J. A. 1977. *The Hynek UFO Report.* Dell Publishing Co., New York.
[7] *Ibid.*, p. 33.
[8] *Ibid.*, p. 244.
[9] *Ibid.*, p. 139.
[10] *Ibid.*, p. 281.
[11] *Ibid.*, pp. 282-287.
[12] *Ibid.*, p. 287.
[13] *Ibid.*, p. 120.
[14] *Ibid.*, p. 189.
[15] Fuller, J. G. 1966. *The Interrupted Journey.* Dell Publishing Co., Inc., New York.
[16] Sachs, M. 1980. *op. cit.*, pp. 295-296.
[17] Vallée, J. 1980. *Messengers of Deception: UFO Contacts and Cults.* Bantam Books, Inc., New York.

18 *Ibid.*, pp. 155-160.
19 *Ibid.*, p. 158.
20 Vallée, J. 1965. *Anatomy of a Phenomenon: UFOs in Space - a Scientific Appraisal.* Ballantine Books, New York. pp. 1-10.
21 _____ 1980. *op. cit.*, pp. 27-29.
22 Good, T. 1988. *Above Top Secret: The Worldwide UFO Cover-Up.* William Morrow and Co., Inc., New York.
23 *Ibid.*, p. 475.
24 *Ibid.*, p. 253.
25 Hynek, J. A. 1977. *op. cit.*, pp. 7-8.
26 Good, T. 1988. *op. cit.*, p. 183.
27 Cooper, M. W. 1991. *Behold a Pale Horse.* Light Technology Publishing. Sedona, Arizona.
28 *Ibid.*, p. 197.
29 For references to the alleged UFO crashes, see Good, T. 1988. *op. cit.*, pp.253-426.
30 Cooper, M. W. 1991. *op. cit.*, p. 202.
31 *Ibid.*
32 *Ibid.*, pp. 203-204.
33 *Ibid.*, pp. 209-211.
34 *Ibid.*, p. 212.
35 *Ibid.*
36 *Ibid.*
37 *Ibid.*, pp. 68-98.
38 Lindemann, M. 1990. "UFOs and the Alien Presence: Time for the Truth." Manuscript from the 2020 Group, Santa Barbara, California.
39 *Ibid.*, p. 7.
40 *Ibid.*, p. 4.
41 *Ibid.*, p. 19-23.
42 Cooper, M. W. 1991. *op. cit.*, p. 168.
43 *Ibid.*, pp. 165-169
44 Zailan, M. 1993. "Dissident gets a voice." *San Francisco Examiner*, "Datebook", April 4, p. 33. See also Herman, E. S. and N. Chomsky 1988. Manufacturing Consent: The Political Economy of the Mass Media. Pantheon Books, New York.
45 *Ibid.*
46 LaSalle, M. 1993. "Film on rabble-rouser Chomsky and media." *San Francisco Chronicle*, April 9, p. D5.
47 *Ibid.*
48 Summers, A. 1980. *Conspiracy.* McGraw-Hill Book Co., New York, p. 14.
49 The 1979 Select Committee on Assassinations report was quoted by, Summers, A. 1980. *op. cit.*, p. 14.
50 A partial list of the members of the Council on Foreign Relations, the

Jason Group, and the Trilateral Commision is provided free of charge by F.R.E.E., P. O. Box 33339, Kerrville, Texas 78029

[51] *USA Today*, November 15, 1993, p. 4A.

[52] Hulen, D. 1992. "Radioactive soil had secret burial in Alaska in 1962." *San Francisco Chronicle*, October 12, p. A5.

[53] Jackson, R. L. and R. J. Ostrow 1994. "Iran-Contra report blames Reagan, Bush". *San Francisco Chronicle*, January 19, p. A1 and p. A13.

[54] Hilts, P. J. 1993. "Battle heating up over veterans' mysterious Gulf War ailments." *San Francisco Chronicle*, November 25, p. A17.

[55] Lippman, T. W. 1993. "Utah survivors remain bitter about Nevada bomb tests." *The Oregonian* (Portland), May 19, p. A13.

[56] Lear, J. 1988. "The UFO Cover-Up." In, Valerian, V. 1991. *op. cit.*, pp. 239-252.

[57] Howe, L. M. 1991. "The 'alien harvest' and beyond". In, *UFOs and the Alien Presence: Six Viewpoints*, M. Lindemann (ed.). The 2020 Group, Santa Barbara, California, pp. 61-84.

[58] *Ibid.*, p. 64.

[59] *Ibid.*, p. 65.

[60] *Ibid.*, p. 69.

[61] Budd Hopkins quoted in, Ware, D. 1991. "The 'Larger Reality' Behind UFOs". In, Lindemann, M.(ed.) 1991. *op. cit.*, pp. 195-218.

[62] *Ibid.*

[63] Valerian, V. 1991. *op. cit.*, p. 151.

[64] Hopkins, B. 1988. *Intruders: The Incredible Visitations at Copley Woods*. Ballantine Books, New York.

[65] *Ibid.*, p. 79 and pp. 236-238.

[66] *Ibid.*

[67] Fowler, R. E. 1991. *The Watchers: The Secret Design behind UFO Abduction*. Bantam Books, New York. p.183.

[68] Marciniak, B. 1992b. *Bringers of the Dawn: Teachings from the Pleiadians*. Bear and Company Publishing, Santa Fe, New Mexico. p. 26.

[69] Fowler, R. E. 1991. *op. cit.*, p. 183.

[70] NORC survey reported in Fowler, R. E. 1991. *op. cit.*, p. 183.

[71] Lazar, B. 1991. "Alien Technology in Government Hands." In, Lindemann(ed) 1991, *op. cit.*, pp. 87-127.

[72] Morgan, M. 1991. *Mutant Message Downunder*, MM Co., Lees Summit, Missouri, pp. 53-56.

[73] Fowler, R. E. 1980. *The Andreasson Affair*. Bantam Books, New York, and Fowler, 1991. *op. cit.*

[74] _____ 1991. *op. cit.*, pp. 355-357.

[75] Valerian, V. 1991. *op. cit.*

[76] *Ibid.*, p. 15.

[77] *Ibid.*, p. 97.

[78] *Ibid.*, p. 15.
[79] *Ibid.*, pp. 188-189.
[80] *Ibid.*, pp. 173A-237.
[81] *Ibid.*, pp. 303-311.
[82] *Ibid.*, p. 364A.
[83] *Ibid.*, p. 30.

> *"We must always keep humility before us,
> so that we may realize this strength
> cannot proceed from any strength of our own."*
> The Autobiography of Saint Teresa of Avila

Chapter 13

Some Biocultural Aspects of Humans and the ET Question

Do you need a break from all of the astonishing news of the ET presence hypotheses? This chapter will provide somewhat of just such a breather, as we will mostly investigate in this chapter the work of human scientists who have no knowledge or interest in the possible ET presence throughout the history of humankind. However, this is far from a total break from the basic material of this book, because I feel this chapter is necessary in order to understand some of the methods in which the regressive ETs have endeavored to control the human species, and some of the biological and cultural reasons they have, to a large extent, been able to do so. Also, this discussion of behavior will help us understand the behavior of the reptoids and the greys when we discuss them in later Chapters.

Culture

Since the latter part of the nineteenth century, American cultural anthropologists have been studying human cultures, mostly those of technologically primitive cultures. They and sociocultural anthropologists of other countries have shown, over the decades, that almost all humans grow up with the attitudes and rules of their culture, and transmit their culture from generation to generation. Cultural anthropologists have demonstrated that much of the behavior of an individual, including his or her belief systems (e.g., religion), is a reflection of the culture in which he or she grew up. For example, a human that grew up in an Islamic culture is almost always of the Islamic faith and tradition as an adult, and one that grew up in a Christian culture almost always adheres to the Christian faith as an adult, etc. A reciprocal concept of culture is the concept of ethnocentrism. Ethnocentrism is the tendency of humans to interpret and evaluate the cultural beliefs and practices of others in terms of one's own cultural beliefs and practices. Usually these evaluations are quite negative. Thus, for example, the people of the Islamic cultures usually believe that the people of Christian cultures are greatly mistaken in their religious beliefs and, likewise, the people of Christian cultures usually believe the people of Islamic cultures are greatly mistaken in their religious beliefs. The sociocultural anthropologists and historians have demonstrated that virtually all humans of any time period had ethnocentric attitudes, often with tragic results. The people of Islamic and Christian cultures have been killing and maiming each other in the name of "God" since shortly after the founding of Islam in the Seventh Century A.D., and this is but one of the thousands of tragic examples of ethnocentrism that could be given.

Since the early decades of this century, it has become a standard doctrine among cultural anthropologists, and later archeologists, that almost all of the behavioral variation among humans throughout the world is due to the fact that they were reared and live in different cultures, and these behavioral variations have nothing to do with biological or genetic differences. In fact the claim has been made that human culture is superorganic (above biological), and that the study of human behavior, especially the study of human culture, should be completely divorced from biology.[1]

Much of the reason that American cultural anthropologists opted for cultural influences at the total exclusion of almost all biological factors in explaining human behavior was a reaction against the biological deterministic theories that were being propagated by American biologists at the turn of the century.[2] And, of course, cultural anthropologists were much impressed with the data they were gathering that indicated a very strong, if not overwhelming, influence on the behavior of humans by the culture in which they were reared. Biological deterministic theories, on the other hand, purport that virtually all of human behavior is determined by the biology of humans, especially their genes. So there has been what is called nature (biology) versus nurture (the environment, which includes human culture) debate which has continued throughout this century in all behavioral sciences. The early biological deterministic theories were shown to be too simplistic and untenable, so behavioral theories that purported that nurture, or the environment, including culture, was primarily responsible for human behavior were prominent in the middle decades of this century. However, most biologically oriented scientists still felt that biology must be important in behavioral determinants, but it wasn't understood how.

Studies of animal behavior in their natural habitat were begun by biologists and anthropologists (on primates) in the 1930's, and these field studies picked up dramatically in the 1960's. Biologists hadn't studied animals in their natural habitat until the 1930's because they were devoted to the methods of the hard sciences and reductionism, which could best take place in a laboratory.[3]

After years of studies of animals in their natural habitat, the biologists had the data they needed to document the biological influences of behavior, and a new sub-discipline of biology, usually called sociobiology, was launched in the 1970s by Harvard ethologist, E. O. Wilson.[4] Sociobiology, the study of the biological basis of behavior, while it has been controversial, nevertheless is a thriving sub-discipline of behavioral sciences today. Many that study animals in their natural habitat are attracted to sociobiological theory, and also many that concentrate on human behavior have applied sociobiology to the human animal. Wilson, for example, states that there is strong evidence that "...there does exist a human biogram, a pattern of potentials built into the heredity of the species as a whole...," which is the case for all animals. He points out that humans share with other mammals a number of traits, such as prolonged maternal care, a sexual division of labor (males and females doing different things), a tendency toward male dominance, and an extended socialization of the young based in good part on social play.[5] While the sociobiologists are accused of overemphasizing the effects of genes on the social behavior of our species, the hundreds of studies of the past few decades of animals in their natural habitats, leave no doubt that there are many characteristics of human behavior that are very similar to the behavior of animals, especially that of mammals, and more especially that of primate mammals.

A recent study that renders strong support for the "nature" side of the nature/nurture debate has been carried out on monozygotic and dizygotic twins that have been separated from infancy, by a group of psychologists at the University of Minnesota.[6] The dizygotic twins are genetically no more similar to each other than brothers and sisters, while monozygotic twins, the products of the splitting of one fertilized ovum (one zygote/egg) of the mother, have an identical genetic make-up. Similar twin studies have been carried out since the 1930s, but the earlier studies often had a very small sample of twins and/or were otherwise flawed. But, the Minnesota study subjected more than one hundred sets of twins reared apart to a week or more of intensive psychological and physiological assessment. The study tried to measure the leisure time interests, the occupational interests, the social attitudes and the temperamental personality traits of the twins. The scientists who conducted this study came up with the surprising conclusion that "...monozygotic twins reared apart are about as similar as are monozygotic twins reared together." [7] In other words, according to this study, identical twins that are separated at birth and reared apart are going to be about as similar to each other when adults as identical twins that are reared in the same family and experience very similar environmental influences throughout their immature years. Despite this twin study, and the burgeoning science of sociobiology and many more studies and disciplines that support the nature side of the nature/nurture debate, the tremendous influence of culture and other environmental influences on human behavior remains. Many scientists involved in this debate recognize that it is impossible to neatly divide the influences on human and animal behavior into those of nature and nurture. Many scientists recognize that nature verses nurture is a false dichotomy, which is a result of the lack of understanding we humans have of ourselves.

Imprinting and Brain Development

Understanding much of how our human cultures become embedded in individuals can be illuminated by examining studies of the imprinting phenomenon and the corresponding studies of early brain development in animals. I will briefly present a hypothesis that demonstrates how early environmental experiences, including exposure to the culture in which one is reared, becomes "wired" in one's brain through the imprinting and early brain development process. This is a hypothesis that is not in any anthropology text. Most cultural anthropologists and archeologists still resist any proposed intrusion of biological aspects into the realm of culture. But this hypothesis helps explain how humans acquire their culture during their immature years and usually fervently hold onto these beliefs for most, if not all, of their lives. This imprinting-during-brain-development hypothesis helps explain the prevalence of ethnocentrism among humans and provides a partial explanation of how the ETs have been able to control much of human social and spiritual development. Finally, this hypothesis helps resolve, to some extent, the decades-old nature-nurture dichotomy.

Before we examine the neurobiological bases of imprinting and its relevancy to humans, let us take a closer look at the concept of culture, Conrad Kottak, in a popular cultural anthropology text, defines culture as that,

> ...which is transmitted through learning, behavior patterns, and modes of thought acquired by humans as members of society. Technology, language, patterns of group organization, and ideology are aspects of culture.[8]

Cultural anthropologists and archeologists emphasize that culture is the exclusive domain of humans because of our ability to symbolize. Language, in Kottak's definition of culture, is a clear example of humans' ability to symbolize. We recall from earlier discussions that there is much evidence to support the notion that earlier forms of humans, such as *Homo erectus* and Neanderthal humans, could not talk, or had a very rudimentary form of language at best, and that it is now believed by many anthropologists and linguists that language did not "evolve" until modern humans "evolved" some thirty-five thousand, to more than one hundred thousand years ago. With regard to symbolizing, Lewis Binford, a prominent American archeologist, claims that there is no solid evidence of human symbolizing, and therefore human culture, before the Upper Paleolithic, which began around thirty-five thousand years ago.[9] Following Binford's claim, none of the thousands of stone tools and the other artifacts associated with pre-modern human hominids of the Middle and Lower Paleolithic, such as those of *Homo erectus*, are part of a cultural tradition because there is no conclusive evidence that pre-modern human hominids had developed a capacity for language or any other form of symbolizing. So there was no culture in *Homo erectus* (~1.8 million years before the present to as recently as 12,000 years before the present), according to Binford, even though they clearly had some sort of traditions.

Similarly, using the definitions of anthropologists, the hundreds of examples of animal traditions passed on from generation to generation are not considered as culture. Most of these local traditions are known only in a species where more than one population has been studied and usually involve items of the animal's diet. For example, it was found that a lemur of Madagascar, known as the Sifaka (*Propithecus verreauxi*), has different eating traditions in the northern and southern populations. Northern groups of this species were found to eat only the flowers of a particular tree (*Rothmannia decaryi*), while southern groups of this same species of lemur fed extensively on the fruit of the same tree.[10] Similarly, it has been observed that populations of a monkey, known as the grey-cheeked mangabey (*Cercocebus*

221

albigena), that live in eastern Uganda eat the flowers, shoots and fruit of a certain tree (*Sapium ellipticum*), while another study of the same monkey that lives in western Uganda and has the same tree available in its habitat, does not use this tree at all for food.[11]

The common chimpanzee (*Pan troglodytes*) of Africa is the non-human primate species that exhibits the most variations in local traditions. In addition to variation in diet from population to population as we saw with the lemur and monkey we just discussed, chimpanzees are the best non-human primates at "tool-use." Chimps, in populations from east to west Africa, have been observed using several items of their environments, such as stones, twigs, bark and leaves, as "tools" to crack open nut shells, "fish" for termites and ants, acquire honey, and "sponge" for water located in hard to reach places. Usually these local traditions are discontinuous. For example, nut-cracking with stones has been observed only in west African populations of chimps, while termite fishing, made known to the general public by Jane Goodall on National Geographic Specials on TV and elsewhere, has only been observed in eastern and central African populations.[12] If a chimp that was born and raised in west Africa were transported to east African populations of chimps, it would not know how to fish for termites because such food acquisition habits had not been part of the food gathering traditions of its natal group. It may sound like a simple procedure, but it takes much skill to find and prepare a proper twig and acquire termites from a termite mound. A young chimp must spend hours of learning the proper procedure of termite fishing, usually by imitating its mother, who learned it from her mother, etc.[13]

I have concentrated on primates because they are the animals I am most familiar with, having been a primatologist for years, but local traditions have been observed in several other types of mammals and birds. Mammals which have been observed having local traditions include elephants, rodents, ungulates, carnivores, whales and dolphins. Most of these local traditions deal with knowledge of the animal's predators or prey, plant foods they may eat, and other details of their particular habitat, such as boundaries of their territories or home ranges and pathways through them. Mammals and birds learn these local behavior patterns by imitating their mothers or other members of their social groups, if they live in social groups during their immature years.[14]

Wild-living populations of African elephants have been observed to vary considerably in the relative tameness or ferocity they display toward humans. Generally, in areas where the elephants have been hunted little or not at all for several decades, the elephants are relatively tame. But in areas where human hunting has been intense, the elephants are relatively ferocious and dangerous to humans. A husband and wife team who have extensively studied African elephants, Iain and Oria Douglas-Hamilton, report a remarkable story of tradition transmission of an elephant group of South Africa in their book *Among the Elephants*. They report that in 1919 a group of neighboring farmers attempted to annihilate a small population of about 140 elephants that had been destroying part of their crops. The farmers gave this task to a single hunter. This hunter shot the elephants one by one. The elephants learned to be active only at night (nocturnal), which is unusual for elephants, and be very wary of humans in order to avoid being killed. The hunter killed most of the elephants, but he had an extremely difficult time finishing the group off because of the wariness of the elephants toward humans. Finally, in 1930 the hunter admitted defeat and the surviving elephants of the group were granted a small sanctuary. However, several decades later, this group is still primarily nocturnal and they behave extremely aggressively toward any humans. It is clear that there can be few, if any, elephants alive today from the group that were shot-at between 1919 and 1930, so it seems that their nocturnal behavior and fear of humans has been transmitted from adults to the young over the decades.[15]

Observed examples of birds passing social traditions from generation to generation are perhaps even more numerous than those of mammals. For example, it is well known that many species of birds migrate hundreds and thousands of miles each year to the exact area in which they were reared to carry out their own reproductive behavior. While it is not understood how the migratory birds make the long migrations, the fact that they are able to find the area of their hatching and rearing, the exact stretch of beach or rocky cliff or whatever, is well documented in some cases. For example, a population of blue geese returns to the same specific location on shores of Hudson Bay, Canada to nest every year, and it has been demonstrated that these geese have learned this through social transmission from generation to generation in these locations.[16] Another example of social tradition in birds comes from song birds. Adolescent male song birds learn the details of the song(s) of their species from adult males. The ability of song birds to sing seems to be innate, but the young males must learn the details of the song of their species and local population in which they are reared by imitating adults.[17] Widespread species of song birds have been shown to have many local dialects similar to dialects of the language of widespread human populations, such as a "western drawl" or a "southern accent" of the United States.

I could give hundreds more examples of studies where scientists of biological orientations have observed incidences of birds and mammals having some sort of social traditions that are passed from generation to generation, but these few examples will suffice in this brief investigation of culture.

Clearly, only animals that are social to some extent are capable of behavioral transmission of information through social traditions. The two groups of animals on Earth where this occurs most frequently are birds and mammals. In a majority of the species of the other major groups of vertebrates (the fishes, the amphibians and the reptiles), the mother lays eggs and neither parent provides any care whatever for their offspring. In birds and mammals, on the other hand, the young are cared for by at least one parent during their immature years. The one parent to which I refer is, of course, the mother. However, the father also is responsible for much of the care and rearing of their offspring in many birds and mammals. It is during the period of immaturity when most behavioral transmission of information from parent(s) to offspring takes place. Much of the learning by the immatures occurs by imitation of their parent(s) or other members of their social group, if they are animals that live in a social group when they are adults, such as many birds, elephants and most of the two hundred plus species of non-human primates.

Because of the widespread, well documented prevalence of social traditions in birds and mammals, I prefer a much simpler definition of culture than those that are traditionally offered by cultural anthropologists. J. T. Bonner in his book, *The Evolution of Culture in Animals*, provides the following definition of culture: The transfer of information by behavioral means.[18] This simple definition is all we need in this brief discussion of culture, and if the biologically oriented scientists had been studying animals in their natural habitat for as long as American cultural anthropologists have been studying technologically primitive cultures, something like Bonner's culture definition would, I feel, be the base definition of culture. Then, I feel, a special definition for human culture with its symbolic communication (language and writing) and extraordinary complexity relative to animal culture would be added. This is basically the way it is today, although cultural anthropologists and archeologists, as we've seen, have a difficult time acknowledging that animals have anything similar to human culture. But who, besides anthropologists and behaviorally oriented biological scientists, cares about these small arguments concerning definitions and words anyway? Here we are trying to understand the importance and relative permanence of our early learning experiences, and how the regressive ETs amongst us might use this to control us, so we'll use Bonner's definition of culture and proceed.

Bonner points out how much faster is cultural transmission of behavior, than is genetic transfer of behavior. As we have seen there is much evidence that some behavior patterns are transferred genetically from parents to offspring. But one individual can transmit behavior genetically only to his or her offspring. However, an individual can transmit behavior culturally to many individuals anytime.[19] Bonner argues that culture didn't suddenly spring from nowhere when modern humans "evolved," but that culture has a long evolutionary history among animals. One of the first steps in this evolution (and creation or genetic manipulation) is when animals on Earth evolved the capability of multiple-choice responses in behavior situations as opposed to single responses (reflex actions). This involved the centralization of the nervous system, especially the evolution of the brain and its subsequent expansion. This is especially seen in the development of the cerebral cortex, the youngest part of the brain as seen through the evolution-creation sequence as observed on the Earth. Figure 13-1 shows the relative portion of cerebral cortex in the brains of fish, reptiles, a small mammal (rabbit) and humans. This expansion of the brain allowed animals, especially mammals, more learning and memory capacity and more flexibility of behavior patterns. Little difference in the plasticity (complexity) of behavior is seen in animals with little cerebral cortex, such as in fish and reptiles, whereas much more plasticity of behavior is seen in animals like mammals, who have a relatively large cerebral cortex.[20]

Imprinting

In our investigation of how information is transmitted through behavioral means (culture) in animals, we now look at the study of imprinting in animals. Imprinting was discovered by an Austrian pioneer of the studies of animal behavior, Konrad Lorenz in the 1930's.[21] Lorenz discovered that newly hatched infant geese, when deprived of contacts with their mother, but allowed to see him, would follow him around as if he were their mother. These infant geese form an imprint during their first few hours on their mother, and they imprinted on Lorenz in the absence of their mother.

Considerable research on imprinting followed Lorenz's discovery. In the early days of research on imprinting, imprinting was thought to be a characteristic of birds only. Many species of birds were shown to imprint on their mothers and sometimes their fathers (filial imprinting) during the first few hours or days of their lives. In addition to filial imprinting, birds have been shown to imprint on their sexual preferences in the first few hours or days of their lives even though it will be months and years before they engage in sexual behavior themselves. In addition to these social imprints, many birds were seen to imprint on food preferences and habitat preferences.

Following Lorenz, early researchers felt that the imprinting process had two crucial aspects:

1. Imprinting was thought to take place only during a restricted time period of the bird's life called the sensitive period (also called the critical period or sensitive phase). These periods or phases occur very early in the life of birds or other animals. After these sensitive periods have passed, imprinting is impossible.
2. Imprinting is irreversible.

As research continued, however, some researchers found that, in some cases, the preferred object(s) (such as food preferences) on which the animal had been imprinted could be changed under certain circumstances. Some researchers pointed out that some types of imprinting are more persistent than others. Sexual imprinting, for instance, is more persistent than an imprint for food preferences. Also, it was seen that some individual birds imprinted more readily than other individual animals. In light of these and other difficulties encountered during research on imprinting, K. Immelmann, a German researcher on animal behavior, has suggested that the term "irreversible" be replaced by the term "persistence" in the imprinting research.[22]

Further research found that mammals show many behavioral patterns that are similar to the imprinting observed with birds. Evidence indicates that mammals, like birds, have filial imprinting, sexual imprinting, ecological imprinting and social imprinting. For example, it has been found that much of a rat pup's sexual imprinting is through the olfactory (smell) sense. In one experiment, male rat pups lived with and suckled female rats whose nipples and vaginal odors were altered with lemon scent. These rat pups were weaned and not exposed again to the lemon scent and females, until they were about 100 days of age. At this time the males were paired in mating tests with a normal sexually receptive female, or with a sexually receptive female that had been treated with lemon scent. The males ejaculated readily when paired with lemon-scented females, but were slow to achieve ejaculation when paired with normal females.[23] Both the young of mountain sheep and of red deer learn their home ranges where they live from their mother and other adults of the social group into which they are born. This must be true of every mammal that lives in a home range or territory. This, plus the learning of their diet and other details of their habitat appear to be a case of imprinting.[24]

Figure 13-1   *A view of the brains (midbrains, or sagittal sections) of a fish, a reptile and two mammals demonstrating the size of the cerebral cortex (black) relative to older parts of the brain. Drawn after Cambell, 1985*

One study of wild rats (*Rattus norvegicus*) indicates that they are imprinted with the social status of their parents and the clan into which they were born. In this study infant rats that were born into a low status class that lived in suboptimal areas remained in the area of their births and became low-status adults.[25] There is much evidence that infant primates of wild-living populations also imprint the status of their mother in the troop or group in which she lives. With few exceptions, monkeys whose mothers were low-status individuals will grow up to be low-status adult monkeys and those that are reared by high-status mothers will grow up to be high-status adults.

Laboratory studies have shown the quality of maternal behavior of a female rhesus macaque monkey is disproportionately influenced by experiences with her own mother. Generally, if the female monkey received poor mothering from her mother when she was an infant, she also will be a poor mother, at least with her first infant.[27] Again, laboratory studies of the rhesus macaque has shown that male sexual competency is highly dependent on experiences during the first three to nine months of life. With the rhesus macaque the male mounts the hind legs of the female during sexual intercourse; if the young infant male does not observe this when growing up, it has a hard time having sexual intercourse itself when it becomes a young adult. The young male raised in isolation in a laboratory is very awkward when placed with a sexually receptive female, because he hasn't the slightest notion of how to mount a female. The young male that had not been exposed to adult copulation as an infant-child will eventually learn how to do it, especially if paired with an older, experienced female, but learning the technique is generally very awkward at first.[28] Hundreds of more examples could be cited showing that the early experience of animals has lasting consequences in their later life. Animal behavior researchers K. Immelmann and S. Suomi believe it is useful to keep and expand the concept of imprinting. These two researchers acknowledging that there is much variability in the sensitive or critical periods proposed by Lorenz, and that imprints are sometimes reversible, have proposed a simple definition of imprinting: an early learning process with a rather stable result.[29] These two researchers argue that virtually all animals, from invertebrates to humans, pass through sensitive phases which corresponds to the animals' biological development, during which rapid learning takes place that has a lasting effect on the individual.

Humans and Imprinting

There is much evidence for some sort of imprinting phenomena in human infancy and childhood, but it must be pointed out that the sensitive phases or "critical periods" are much longer and much more variable in humans, because we are much longer-living animals than most of the animals we have been considering.

There is a widespread notion that there is an early period of human development when there is imprinting for a primary socialization pattern (more love, less love, no love, etc.) that will be with the child the rest of its life. The exact period (sensitive phase) proposed for this primary socialization varies from the first six months of life to up to the first eighteen months of a child's life. However, there is much support that a human learns its primary socialization imprints in the first few months of its life.[30]

Indeed, theories postulating that young children are extremely susceptible to events that may have a very strong influence on one's adult behavior have been around since Sigmund Freud proposed such a theory in the first decade of this century.[31] Freud proposed a number of developmental "stages" a human passes through from infancy to adolescence. Some of Freud's stages are the (1) oral stage, from birth to age 1½ years; (2) the anal stage, from 1½ years to three years; and so on through the genital stage of adolescence. Freud believed that these stages corresponded to biological developmental periods.

The idea that stages or sensitive periods correspond to biological development have been popular among many psychologists and psychiatrists since Freud. While everybody cannot agree as to how many "stages," what they should be called and exactly what years stages begin and end, several have been proposed over the years. One such proposal was made by psychologist John Bowlby, who did research about childhood sensitivity for decades. Bowlby contends that periods of sensitivity correspond to structural changes in the developing child. These structural changes occur principally in the physiology and anatomy of the brain as maturation proceeds.[32]

Another proposed sensitive period that is clearly related to brain development is the theory of language acquisition proposed by linguist Eric Lenneberg.[33] According to Lenneberg, the critical period (his terminology) begins in most children at about two years of age and ends around the time of puberty (c. 12 years). According to Lenneberg, language acquisition is associated with neurological growth. He points out that the human brain weighs 24% of its adult weight at birth. The "critical period" for language acquisition begins at about age two when the child's brain is about 60% of its adult weight, and ends at puberty when the brain is near 100% of its adult size. Most children learn their native tongue with great proficiency during this period. After this period is past, people have a much more difficult time learning a foreign language, and they will almost always speak a language learned after this "critical period" with an accent. This corresponds to the decrease in learning capability or loss of plasticity after sensitive phases are past, observed in many imprinting experiments on animals. Another rather striking example of this loss of plasticity is found in the fact that Japanese speakers who have not been exposed to English during the "critical periods" for language acquisition have a very difficult time learning to pronounce the English "r" or "l".[34] On the other hand, Americans of Japanese descent or the Japanese themselves who were exposed to English during their childhood have no trouble pronouncing the English "r" or "l" whatsoever.

Probably nobody has written as much about proposed cognitive-developmental periods or stages as Swiss psychologist, Jean Piaget.[35] The ages which Piaget has given for the periods and stages of intellectual growth are not absolute, but are average ages during which they appear in normal development.

Piaget's theory of intelligence is so complex that it is not all encompassing and it is not surprising that it is not without its critics. Psychologist R. J. Sternberg, for example, states that Piaget overestimated the age at which children can accomplish many intellectual tasks, and that Piaget's concept of a step-wise development of intelligence was probably overly formalistic and logically based.[37] This critic states that the whole concept of stages of growth or stages of intellectual development has come under question. He says that many developmental psychologists have abandoned the notion of stages of intellectual development altogether, believing it creates more smoke than fire in helping to understand how intelligence develops.

Still, evidence for stages of intellectual development in humans continues to surface. In a recent study of brain development, 577 children were monitored using an electroencephalograph machine. Using this machine the researchers were able to detect discrete growth spurts that appeared in specific locations of the brains of the children at specific postnatal periods. The scientists conducting this study state that the growth spurts of children's brains overlap the timing of the major developmental stages described by Piaget.[39] Similarly, an older study that used techniques of less resolution, also found evidence of brain growth spurts in specific regions of the cerebral cortex of children at ages that overlapped the Piaget periods of cognitive development.[40]

The recent and ongoing discoveries of the correlation of periods of rapid learning corresponding to the growth spurts of the brain may be the key to understanding all of the above research on animals and humans. It is probable that all of the sensitive period-imprinting-cognitive development stages observations of animals and humans that scientists have been making for decades are reflections of growth-spurts in the brains of animals and humans.

Early-Life Experiences and Brain Development

In the human species the brain of a newborn baby weighs about 330 grams. The human brain grows extremely rapidly during the first two years of life. By the time a child is two years old its brain weight has tripled to about 1000 grams. Human brain growth slows down after the child is two years of age, but it still increases its weight substantially. A child's brain weighs about 1,250 grams by age seven and 1,350 grams by the age of twelve. At roughly the age of fourteen, the brain reaches its adult weight, which is approximately 1,380 grams.[41]

The human brain consists of billions and billions of microscopic specialized cells known as neurons. All of the billions of neurons that a human is going to have are present at birth. These neurons increase a bit in size and so do a number of cells that support brain function (e.g., glial cells), but most of the extraordinary brain growth of young humans is accounted for in the formation of billions and trillions of small connections between the neurons that occurs in one's younger years.

In order to understand a bit of how this happens, we must look at some of what neuroscientists have learned about the brain in the past few decades. We see in Figure 13-2 the depiction of a much magnified neuron of the brain. Not all of the neurons of the brains look like this, but the figure represents a "typical" brain cell enough for us to learn a bit of how the brain develops.

The billions of neurons of the brain connect and communicate with each other through their small branch-like extensions, known as dendrites, and the longer extension known as an axon. The dendrites and axon of a neuron connect with dendrites and the cell body of other neurons at junctions called synapses. At these synaptic junctions between the fibers of 2 neurons, a nerve impulse triggers the release of chemicals known as neurotransmitters. These neurotransmitters modify and affect the electrochemical state of the "target neuron." Each individual neuron usually ends up with thousands and thousands of synaptic connections with other neurons. A neuron's primary output to other neurons is conducted through a neuron's axon.

Before and after the birth of animals, including humans, many of the dendrites and especially the axon, grow automatically and form synaptic junctions with other neurons, often some distance away. How these fibers of a neuron grow automatically, or innately, and "find" other neurons in a particular region of the brain is not understood by human scientists. It is assumed that these automatic, innate connections are the result of millions of years of evolution. Whatever the evolutionary-creation history that is responsible for these automatic connections, much research with animals has shown that the continued existence and refinement of these automatic connections of brain cells made shortly before and after the birth of an animal depends upon the experiences of the infant animal.

Perhaps this has best been demonstrated by experiments centered around the development of sight in domestic cats. A new born kitten first opens its eyes about ten to fourteen days after it is born. At this time, some "automatic" connections from the eye of the kitten to the visual cortex in the rear of its brain have already been made. Then, for the next several weeks the visual system of the kitten develops very rapidly. The connection between the eye and the visual cortex of the brain of the kitten made during the first four weeks of the kitten's existence are believed to be genetically determined. Thereafter, follows a "critical period" from the 4th to about the 12th week when the actual visual experience of the kitten in the world is crucial to the development of the eyesight of the kitten.[42] Under normal circumstances the kitten's visual experience from week four to week twelfth and beyond is centered around its mother and litter mates, and it will begin to know the details of its environment, including the type of animals cats eat. However, scientists in laboratories have altered many kittens' visual experiences in order to learn about the development of vision in animals.

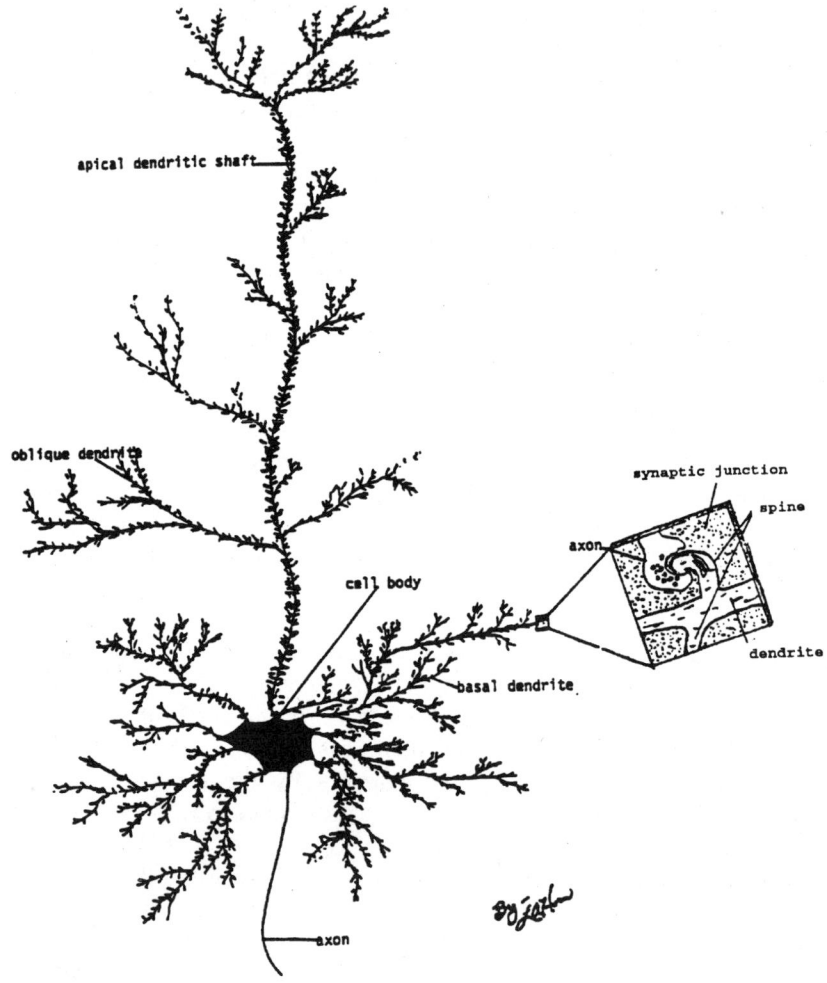

Figure 13-2 A neuron. Drawn from Greenough, 1975, and Greenough, Black and Wallace, 1987

In one experiment, one eye of kittens was closed throughout its "critical period" of visual experience from age four weeks to twelfth weeks. Then the researchers opened the eye that had been shut and covered the kitten's eye that had developed normal vision. Under these conditions, the kittens behaved as though they were blind, as they bumped into walls and other objects. The connections from the eye that had been closed to the visual cortex that had automatically formed in the first four weeks of the lives of the kittens had not been reinforced and modified during the critical period of a kitten's visual experience. Therefore, the kittens could not see out of the eye that had been covered. The researchers then tested the kittens in a way (electrophysiologically), whereby they could measure the responses of neurons in the visual cortex. They found that almost none of the neurons of the kittens' visual cortexes responded to an input from the eye that had been covered during the critical period for normal visual development. The connections from the eyes to the visual cortex of the kittens had changed so that virtually all of the connections came from the eye that had been allowed normal vision, and no connections came from the eye that had been covered.[43]

In another experiment, kittens were exposed to vertical or horizontal lines during their critical periods of visual development. These kittens were reared mostly in the dark. But, for five hours each day, the kittens of this experiment were visually exposed to lines; one group was exposed only to horizontal lines and another group was exposed only to vertical lines. In other words, the entire visual experience of these kittens during this critical period of visual development was either vertical or horizontal lines. After five months (well past the "critical period" of visual development) the kittens behaved as if they could not see lines oriented in the direction in which they had no visual experience. For example, a kitten exposed only to horizontal lines ignored a rod held vertically and shaken, but immediately ran to the rod and played with it when it was held horizontally.[44]

In a recent study, Dr. E. Knudsen investigated the connections of hearing and brain development in barn owls.[45] Dr. Knudsen found that the basic brain connections of the barn owl's hearing develops automatically according to genetic instructions. However, these early automatic brain connections in the development of a young barn owl's hearing are still quite flexible. During the young owl's early life, its hearing experiences help refine and solidify the connections already made. Knudsen concludes that in the course of normal development of hearing, that the developmental flexibility of the hearing system of the animal allows it the ability to adjust. This flexibility enables the owls to fine-tune its connections in the hearing regions of the brain and respond to changes in hearing brought about by the growth of its head and ears. Once the owl reaches adult size, however, the flexibility of the hearing system decreases substantially.

These experiments with domestic cats, owls and on many more animals have given us a clear picture of the neural development of the sensory systems of animals. First, the brain circuits of the sensory systems are roughly formed through automatic synaptogenesis from genetic instructions. However, there remains sensitive phases of development during which the final connections of the neurons are made. These final connections depend upon the actual experience of the animals, and they correspond to spurts of brain growth. Many of the sensitive phases occur automatically during the infant-childhood years of the animals, while other sensitive phases will occur only after other sensitive phases have occurred. During the early automatic period of connection formation, there is an overproduction of synapses. As development proceeds, the extra, surplus connections are lost, depending upon which ones are used. Those that are not used will be eliminated. This selective retention of connections determine the final "wiring diagrams" of the animal's sensory systems. And finally, there is a decrease in brain flexibility as the animal advances in age.[46]

It would be hard to overestimate the number of connections (synapses) in the final "wiring diagram" of the sensory systems. One brain cell (neuron) in the human visual cortex can receive as many as 10,000 to 30,000 input connections from other cells and there are as many as 200,000,000,000 neurons in the human brain.[47]

It is reasoned that the flexibility in the development of the sensory systems is to take into account the natural variations in the rates of growth in animals - growth of the head, ears, eyes, etc. The sensory systems could not stay fine-tuned and account for the natural growth of the animal if these sensory systems were under rigid genetic control. However, this elegant self-tuning process of brain development does leave an animal highly susceptible to false and distorted information that would probably have a lasting negative effect on the adult animal, as demonstrated by the kittens subjected to dark rearing conditions in scientific experiments.

We now have very good information concerning how sensory systems develop in we animals. What about learning behavior, such as when we humans learn a language? Would it not be reasonable to assume that there is a similar interaction of the brain of the learning child, such as the child that is learning a human language, and his or her actual learning experience? Is it not a reasonable hypothesis that there is much automatic "wiring" of the brain with considerable flexibility in the earlier years of the learning of almost anything, such as a language, but less flexibility after the passing of a susceptible period?

Isn't it reasonable to assume that at least part of the reason why Japanese speakers who have not been exposed to English between the ages of two to twelve usually have a hard time pronouncing the English "r" and "l", can be found in the neuron connections made and not made during this period? While human scientists don't understand exactly how brain development and the language learner interact, I have borrowed an electrical term and suggest that these Japanese speakers are "wired" in such a way as to make pronunciation of the English "r" and "l" very difficult.

Most neurobiologically oriented scientists believe that learning behavior development and sensory system development occur in a very similar manner.[48] But, there hasn't been as much research done on the biological aspects of learning as there has concerning the biological aspects of the development of sensory systems. However, there are a number of experiments that have been performed on laboratory rats and mice for years that support the idea that there is much flexibility in brain development during the learning process.[49]

For example, there have been many experiments performed that compare rats or mice that are raised in more complex environments, usually with other rats or mice, to those raised in considerably less complex environments, such as in individual cages.

In behavior and learning tests, such as learning mazes, it is not surprising that mice and rats that were reared in more complex environments did better than those that were reared in less complex environments. Nor is it surprising that when the brains of the rats or mice were examined, the brains of the rats or mice that were reared in more complex environments had up to 25% more connections (synapses per brain cell) than those reared in single cages.

In one experiment the researchers tried to determine if actual learning could be correlated to brain development. These researchers trained young adult rats on a maze of changing patterns for twenty-five days. Then they compared the visual cortex of brains of the trained rats to those of untrained rats. The scientists found that two types of neurons (cells) of the visual cortex had many more connections (dendrites) in them than those same cells of untrained animals. A third type of cell in the visual cortex was about the same in the two groups of rats. This shows that the effects of training and learning on the developing brain are more localized and specific than the effects of simply exposing an animal to a more complex environment.[50] These types of result are what one would expect to see if there was much modification of the physical brain during early learning experiences.

Let us recall the nature/nurture debate. Most who approach this debate know the nature-nurture dichotomy is a false dichotomy. Much of this dichotomy would disappear if behavioral science had as its cornerstone the interaction between an animal's environment and its biological "vehicle", especially the interaction between the environment and the animal's developing brain during its immature years. Our investigation has made clear the importance of the automatic, genetically programmed plan for the development of the animal (nature), plus the importance of receiving the proper environmental stimuli (nurture) for the proper refinement of neuronal development. Both nature and nurture are crucial to the process. If nature breaks down, say, because of genetic defects, the animal may not develop into a functioning member of its species. Similarly, if an animal is reared under environmentally deprived circumstances, it will have a difficult time becoming a functioning member of its species.

Evidence indicates that a tremendous amount of what we have come to call a person's culture becomes biologically based. Evidence indicates that the acquisition of culture in animals, including the human animal, during the immature stages of the animal's life is largely through a process known as imprinting which corresponds to periods of rapid biological development, especially in the brain. The imprinting evidence indicates that one's culture becomes wired in one's brain by the time one reaches adulthood. Of course, adult learning occurs. Evidence indicates that new synapses (connections) arise in the brain of a rat throughout much, if not all of its life[51], and presumably in other mammals, such as humans, as well. But, here we have attempted to emphasize how inordinately important early life experiences are in the formation of one's culture and mind-set. Given one of the characteristics of imprinting, namely the persistence of what an animal learns during its immature years into later life, I feel it is probable that human imprinting is a major factor in the widely observed phenomenon of human ethnocentrism. If one's culture is "wired" in one's brain through a process that began at birth, one "knows" that his or her way of life is the best, and that his or her culture represents the proper reality system, and looks askance at other cultural beliefs.

The Regressive ETs and Human Imprinting

Historical evidence indicates that this pattern of ethnocentrism has been understood and used by the ETs to keep humans controlled and separated from the very beginning of human civilization. The quoted passages of Genesis concerning the attempt of humans to build a Tower in Babel, probably tell in mythic-historical form what the "gods" have done to keep humankind divided and "...scattered abroad upon the face of the whole Earth." Right from the start, the Sumerian "gods" gave humans a strongly hierarchical governmental system designed so that a few "elite" at the top of society could control the masses at the bottom of society. In addition, the ET founders of our first civilizations taught humans how to hate and wage war with their neighbors, through the ETs own wars centered around greed and power struggles, interwoven with different beliefs systems. What terrible "imprints" of how to run a civilization we humans were provided! As early human civilizations fragmented and different religions and belief systems proliferated, the imprinting process through which humans learned mostly positive traits of their own culture and mostly negative traits of other cultures, served to reinforce the divisions of humankind. No wonder bloodshed, separation and wars have characterized human civilizations for more than 5000 years. The ETs have almost certainly had as much and more knowledge of the profound influences of the early-life experiences of humans as modern science has at present, and have used such knowledge to divide and control humans. And, we have seen evidence that regressive ETs have continued to divide humans throughout history by tampering with the Truth, especially with regard to spiritual knowledge. In the late twentieth century, in a large nation, such as the United States, the population is not only separated from other nations, it is divided internally into sub-cultures and into sub-sub-cultures. In the United States, we are divided along racial and ethnic lines (Blacks, Whites, Hispanics, Asians, Indians, etc.). Americans are divided along monied class lines, those few with much money versus the vast majority who have just enough money to survive (and even take the kids on vacation once in a while), if they work hard all their lives, plus the thousands that have no money at all and live homeless on the streets of the cities. Americans are divided by religions and politics (Republicans versus Democrats) and a host of other "differences," such as those who do or do not believe in abortion, and so on. We are living in an alienated (oops,.... there they are again!), divided, and controlled society.

The regressive ETs undoubtedly use much more than knowledge of the inordinate influences of early life learning experiences and other "three-dimensional" biological knowledge to promote and maintain the control and divisions of humankind. There is some evidence, for example, that they may be able to negatively influence humanity through the collective unconscious. Although the concept of the collective unconscious was introduced by the Swiss psychologist, Dr. Carl Jung, decades ago, most American psychologists give it little or no support because of the non-materiality of the theory. Basically,

Jung claimed that all of the thoughts and actions of humans are preserved in a non-material manner he referred to as the collective unconscious. Jung's research and reasoning lead him to believe that, despite the non-material nature of his proposed collective unconscious, humans have unconscious access to it, and it has a profound influence on the mental evolution of humankind.[54] Recently a similar idea has been proposed by English biologist, Rupert Sheldrake, which he calls the *hypothesis of formative causation*. Sheldrake has proposed that the growth and development of all biological organisms is not directed only by biochemical and genetic mechanisms, as the current biological paradigm claims, but also by a "morphic field" that exists outside the organism. These morphic fields, Sheldrake argues, have an influence over space and time, carry a collective memory of all previous organisms of each kind, and are constantly evolving.[55] While quite radical to materialist biological scientists and largely ignored, Sheldrake's hypothesis has received some experimental support in recent years.[56] If the Jung and Sheldrake hypotheses are even somewhat correct, negatively-oriented ETs might be able to influence the human collective unconscious or morphogenetic field.

We have already reviewed probable examples of the greys using holograms to found new religions, new divisions, and spread fear through doomsday prophecies. Judging from the accounts of abductees, the greys have advanced technology to cause most abductees to have no memory of their usually frightening and traumatic abduction. We briefly noted sources that indicate that the universe in which we live is a multidimensional universe, and countless metaphysical sources claim that we humans are multidimensional beings. Our physical bodies are surrounded by our etheric "bodies" which in turn are surrounded by our astral bodies, claim many metaphysical sources. Our human biological science has no belief in- and no knowledge that such bodies beyond the three dimensions exist. Regardless of how you feel about it, consider for the sake of argument that these non-three-dimensional bodies do exist and that the ETs know all about them and how they relate to our physical bodies. How much easier would it be for the aliens to manipulate and control us, as abductees report that they do?[52]

Despite our ignorance, we do not have to continue to agree to the terms of the divided, controlled societies which we ultimately choose to honor and support. We can work within and without ourselves to change our reality in order to play a more positive, creatively harmonious role, rather than that of victim which we have played and chosen throughout most of our history. We can turn to a spiritual path based on love of the Divine, love of ourselves, and love and respect for all life. Despite the imprints from our early lives that tell us that we humans are divided along racial, ethnic, and national lines, and a host of other ways, one body of evidence suggests that all entities, even the regressive ETs, are all extensions of the One Great Prime Creator, and that we are One in Spirit. Despite the lack of spiritual progress on the Earth, many esoteric sources inform us that we can evolve physically, culturally and spiritually to be more intelligent, light-filled beings very rapidly, indeed, within this lifetime. There is evidence that the next fifteen or twenty years is a time for potentially very rapid conscious and spiritual evolution for humankind. This is the main "message" I/WE have to offer and support in this book, and will be discussed in later chapters.

In the meantime, a large part of our conscious and spiritual evolution is learning who we are, the truth about our origins and existence and all we can about our biological beings. We can come to understand how the learning experiences we have early in life can and do have an inordinately strong influence on our attitudes as adults. Nevertheless, we will learn that we are far from victims and/or slaves to our social and cultural imprints. All of us should know that our imprints can be overcome. In the first place, there are many therapy-oriented treatments designed to allow one to move past ones early-life experiences, no matter how traumatic, and become a more intelligent, integrated, entity. And other techniques of "rewiring" our brains are rapidly being developed. Some of these new techniques for correcting or modifying our personal imprints, such as those using drugs or regression therapies that take people into their past lives, might seem quite radical to us now. But, as we strive, individually and collectively, to consciously move toward the light, new, as well as older techniques of "rewiring" our brains will become more and more prominent in our society. We humans do not have to behave in egoic, stupid ways, emphasizing the divisions of humankind, as the regressive ETs wish us to behave.

We humans who chose, with firm conviction, to follow the paths of the Truth, will come to understand much more than we do at present about the development of our biological bodies, about biological evolution, about our spiritual existence, about our spiritual evolution, and about much more light-inspiring information. The more we learn about our existence, no matter how "unbelievable" it seems when we first begin to learn some of the shocking details of the Truth, the more we will understand our evolutionary capabilities, and be able to consciously evolve away from the dark machinations of the secret government-ET coalition, toward the Divine. But, alas, any movement we humans may have had toward the Divine has been deliberately hindered by older species of this universe, who get their kicks out of dominating and controlling younger and more primitive species, like us. Let us take a closer look at two specie of regressive ETs that have been dominating and controlling much of humankind for centuries.

[1] Kroeber, A. L. 1917. "The Superorganic". American Anthropologist, *19*: pp. 163-213; and White, L. A. 1973. *The Concept of Culture*. Burges, Minneapolis.

[2] Debates between biological determinists and cultural anthropologists of the first decades of the 20th century are reviewed in, Cravens, H. 1978. *op. cit.*

[3] Augros, R. and G. Stanciu 1988. op. cit., p. 230; and Cravens 1978. *op. cit.*

[4] Wilson, E. O. 1975. *Sociobiology, the New Synthesis*. Harvard University Press, Cambridge.

[5] _____ 1977. Foreword of *Sociobiology and Behavior* by D. P. Barash; Elsevier North-Holland, Inc., New York, pp. xiii-xv.

[6] Bauchard Jr., T. J., D. T. Lykken, M. McGue, N. L. Segal and A. Tellegen 1990. "Sources of human Psychological differences: the Minnesota study of twins reared apart. " Science *250*, pp. 223-228.

[7] *Ibid.*, p. 223.

[8] Kottack, C. P. 1979. *Cultural Anthropology, 2nd Ed.* Random House, New York.

[9] Binford, L. R. 1982. Comment on: "Rethinking the Middle/Upper Paleolithic transition", by R. White; Current Anthropology *23*, pp. 177-181.

[10] Richard, A. 1977. "The feeding behavior of *Propithecus verreauxi*." In *Primate Ecology*, T. H.Clutton-Brock (ed.), Academic Press, London, pp. 72-96.

[11] Chalmers, N. R. 1968. "Group composition, ecology and daily activities of free living mangabeys in Uganda." *Folia Primatologica 8*: pp. 247-262. And, Waser, P. 1977. "Feeding, ranging and group size in the mangabey, *Cercocebus albigena*," In, *Primate Ecology: Studies of Feeding and Ranging Behavior in Lemurs, Monkeys and Apes*. T. Clutton-Brock (ed.), Academic Press, London, pp.183-222

[12] A summary of the observed variation in cultural traditions of local chimpanzee populations is found in: McGrew, W. C., C. E. G. Tutin, and P. J. Baldwin 1979. "Chimpanzees, tools, and termites: cross-cultural comparisons of Senegal, Tanzania, and Rio Muni." Man, *14*: pp. 185-214.

[13] The intricacies of "termite fishing" are discussed in, Teleki, G. 1974. "Chimpanzee subsistence technology: materials and skills." Journal of Human Evolution, *3*: pp. 575-594.

[14] Mainardi, D. 1980. "Tradition and social transmission of behavior in animals." In: *Sociobiology: Beyond Nature/Nurture?* G. Barlow and J. Silverberg (eds.),Westview Press, Boulder, Colorado, pp. 227-255.

[15] Douglas-Hamilton, I., and O. Douglas-Hamilton 1975. *Among the Elephants*. Viking Press, New York.

[16] Healey, R. F., F. Cooke and P. W. Colgan 1980. "Demographic consequences of snow goose brood rearing traditions." Journal of Wildlife Management, *44*: pp. 900-905.

[17] Marler, P. 1984. "Song learning: innate species differences in the learning process." In: *The Biology of Learning*, P. Marler and H. S. Terrace (eds.). Springer-Verlag, Berlin, pp. 289-309.

[18] Bonner, J. T. 1980. *The Evolution of Culture in Animals*. Princeton University Press, Princeton.

[19] *Ibid.*

[20] Stryker, M. P., J. Allman, C. Blakemore, J. M. Greuel, J. H. Kaas, M. M. Merzenich, P. Radic, W. Singer, G. S. Stent, H. Vanderloes, and T. N. Wiesel 1988. "Group report: principles of cortical self-organization." In: Neurobiology of Neocortex, P. Radic and W. Singer (eds.), John Wiley and Sons, Limited, New York, pp. 115-136.

[21] Lorenz, K. 1935. "Der Kampan in der Umwelt des Vogels." Journal of Ornithology, *83*: pp. 137-213, 289-413.

[22] Immelmann, K. 1972. "Sexual and other long-term aspects of imprinting in birds and other species." Advance Study of Behavior, *4:* pp. 147-174; and, Immelmann, K. 1975. "Ecological significance of imprinting and early learning." In: Annual Review of Ecology and Systematics, *6*, R. F. Johnston, P. W. Frank and C. D. Michener (eds.), Annual Reviews Inc., Palo Alto, pp. 15-37.

[23] Fillion, T. J. and E. M. Blass. 1986. "Infantile experience with suckling odors determine adult sexual behavior in male rats." Science *231:* pp. 729-731.

[24] Geist, V. 1971. *Mountain Sheep*. University of Chicago Press, Chicago. And for red deer: Schloeth, R. and D. Burchardt 1961. "Die Wanderung des Rotwildes (*Cervus elaphus L.*)" Rev. Suisse Zool. *68*: pp. 146-155.

[25] Calhoun, J. B. 1962. *The Ecolgy and Sociology of the Norway Rat*. U. S. Department of Health, Education, and Welfare, Bethesda, Maryland.

[26] For example, see: Housfater, G. J. Altman and S. Altman 1982. "Long-term consistency of dominance relations among female baboons (*Papio cynocephalus*)." Science *217:* pp. 752-755

[27] Ruppenthal, G. C., G. L. Arling, H. F. Harlow, G. P. Sacket, and S. J. Suomi 1976. "A ten-year perspective of motherless-mother monkey behavior. Journal of Abnormal Psychology *85*: pp. 341-348.

[28] Goy, R. W. K., K. Wallen and D. A. Goldfoot 1974. "Social factors affecting the development of mounting behavior in male rhesus monkeys." In: *Reproductive Behavior*, W. Montagna and W. Sadler (eds.). Plenum Press, New York.

[29] Immelmann, K. and S. J. Suomi 1981. "Sensitive phases in development."

In: *Behavioral Development*, K. Immelmann, G. W. Barlow, L. Perinovich and M. Main (eds.). Cambridge University Press, Cambridge, pp. 395-431.

[30] Review in: Hess, E. H. 1973. *Imprinting: Early Experience and the Developmental Psychobiology of Attachment*. Van Nostrand Reinhold Co., New York.

[31] Freud, S. 1953. "Three essays on the theory of sexuality." In: *The Standard Edition of the Complete Psychological Work of Sigmund Freud, Vol. 7*. Hogarth, London, pp. 125-248. (Essays originally published in 1905.)

[32] Bowlby, J. A. 1969. *Attachment and Loss, Vol. I*. Basic Books, New York.

[33] Lenneberg, E. 1987. *Biological Foundation of Language*. John Wily and Sons, New York.

[34] Goto, H. 1971. "Auditory perception by normal Japanese adults of the sounds of 'L' and 'R'." Neuropsychologica *9*: pp. 317-323.

[35] Piaget's proposed cognitive-developmental stages summarized in: Ginsburg, H. and S. Opper 1979. *Piaget's Theory of Intellectual Development (2nd ed.)* Prentice-Hall, Englewood Cliffs, New Jersey.

[36] Horn, G. 1985. *Memory, Imprinting and the Brain*. Carendon Press, Oxford.

[37] Sternberg, R. J. 1985. "Human intelligence: the model is the message." Science *230*: pp. 1111-1117.

[38] DeLoach, J. S. 1987. "Rapid change in the symbolic functioning of very young children." Science *238*: pp. 1556-1557.

[39] Thatcher, R. W., R. A. Walker and S. Guidice 1987. "Human cerebral hemispheres develop at different rates and ages. Science *236*: pp. 1110-1113.

[40] Matousek, M. and I.. Peterson 1973. "Frequency analysis of the EEG in normal children and adolescents." In: *Automation of Clinical Electroencephalography*, P. Kellaway and I. Peterson (eds.). Raven, New York, pp. 75-102.

[41] Winson, J. 1985. *Brain and Psych*. Anchor Press/Doubleday, New York.

[42] Hubel, D. H. and T. N. Wiesel 1963a. "Receptive field of cells in striate cortex of very young, visually inexperienced kittens." Journal of Neurophysiology *26*: pp. 994-1102.

[43] _____ 1963b. "Single-cell responses in striate cortex of kittens deprived of vision in one eye." Journal of Neurophysiology *26:* pp. 1003-1017.

[44] Blakemore, C. and G. F. Cooper 1970. "Development of the brain depends on the visual environment." Nature *228*: pp. 447-448.

[45] Knudsen, E. I. 1987. "Early experience shapes auditory localization behavior and the spatial runing of auditory units in the barn owl." In: *Imprinting and Cortical Plasticity*, J. P. Rauschecker and P. Marler (eds.), John Wiley and Sons, New York, pp. 7-21.

[46] Greenough, W. T., J. E. Black and C. S. Wallace 1987. "Experience and brain development." Child Development *58*: pp 539-559.

[47] *Ibid.*

[48] Rauschecker, J. P. and P. Marler 1987. *op. cit.*, p. 361.
[49] Studies summarized in, Greenough et al. 1987, *op. cit.*
[50] *Ibid.*, p. 548.
[51] *Ibid.*
[52] Crowell, C. 1990. "The government factor." In: Valerian, V. 1991. *op. cit.*, pp. 129-132.
[53] Marciniak, B. 1992b. *op. cit.*, p. 62.
[54] Jung, C. G. 1971. *The Portable Jung*. The Viking Press, New York, p. 60.
[55] Sheldrake, R. 1982. *A New Science of Life*. Thacher, Tiburon, California.
[56] eg., Bulletin of the Institute of Noetic Sciences, 1991. *Volume VI/3*, Sausalito, California.

> "*Living according to these myths can mean living in ignorance. For example, the only way to maintain the myth of knowing and understanding everything is to ignore a whole universe of other information. When one clings to the myth of innate superiority, one must constantly overlook the virtues and abilities of others.*"
> Anne Wilson Schaef, *Women's Reality*, 1981

Chapter 14

The Reptoids

We have reviewed much about the reptoids, possibly the Anunnaki of the ancient Sumerians, but let us now take a closer look at the nature of our creator gods. The Pleiadian Plus group, as we have seen, refers to our "creator gods" and founders of the ancient Sumerian civilization, as the "lizzies" in order to inject some humor into a situation that might be considered terrifying to humans whose consciousness is encountering them for the first time. These creatures are said to be part human and part reptile. They are said not to exist in time as we know it, but mostly in other dimensions where they are supposed to live for thousands of years. There are said to be benevolent beings amongst the lizzie population, but a vast majority of our creator gods were tyrannical and did not honor life or understand their connection to life, even that of their own creations. Furthermore, and most importantly, we should know something about the reptoids because they are reportedly soon going to re-enter our dimension in order to try to continue to control us.[1]

Let me remind the readers that, while the reptoids did participate in at least one of the genetic manipulations that created we modern humans, apparently, and, therefore, are properly called our "creator gods," our creation is much more complicated. The ultimate human creation stems from what is called the Prime Creator of this universe. Furthermore, it appears that most of the entities between humans and the Prime Creator involved in our creation were positive entities

The Reptoids

who had the highest intentions of Love and light in our creation. Despite the endeavors of the reptoids and other regressive ETs to control us, all humans, by accepting their own dark sides and those of others, can turn toward love and light and escape the negativity of the reptoids and others. Part of the procedure of spiritually evolving beyond the lizzies and other regressive ETs is to understand their natures and motivations.

The descriptions of the lizzies by the Pleiadian Plus group depict them as truly amazing creatures. But Pleiadian Plus is a very esoteric group by Earth standards; so what "earthly evidence" for the existence of the lizzies do we have?

I studied all of the cylinder seal illustrations of the ancient Mesopotamian civilization that I have around my study, to see if anything like a reptoid is depicted. Most of the "gods" of the Sumerians and later Mesopotamians, when depicted, look very human, some of them look bird-like, but I found none that looked like it may have been a lizzie. Sitchin states and implies that the Mesopotamian "gods" were a lot like we human primates in his books that cover Mesopotamian history. In all of Sitchin's excellent books that cover Mesopotamian history, there is not one reference to anything like a reptoid.[2] The reader will recall that most of the Sumerian accounts of history were written on clay tablets at around 2,500 B.C. Many of the history stories, usually written in what we would call mythic form, are about events that took place prior to (or at the time of) the flood, especially the stories of the creation of humans and of the Great Flood itself.

The first six chapters of the first book of the Bible also deal with the period of time before the Great Flood. This includes the Adam and Eve story, the references to the descendants of Adam and Eve, and the story of Noah and the Ark. The reader will recall that the evidence is that these portions of the Bible, written around 1,000 BC, were borrowed and edited selectively from existing accounts of history, mostly from the texts of Sumer and later Mesopotamian civilizations.

As we've seen, there is a large body of religious literature not in the Bible that also deals with the period before the Great Flood. Most of these works have been lost, hidden and/or an attempt was made to destroy them, or they have been selectively excluded from the Bible or otherwise suppressed. This is part of the giant cover-up, part of the duping of humankind. Certainly the ETs who have endeavored to enslave and control humankind do not want humankind to know of their ancient, pre-Deluge past. These creator gods certainly did not want humankind to know of the true story of human "creation" (that we were created to be slaves and dupes) and the true nature of one of

The Reptoids

our creator gods, the reptoids.

We have already reviewed some of the work of R. A. Boulay and his book, *Flying Serpents and Dragons: The Story of Mankind's Reptilian Past*, in Chapter 5. We saw that Boulay lists several historical sources of information about the lizzies which were excluded from the Bible. We recall that Boulay uses such works as the *Book of Jubilees*, three books of *Enoch*, the *Haggadah*, or the oral tradition of the Jews, the recently discovered Dead Sea Scrolls, and other works to document the existence of the lizzies and their behavior toward humans.[3]

The primary, present-day, "Earth data" we have to verify the existence of the reptoids is that they have been seen several times, usually by modern-day abductees. In a majority of abductee cases they are not seen at all, but they are sometimes observed standing aside and simply observing what is happening. The relationship between the reptilians and the grey species is usually not known to the abductees. Information that Valerian has gathered makes it clear that most of the greys "work" for the lizzies, or are themselves slaves of the lizzies. According to Valerian, the reptoids apparently left the grey species in charge of the Earth when most of them left the Earth several thousand years ago.[6]

It is not only abductees, however, who observe the lizzies. Linda Howe tells of a "cow abduction" case in Missouri where a reptoid was sighted. A married couple watched the apparent cow abduction from the porch of their farm through binoculars. While two, small, white-skinned, large-headed beings worked with the cow, a few feet away a "lizard-guy" watched the proceedings. The human couple described the "lizard-guy" as a humanoid creature that stood about six feet tall, had greenish, scaly skin like a reptile, and large pale green eyes with vertically slit pupils.[7]

Alice Bryant and Linda Seebach report two apparent abductions of humans by reptilian beings in their book *Healing Shattered Reality: Understand Contactee Trauma*. Two men who were not acquainted at the time were abducted and given rough physical examinations on board a craft by seven foot tall lizard-like beings.[8] Both men are reported to have lived lives of terror and secret shame following the abduction.

In another case, a rancher was working with his cattle on his ranch in New Mexico, when suddenly his herd spooked and scattered. He looked up and saw the reason his cows had become frightened - a UFO craft was hovering over him. His first reaction was one of anger at this craft for having scattered his cattle, and he fired two shots with a 30/30 rifle he carried. His horse then threw him and he was dazed. The next thing this rancher knew he was taken into the craft, pushed onto an examining table and undressed. He was given a very rough examination by five reptilian-like beings, three males and two females. He described these creatures as having "...humanoid bodies with the exception of scaly fish features and tails like reptiles. They wore few clothes. The female sexual parts were similar to *Homo sapiens* (sic). Their legs and arms were somewhat lizard-like. Their facial features resembled frogs, yet were human-like at the same time. The eyes were large, and protuberant. They had three fingers and three toes..."[9]

Sometimes human encounters with a lizzie have been more pleasant. Denise and Bert Twigs, in their book *Secret Vows*, claim to have an ongoing relationship with a couple that claims to be from the Andromeda constellation. Both claim to have conceived several half-alien, half human children, both in their home in Oregon and on the spacecraft on which the Andromeda couple are now residing. On one of their extended visits to the "mother ship" of the alleged Andromes, they met the security man:

> The top security man for the Andromes was not an Androme. He looked like a reptile, like an upright alligator, although I don't believe his mouth was as long as an alligator's, and he did speak our language. His smile lit up his whole face... Bert and I both remembered this on the same night, and since then we have fondly referred to our new acquaintance as "the alligator man."[10]

Perhaps, the Twigs met an evolved lizzie, one that would not harm humans. The details of this "alligator man" given here are too scant to tell, but at least we know that some of them are capable of smiling.

There was a recently reported abduction case where the reptilian creatures demonstrated how they could change form. A woman in Seattle was abducted (1990) by beings that looked human. They told her, however, they were not really human and allowed her to see their true form. They were in fact a reptilian species.[11]

Another example of the "shape-shifting" capability of the lizzies is found in the book *Close Extraterrestrial Encounters* by Dr. Richard Boylan with Lee Boylan. Dr. Boylan is a clinical psychologist and ET researcher, and has treated several abductees, often using regressive hypnosis, mostly in the Sacramento, California area. Dr. Boylan's patients report contact with greys and reptoids and "jawas" ("praying mantis" types) species, but not with "blonds" or "Nordic" ETs. Boylan reports that several of his abductee patients, which he calls "experiencers," saw ETs they thought were human, but in every case,

> ...this turned out to be an ET-imposed mental visualization in the mind of the experiencer, a so-called "screen memory". Upon closer examination, the experiencer was able to see the actual non-human face of the ET behind the mentally imposed "human" mask. One experiencer believed she was encountering a human "spaceman." Because of certain details of the account, I believed the "human" appearance was not the way the being actually looked. I invited the experiencer to look closely and carefully at the face of the "human" spaceman. When she did so, she suddenly was startled. "Oh my!" she said, "It's not a human after all. It's one of those Greys." Other details made me suspect we were still not to the bottom of the matter. I suggested she study closely the face of her "Grey" visitor. Again she soon startled. Now she could see that it was a "reptoid" who had previously cloaked himself mentally as a "Grey." [12]

This "shape-shifting" ability of the reptoids might be one of the reasons I could not find any lizzie-type creature depicted on the cylinder seal drawings of the ancient Sumerians. Evidently, reptoids are able to change form and appear to be what they aren't. The observations from these abductions suggest that reptoids can appear to be human if they want to shift shapes. The Boylans also report several cases of this "shape-shifting" capability with the greys.[13]

We recall that the Pleiadian groups tell us that some of their ancient ancestors sometimes worked with the reptilian species, and some renegade Pleiadians still do. The reader will recall that Pleiadians supposedly are primates and, in fact, many of their genes were inserted into the new Earth species by Enki to allow them to be fertile and make humans more primate-like. Perhaps some of the Sumerian depictions of their gods were, in fact, renegade Pleiadians or some other primate species who were working with the reptoids. Human-looking "blondes" have been seen in about 60% of the abduction cases that are happening today.[14] These human-looking ETs are observed working along with the greys, so probably human-looking ETs worked along with the reptoids in the ancient past.

The Reptoids

Jason Bishop, an ET investigator, reports that the ET reptilians average from six to seven feet in height. They are cold-blooded like the reptiles that live on Earth. This means their body temperature and metabolism depend on the temperature of the environment. Bishop notes that the reptilians are well-suited for space travel due to their ability to hibernate. The reptilians have scales all over their body and have no sweat glands, according to Bishop (Figure 14-1). They have three fingers, with an opposing thumb. Their eyes are catlike and they have a short stubby muzzle. They are mostly meat eaters, according to Bishop.[15]

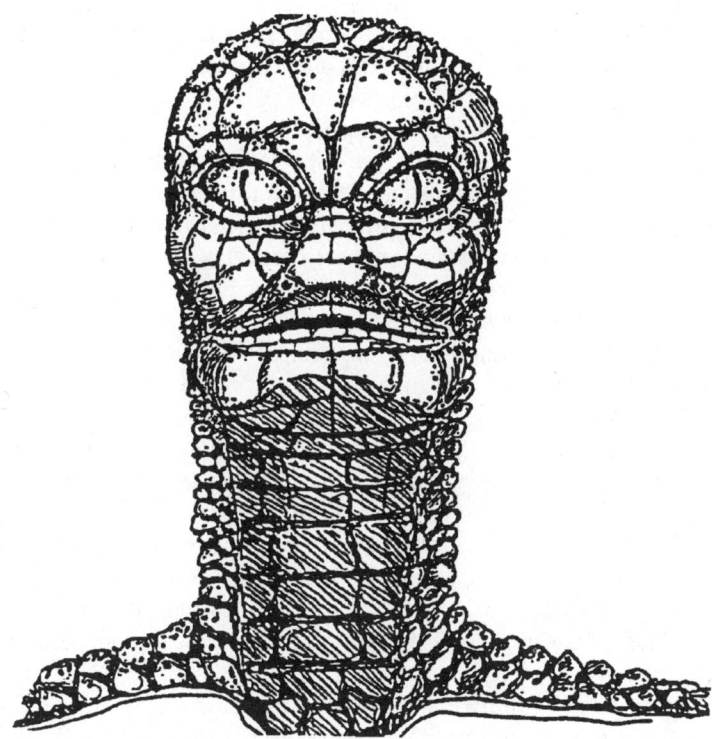

Figure 14-1 A "lizzie", or reptoid species drawn from abductee eyewitness accounts. From Revelations of Awareness, Issue 92-4, Cosmic Awareness Communications, P.O. Box 115, Olympia, Washington 98507. With permission.

Unlike the reptiles that live here on Earth, however, these humanoid reptoids have different types of teeth, as we humans and all mammals do. Bishop reports that these reptilians have incisors, canines, and molars. All Earth mammals but absolutely no reptile native to the Earth have differentiated teeth. If Bishop's information is correct, this

The Reptoids

must be some of the "part humanoid" or "part mammal" aspect of the lizzies.

Bishop informs us that the elite leaders of the reptoids are from the planetary system in the constellation of Draco. The very elite have wings that can be folded back against their body. However, the "soldier class" and the "scientists," or a vast majority of the members of the reptilian species, do not have wings.[16]

The rancher in New Mexico mentioned earlier, who suffered such a rough and traumatic abduction at the hands of five reptoids, saw eggs, large eggs in the space craft where he was examined.[17] These eggs were as big as watermelons and were kept in large jars. All but a handful of the thousands of species of reptiles that live here on Earth hatch from an egg. As we reviewed earlier, usually the females of the reptilian species of Earth lay several eggs, hide them in some way, then abandon them. The infant reptiles are usually on their own from the time they hatch, as reptile parents do not generally provide any care for their young. Virtually no female mother of any Earth species of reptile forms a bond with her offspring, as is characteristic with birds and mammal mothers. An herpetologist recently told me that crocodile, python and cobra snake mothers provide care for their offspring, but no mothering behavior and bonding are known from the hundreds of other species of reptile on Earth.

Esoteric sources suggest that there is no bond formed between the mother and her offspring in the ET reptilian race.[18] As we observed in the previous chapter, all Earth species of bird and mammal mothers, and many fathers as well, imprint and form bonds with their offspring, and also provide care for them during their immature years. Several women and a few fathers have communicated to me that they felt a very strong bonding emotion with their infant(s) soon after they were born. This bonding emotion, we feel, holds one of the keys to understanding the relationship between the reptoids and one of their "creations", we humans.

Esoteric sources inform us that our emotions connect us with the spiritual body, and are the key to spiritual evolution.[19] On Earth, birds and especially mammals, exhibit strong emotions in their behavior. In fact, mammals have a specialized portion of their brain that generates and processes emotion, the limbic system, which the reptile does not have. The mammal has the basic brain that a reptile has, plus other parts, especially the cerebral cortex, also known as the neocortex (see Figure 13-1). Within this neocortex is the limbic system (Figure 14-2), which is quite large in proportion to the rest of the brain. The limbic system itself has many connections and "loops" between the neocortex and with "lower" areas of the brain.[20] When confronted with the threat of a possible predator, for example, the limbic system generates the strong emotions that prepare the animal to behave in the appropriate way to avoid the danger - usually fleeing. The Earth reptiles also, of course, behave in a manner that will allow them to avoid a predator. But the range of response of Earth reptiles seems quite limited relative to the responses available to Earth mammals in similar situations; the response of Earth reptiles to a predator is limited to fight (attack) or flight, nothing in between these extremes, such as fight until the animal maneuvers to gain access to a fast route of escape. And, Earth reptiles do not exhibit the physiological responses that are characteristic of emotions in Earth mammals, such as an increase in the blood sugar level and rise in the heart rate.[21]

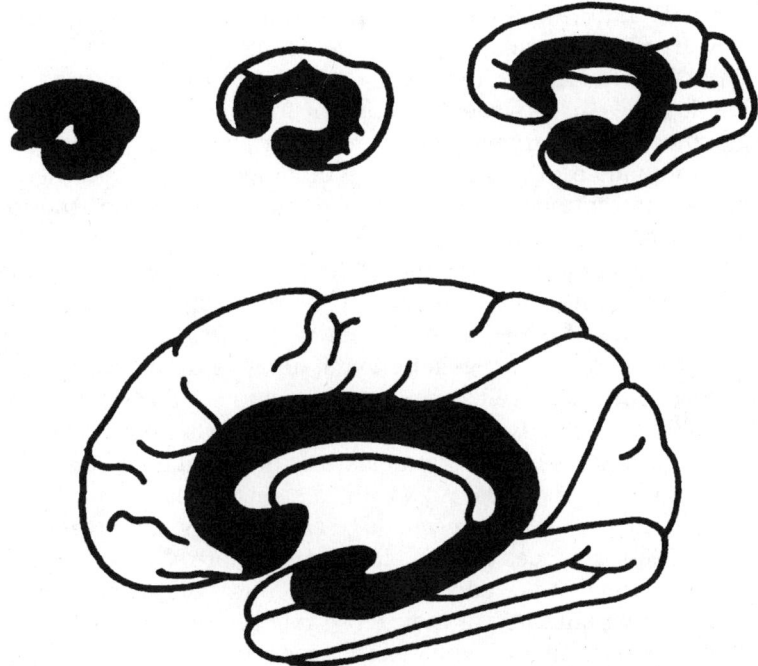

Figure. 14.2   *Sagittal views of the brains of four mammals showing the extent of the limbic system relative to the cerebral cortex.*

The amount of care a mammalian mother gives her young can find no parallel among the Earth reptiles. Many reptile parent-offspring relationships follow the example of the huge Komodo Dragon that today lives only on remote islands of Indonesia, the island country in southeast Asia. This giant ten-foot lizard has programmed in its genes a unique behavior pattern that allows its young to survive. As soon as the eggs hatch, the young lizards are genetically programmed to climb up the nearest tree as fast as they can, where they will remain for a period of several months. The reason these infant lizards are programmed to rapidly climb trees and remain there during their early development is to avoid being eaten on the ground by predators, especially their own mother. Apparently, the mother Komodo Dragon has no means of distinguishing her own progeny from all the other small animals normally constituting her prey.[22]

Of course, the humanoid reptoid would not be expected to be so totally devoid of emotions. The physical descriptions given by human abductees of the reptoids indicate that large parts of them are mammalian. The differentiated type of teeth, their "human-like" female sexual parts, and their somewhat "humanoid" appearance indicate that they are part mammalian. Are these beings the result of a crossbreeding of reptile and primate genes? The fact that they are cold-blooded, lay eggs, have scales, and do not form bonds with their offspring, indicates that largely they are more reptilian than mammalian. This predominance of reptile genes would seem, on the whole, to have made the reptoids more aggressive and more limited emotionally than the average human primate.

We have reviewed evidence that indicated that when the reptoids combined their genes and those of a primitive Earth humanoid primate, something close to *Homo erectus*, the first generations of humans were sterile. Later genetic manipulations by the Sirian Enki and possibly others, as well as sexual intercourse of the Pleiadian ET "sons of god" and the daughters of men at a later time, provided an infusion of primate genes into humans that made humans fertile and much less like the reptoids. As we have seen, Boulay presents evidence using only "Earth sources," that early humans had proportionately more reptilian genes than the later, much more short-lived, more primate humans. These early humans, Boulay believes, became ashamed of any reminders of their close relationship to their reptoid genetic creators. Boulay speculates that the story in the Bible of the angry reaction of Noah when he is seen naked by his youngest son indicates that Noah wanted to hide something from his son. The story in the Ninth Chapter of Genesis has Noah becoming drunk with the wine from grapes of his vineyard, and,

> he was uncovered within his tent. And Ham, the father of Canaan, saw the nakedness of his father, and told his two brethren without. And Shem and

The Reptoids

> Japheth took a garment, and laid it upon both their shoulders, and went backward, and covered the nakedness of their father; and their faces were backward, and they saw not their father's nakedness. And Noah awoke from his wine, and knew what his younger son had done unto him. (King James, Genesis 9: 21-24)

And Noah condemned the offspring of his youngest son to be lowly servants of his older brothers and their families. This is quite cruel punishment for seeing a parent of the same sex naked. The whole story seems somehow irrational, as if a crucial part had been cut out. Boulay thinks Noah's totally irrational reaction to his son seeing him naked was due to the fact that Noah was sensitive (ashamed?) of traces of reptilian ancestry, such as patches of scaly hide that appeared on his body.[23]

The Pleiadian Plus group assures us that there are benevolent lizzies, although they seem to indicate that most of the ones humans have and will encounter do not have the best interests of humans in mind. So, the lizzies must have some emotions similar to the love that human mothers can feel for their newborns, and the love that all humans are capable of expressing. If some reptoids are able to evolve spiritually, what has stopped the spiritual evolution of most of them?

Clearly, they must have existed long before humans if they genetically engineered a version of modern *Homo sapiens*. Yet they seem to have evolved so little spiritually, as is evidenced by their lack of expressing love, which is the emotion the spiritual traditions emphasize as the key emotion of higher spiritual consciousness. The Awareness consciousness states that a vast majority of reptoids not only do not love humans, but have contempt for them and will show them no mercy or compassion.[24] Certainly, the Sumerian historical records tell us as much, assuming Boulay's interpretation of ancient historical records and esoteric sources that state that the Anunnaki were reptoids, are correct. It is probable that the lack of compassion demonstrated by the reptoids can be at least partially explained by their biological nature - namely the preponderance of reptilian genes in their genetic makeup.

Let us review a few aspects of these creatures that we have learned through historical and esoteric sources. They are very intelligent, and they have a marvelous technology. They can travel between different dimensions and their life spans number in the thousands of years. They have space travel technology and, in general, their technology is way beyond what we humans are capable of mechanically performing. They did rapidly advance our biological evolution through genetic manipulation, and taught the ancient Sumerians about civilization, including the art of food production necessary for civilization.

Highly developed spiritual teachings indicate that we are not to judge others, for all entities are evolving toward the Prime Creator at their own rates. All esoteric teachings indicate that we should not think of the reptoids as evil beings, but simply as beings who, for whatever higher purpose, are not able to feel compassion for others, especially beings such as ourselves. Perhaps the attitude of the reptoids toward us can be compared to the average American's attitude toward cattle - that cattle are dumb animals which can be exploited and controlled without compassion, and are good to eat! Apparently, the reptoids are enamored with power, and get their thrills from conquering and dominating others. Humans have learned this program from their reptoid creators very well, as indicated by the extant historical records since the founding of civilization, and the present state of global affairs where humankind continues to struggle to resolve its critically horrendous social, political and ecological problems. Aggressive dominance, through greed and power over the weak seems to be the order of the day, and throughout human history. Humans have agreed to live with the program of fear and chaos the reptoids and their allies have perpetrated throughout the ages. As Pleiadian Plus states, it is hard for us to understand how entities can feed on consciousness, how lizzies might feed on fear, but they clearly state that "Consciousness feeds consciousness." [25] So it is imperative for us not to feed the reptoids our fear. Instead, we must hold the higher frequency of love, and live on the frequency of unconditional love at all times. We are told that we are receiving assistance from benign entities all over the universe as we learn to hold these higher frequencies. When we learn to hold the higher frequencies of love, we are told, we will be prepared to meet these "gods," the lizzie gods from our ancient history as they re-enter our reality in the next few years.

The allies of the reptoids in their domination and control agenda, the greys, also exhibit many unusual and extraordinary (to us) characteristics. A closer look at the greys is in order, especially since these are the ETs living humans have come into contact with the most.

[1] Marciniak, B. 1992b. op. cit., pp.23-38
[2] Sitchin, Z. 1976, 1980, 1985, 1990, 1991. ops.cit.
[3] Boulay, R. A. 1990. op. cit., p. 61.

[4] Ibid., p. 1.
[5] Ibid.
[6] Valerian, V. 1991. op. cit., pp. 100-100D.
[7] Howe, L. M. 1991. op. cit., p. 73.
[8] Bryant, A., and L. Seebach 1991. Healing Shattered Reality: Understanding Contactee Trauma. Wild Flower Press, Tigard, Oregon.
[9] Ibid, prologue.
[10] Twigs, D. R., and B. Twigs. 1992. Secret Vows: Our Lives with Extraterrestrials. Wildflower Press, Tigard, Oregon, p. 141
[11] Krill, O. H. No Date. "Orion Based Technology, Mind Control and other Secret Projects". Arcturus Book Service, Stone Mountain, Georgia, p. 29.
[12] Boylan, R. J. and L. K. Boylan 1994. Close Extraterrestrial Encounters: Positive Experiences with Mysterious Visitors. Wild Flower Press, Tigard, Oregon, p. 155.
[13] Ibid.
[14] Valerian, V. 1991. op. cit., pp. 114-115.
[15] Bishop III, J. 1991. "Reptilian-Grey Data". In, Valerian, V. 1991. op. cit., p. 96.
[16] Ibid.
[17] Bryant, A. and L. Seebach 1991. op. cit., prologue.
[18] Revelations of Awareness 1992. Issue 92-4, p. 10. Olympia, Washington.
[19] Marciniak, B. 1992b. op. cit., p. 32.
[20] Winson, J. 1985. op. cit., p. 186.
[21] Campbell, B. 1985. op. cit., pp. 52-54.
[22] Affenberg, 1972. "Komodo Dragons." Natural History 81, pp. 52-59. Quoted in Campbell, B. 1985. op. cit., p. 54.
[23] Boulay, R. A. 1990. op. cit., p. 293.
[24] Revelations of Awareness 1992. Issue 92-4, p. 8.
[25] Marciniak, B. 1992b. op. cit., p. 26.

"I remember when I used to dismiss the bumper sticker "Pray for Peace." I realize now that I did not understand it, since I also did not understand prayer; which I know now to be the active affirmation in the physical world of our inseparableness from the divine; and everything, especially the physical world, is divine. War will stop when we no longer praise it, or give it any attention at all. Peace will come wherever it is sincerely invited. Love will overflow every sanctuary given it. Truth will grow where the fertilizer that nourishes it is also truth. Faith will be its own reward. Believing this, which I learned from my experience with the animals and the wild flowers, I have found that my fear of nuclear destruction has been to a degree lessened. I know perfectly well that we may all die, and relatively soon, in a global holocaust, which was first imprinted, probably against their wishes, on the hearts of the scientist fathers of the atomic bomb, no doubt deeply wounded and frightened human beings; but I also know we have the power, as all the Earth's people, to conjure up the healing rain imprinted on Black Elk's heart. Our death is in our hands. Knock and the door shall be opened. Ask and you shall receive. Whatsoever you do to the least of these, you do also unto me and to yourself. For we are one. "God" answers prayers. Which is another way of saying, "the universe responds." We are indeed the world. Only if we have reason to fear what is in our own hearts need we fear for the planet. Teach yourself peace. Pass it on."
Alice Walker, *Living By The Word*, 1987

Chapter 15

The Greys

Let us now investigate the ET group we have been calling the "greys," who are seen by some of the more recent authors of the ET literature, (which now includes me) as being at least partially responsible for the generally miserable condition most of humanity has been wallowing in throughout its history. Who are these ETs who apparently told members of the secret government that they had been for a

The Greys

long time and were at the present time "...manipulating masses of people through secret societies, witchcraft, magic, the occult, and religion?"[1] Who are these apparently multidimensional beings that seem to have had more contact with surface humans in the past few years than any other ET group? As we have seen, the greys seem almost certainly responsible for thousands and millions of cattle mutilations, human mutilations, and human abductions that have been reported since the early 1960s in literally hundreds of worldwide publications.

We have already reviewed many of the methods used by the greys to manipulate and control human societies. In this chapter I wish to briefly investigate the general nature of these ETs, mainly concentrating on their biological, psychological, and spiritual aspects. As we have seen, the evidence is that the greys are clearly far superior to the human species in technological and metaphysical ability. Why does this older species that has clearly evolved further than humans in some respects, wish to inhibit the spiritual evolution of a relatively new, less knowledgeable species? Why have the greys not evolved spiritually to the point where they would have consciously turned toward the light, and possess nothing but love and compassion towards all of creation, as the more mystical traditions of our religions and many esoteric sources indicate should be the goal of humankind and of all conscious entities?

There are, by now, hundreds of books and pamphlets that refer to the greys. The best single source about the greys by an author who has, in my opinion, come to understand the truth about them is Valdamar Valerian's *Matrix II*.[2] Valerian has amassed data about the greys from UFO researchers who, at one time in their lives had access to some "above top secret" files of one of the United States security agencies, and from other human researchers, as well as from esoteric sources. Other esoteric sources will be used, as well, in this brief attempt to understand what this "alien thing" is all about.

There are several types of entities that are known by humans as "greys." There are said to be twenty-two subspecies of greys.[3] There are tall greys that are about 7 feet tall (Figure 15-1), and there are very short greys that are from 3 to 5 feet tall (see Figure 12-2). There are said to be some greys that have a benign attitude towards humans, although there are few human reports of contacts with such greys.

The United States Government performed autopsies on a number of greys (all of the very short type) that came into their hands mostly through saucer crashes. These greys have small thin bodies with large egg-shaped heads, which are proportionally much larger relative to body size than the human head. These greys have slanted, large, tear-shaped eyes. They have three or four claw-like fingers, and tough grey skin of reptilian nature. Their internal organs are similar to those of humans, but unknown to human biological science and have "...obviously developed according to a different evolutionary process."[4] Their brains are absolutely larger than those of humans, despite their small body size. They have a non-functioning digestive system. Greys apparently subsist on human blood and other biological substances which they absorb while soaking in large vats. Waste products are excreted through their skin. These somewhat reptilian-like greys also have some plant characteristics. They do ingest some sort of food which is converted into energy by their bodies by a process that is similar to the process of photosynthesis used by Earth plants. The autopsied greys had chlorophyll in their bodies for this plant-like function. Their blood is greenish. Autopsies of their large brains have revealed that they actually have two separate brains, isolated from each other by a thin bone. There is no physical connection between the "two brains." Some of the autopsies have revealed a crystalline network in their brains which is thought to give them telepathic capabilities.[5]

Most of the greys on Earth (in underground dwellings) are believed to be from, or have ancestors from one of the planets of the Orion constellation of stars. The "Orions" have a very negative reputation in much of the esoteric UFO/ET/alien literature because they have engaged in many conflicts and destructive wars with other ETs of this universe, as we saw in the book we used as an introduction to human galactic heritage, *The Prism of Lyra*.[6] Other esoteric sources inform us that the greys that have perhaps had the most contact with humans, through abductions, in the past 30 years, are from a star system neighboring the Orion constellation, known as Zeta Reticuli. While evidence indicates that many of the greys have been on Earth for centuries manipulating human societies, the Zetas are believed to have arrived on the Earth in the past 40 years, having made treaties with the secret government of the United States and probably other Earth governments. The Zetas apparently have a genetic base that is more insectoidal than that of the more reptilian greys described above. All of the greys almost certainly have their own hierarchy amongst themselves. Most of the greys are themselves servants, or mercenaries, of the reptoids, or the reptilian race.[7] The greys, or most of them, make up what is known as the Orion Empire. The reptoids are members of what they call the Federation of the Draconian Reptoids.[8] The Orion Empire and the Federation of Draconian Reptoids work together and with other regressive ETs to try to domi-

The Greys

nate and control several planets in this universe, including Earth.[9]

The short greys are overseen within their own ranks by the much taller 7 to 8 foot greys (Figure 15-1). These tall greys are the ones that actually carry out "diplomatic" missions, such as secretly negotiating treaties with heads of human governments. As mentioned, the greys in general, and the small 3 to 5 foot greys in particular, have been likened to mercenaries.[10]

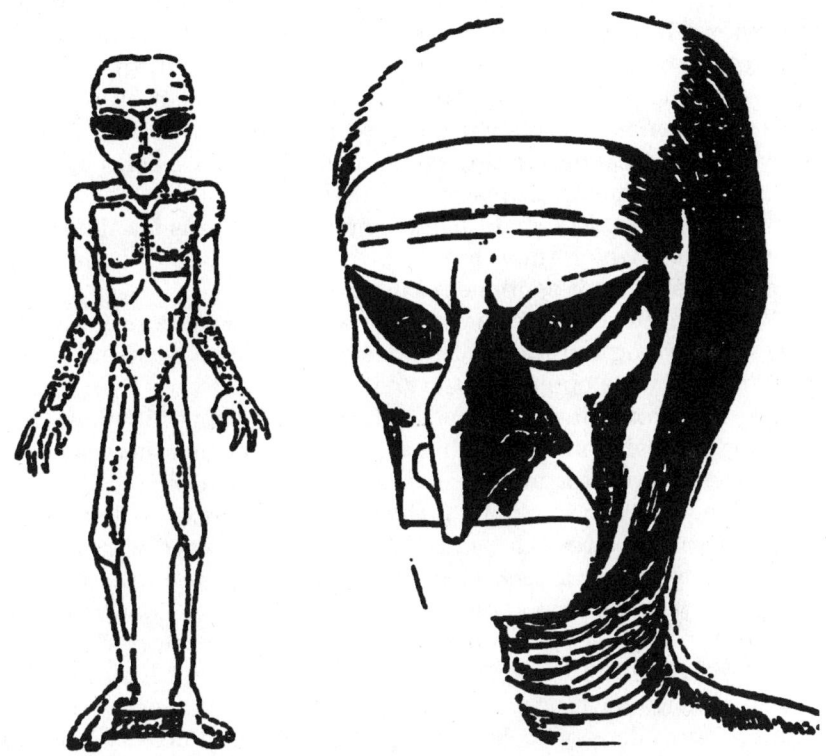

Figure 15-1   *A drawing of a tall grey from a particular subspecies (or species?) of grey (the Orion grey, type 1). From "Matrix II: The Abduction and Manipulation of Human Beings Using Advanced Technology", 1989/1990, by Valdamar Valerian, Leading Edge Research Group, P.O. Box 481-MU58, Yelm, Washington 98597. With permission.*

The greys, as a group, are reputed to be rather weak genetically, although there are reports that this is misinformation. Most of them are reported to have no genitals, and are unable to sexually reproduce. Reproduction among the greys takes place mostly in the laboratory through a cloning process. Remember our own Earth scientists can "manufacture" a simple organism, such as a frog, through cloning. It

The Greys

is, therefore, easy to imagine an older species being able to clone much more complex beings, such as themselves. However, they have repeatedly weakened themselves genetically over the course of their history because of nuclear wars in their past,[11] genetic engineering done on their species in the past, through the deleterious effects of thousands and millions of years of cloning,[12] and other reasons of which we or they may not even be aware.

With this much understanding, we can begin to comprehend their behavior a little better. They need the cattle and humans they mutilate for the blood they depend on for survival. One of the primary reasons for millions of human abductions is the acquisition of human genetic material. Reports suggest they are trying to correct some of their genetic weaknesses through creating a hybrid species by crossing grey and human genes. As we saw in Chapter 12, they often impregnate a human female and let her carry the fetus for about three months, after which time they take the fetus to one of their underground labs where it continues to develop.

We learn more about the genetic and other weaknesses of the greys by turning the focus of our investigation to the metaphysical and spiritual development of the greys. We have already seen that the greys often exhibit marvelous metaphysical capabilities, such as walking through walls and mental telepathy, although there seems to be a question of how much these capabilities are a result of natural evolution, and how much of it is the result of advanced technology.

In Valerian, we read that many of the greys involved with humans, like their reptoid masters, exist in dimensions beyond the three dimensions that humans know.[13] What we call "dimensions" are often referred to as "densities" in the metaphysical literature. There are many fifth and fourth density greys interacting with humans. Fifth density beings have no bodies, but fourth density beings, especially "lower fourth density" beings might appear as third density beings to humans. In fact, fourth density greys can transit between the third and fourth density.

To we third density beings, anything higher than our density would seem to be marvelous and unbelievable. As we have noted, there is a general perception among humans that entities that do not have a third density body like humans, must be both superior and good. Apparently this is not the case. Valerian states that the lower fourth density is always subject to tyranny and that the fifth density is the last density above humans that can contain negativity.[14] If this is true, we can garner a perspective of how primitive the human animal is, relative to other galactic beings. We have been told by the Pleiadian Plus group and others that we humans can and will soon evolve higher density capabilities than we have now. The proportion of the greys that can appear in our density, relative to those of a higher density is unknown. With regard to the spirituality of the greys, let us first turn to the subject of their emotions, or lack of emotions. Several sources give evidence that the greys simply do not have any emotions.[15] The Pleiadian Plus group state that

> emotion is your road or bridge or ticket to the spiritual self. When people deny the emotional self, they can't get into the spiritual realms...[16]

This would seem to be, if true, a serious weakness of the greys. One esoteric source claims, however, that the greys' lack of emotions is actually of their own doing. Through engineering (genetic) and inbreeding, they have removed emotions from their beings.[17] Without emotions they have come to a physical and spiritual dead end. In the past they wished to make themselves purely intellectual, and this, they believed, could be done by eliminating the emotions from their beings. So, they eliminated emotions and thereby, greatly hindered, or actually eliminated their ability to get in touch with their spirituality. It is hard to imagine how an intelligent species could have so damaged themselves. Apparently, in addition to their spirituality, they also have lost the ability to reproduce through sexual union. When they eliminated their own emotions, they eliminated the emotions necessary for "normal" biological reproduction, such as love and passion.[18]

When viewed from this perspective, the greys seem to be almost a sad and pathetic race, despite the atrocities they are known to have committed on humankind. The non-corporeal entity, the Awareness, through his channel, states:

> They (the greys) are essentially a dying race. Their only hope is through intermingling or interbreeding or intergenetic relationship with humans.[19]

We might conclude that they actually need us, yet some of them treat us so cruelly. Cruelty is perhaps the wrong word to use, if the above information about the greys is mostly correct. It would be more accurate to say that the greys lack compassion for humans, rather than that they are cruel to us. They do not have the genetic capability to be compassionate and benevolent. They treat us like Western societies treat beef cattle. The beef cattle are raised without compassion, hauled off to market, slaughtered, then eaten. Humans of "beef cattle cultures" don't wish cruelty and evil for the beef cows, they are simply controlling and exploiting a lower species of "dumb" animals. Similarly, the greys are simply dominating and exploiting a lower, less intelligent ("dumb") species. The difference is that the species the greys are trying to control and exploit is a species which possesses a higher consciousness, a spiritual capability that gives it the means of evolving beyond the silly domination-victim game of the greys and their masters, the reptoids. No wonder they have endeavored to destroy and distort spiritual teachings that have been presented to humankind through benign entities. The Awareness informs us that though the greys are fascinated by the spiritual capabilities of humans, they are actually frightened by human spirituality and avoid abducting humans that exhibit many of the human spiritual capabilities. The Awareness makes it clear that it is not referring to religion when it refers to spiritual energy:

> Spiritual is what this Awareness speaks of, and spiritual energy is enhanced by compassion and the caring of the soul for others, and the love of the Divine and the soul's association and integration with those higher Divine Forces. It does not mean the mouthing of slogans of a religious nature or the repetition of certain phrases or names associated with a religious belief. This is not the same as spirituality.[20]

So, this is apparently the key to enhancing one's chances of avoiding abduction, mutilation, mind control games, implants, or any direct contact with these regressive intruders/visitors to our planet - the development of one's higher consciousness, spiritual awareness and powers!

As we've seen, the greys do not like to be seen for what they are, but prefer to be worshipped as if they were gods. The greys feel that they can get humans who worship them to do almost anything for them. We have referred to the implants the greys usually leave in abducted humans. Secret U.S. government documents report that the greys implant small devices, usually near the human brain, which potentially give them control and monitoring capability of the humans they implant.[21] We've seen that millions of people throughout the world have been abducted and implanted with one of these small devices. An esoteric source estimates that approximately 30 to 40 million people have been implanted by the greys in the United States alone.[22] We might well ask why the greys wish to try directly to monitor and control humans through mechanical devices. Hasn't their nonmechanical manipulating of human religions and other aspects of human societies been sufficient to control enough people enough of the time? As we have seen, Bramley, Cooper, Valerian and others have presented evidence that the greys have been able to manipulate human cultures and institutions through interfacing with human "secret societies," and now secret governments. One esoteric source who maintains that it has the best interest of humans in mind, claims that the greys have developed many "channels" of Earth humans in order to promote their own interests here on Earth.[24] And now they are implanting millions of people on this planet. For what purpose are they doing this? What is their agenda here on Earth, anyway? Whatever it is, judging from the prophecies and predictions of major changes that are soon to occur here on Earth, the greys might soon be prepared to make some dramatic moves with the human element with whom they have secretly been interfacing.

One thing seems very clear: we humans should all be working intensely on our spiritual natures. We should all be striving to evolve spiritually toward the Light, or Truth. In this way, exoteric and esoteric sources tell us, we can largely escape the agenda of the greys and other negatively-oriented ETs, and all negative forces in this universe. Simply being aware of the greys, who they really are and how they have influenced the history of humankind, is part of our consciousness and spiritual evolution. We should not fear the greys or become engrossed in the facts of their present and past presence on Earth, but concentrate on developing our own awareness, emotions and spirituality. Despite being technologically and intellectually young and primitive relative to the greys, we humans may have innate spiritual capabilities that are apparently beyond those of these ETs.

Esoteric sources and mystical traditions here on Earth suggest that humans should develop a society based on love and compassion for each other and all of Creation in order for us to reach our fullest spiritual potential. This love and compassion should include the greys, who, as we have seen, are a rather pathetic, dying species, having genetically engineered emotions out of their physical entities. This does not mean we have to tolerate their diabolical behavior. Nor does it mean that we are responsible for saving them from their sad physical and spiritual state.[25]

There is much more information out about the greys that I have not covered here, since we must not dwell on them too much, unless we need to pull them into our lives, and I do not! I have found the areas I wished to cover, especially in trying to answer the question: why has an older, more advanced species dwelled in and propagated darkness when they could have been moving toward the light themselves?

In this next section, let us investigate how regressive ETs may have distorted and otherwise influenced the major world religions.

---

[1] Cooper, W. 1991. *op. cit.*, p. 209.
[2] Valerian, V. 1991. *op. cit.*
[3] *Ibid.*, p. 102.
[4] *Ibid.*, p. 88C.
[5] *Ibid.*
[6] Royal, L. and K. Priest 1981. *op. cit.*
[7] *Revelations of Awareness* 1990. No. 90-9, Issue 326, pp. 2-4.
[8] _____ 1992. No. 92-11, issue 403, p. 10.
[9] _____ (see reference 7, this chapter).
[10] Valerian, V. 1991. *op. cit.*, p. 89.
[11] *Ibid.*, p. 92.
[12] *Ibid.*, p. 100A.
[13] *Ibid.*, p. 95.
[14] *Ibid.*, p. 95 and p. 103.
[15] eg., *Ibid.*, p. 90.
[16] Marciniak, B. 1992b. *op. cit.*, p. 64.
[17] *Revelations of Awareness* 1990. No. 90-14, Issue 367, p. 18.
[18] _____ 1993. No 93-5, Issue 414, p. 12.
[19] _____ 1990. No. 90-14, Issue 367, p. 18.
[20] _____ 1992. No. 92-11, Issue 403, p. 10.
[21] Valerian, V. 1991. *op. cit.*, p. 88 C.
[22] *Revelations of Awareness* 1993. No. 93-8, Issue 417, p. 3.
[23] Valerian, 1991, p. 97.
[24] *Revelations of Awareness* 1991. No. 91-16, Issue 391, p. 5.
[25] _____ 1993. No. 93-7, Issue 416, p. 12.

"Our ideal is seeking us. Open your eyes, it is here, in your home, in the multitudinous acts of mutual love and sacrifice, in the exalted experience of friendship, in shop, store and office, in your community, in social work, in civic work, in religious work, in the humblest and highest task it is there."
Rabbi Abba Hillel Silver, from a sermon, *The Vision Splendid*

Chapter 16

Jehovah and the Ancient Hebrews

In my childhood and adolescent years, as a young Christian of the Episcopalian variety, I, of course, became much more familiar with the New Testament and the story of Jesus, than with the Old Testament. However, the Old Testament was part of the Bible, part of the story of God Almighty. Therefore, I did become somewhat familiar with the Old Testament, primarily through stories from Sunday school printed by the church for the consumption of children like me. Also, an occasional Hollywood movie would come through town that told a story of the Old Testament, such as "The Ten Commandments." So, I was somewhat acquainted with some of the stories of the Old Testament. I, of course, knew the story of Adam and Eve in the Garden of Eden. The story of Abraham where he was commanded to sacrifice his son to "the Lord," but was relieved of this tragic "duty" after he had made preparations for the sacrifice, was familiar to me. I memorized the Ten Commandments, all of the names of the books of the Old and New Testaments, and the Twenty-third Psalm for my Boy Scout "God and Country Award." But, I never actually read any of the Old Testament myself. When I learned in recent years, however, that much of the Old Testament contains much historical information, I read most of the Old Testament.

I accept Sitchin's contention that Genesis is mostly an edited, Hebrewized copy of much more ancient Mesopotamian texts. This includes the stories of Abraham, the supposed grand patriarch of the ancient Hebrews, with whom Jehovah (also known as Yahweh) made the pact to provide the ancient Hebrews with the land which was then known as Canaan. Sitchin believes that, in fact, the story of the interaction of Abraham and Jehovah is a much Hebrewized version of an ancient Sumerian story, even though such a Sumerian story has never been found.[1] According to Sitchin's analysis of the fragmentary remains of Mesopotamian tablets that do exist, Abraham was the son of a high Sumerian priest and became a priest himself at the Sumerian city of Ur. His mission when he was located in the part of the world then known as Canaan, later to become Israel and Palestine, was to lead a military group that was to help defend the spaceport of the gods located in the Sinai Peninsula.[2]

In any case, we have reviewed Sitchin's arguments that the Old Testament is a historical document, including the books of the Old Testament that succeed Genesis, beginning with Exodus, which record the interaction of the entity Jehovah with the ancient Hebrews, and their interactions with other people of the ancient Middle East. When I first began to read the Old Testament after Genesis, I was astounded by the violence, the savagery, and the brutality the Lord God Jehovah demonstrated, both to the enemies of Hebrews and, on occasion, to the Hebrews themselves.

Allow me to observe my own evolution of my understanding of "God." As a child and adolescent the one "God" of the (only) universe was an ethereal entity, a nonphysical being that was incomprehensibly higher, more holy than we humans. I was 20 years of age before I had a conversation with a friend that began to awaken me, if ever so slightly, to the fact that the "Old Testament God" was much more severe and more unforgiving than the "New Testament God." What was this? I had been taught that the "Old Testament God" was identical to the "New Testament God." This conversation marked the first time in my life that I began to think of "God" as being more than one entity, or an entity that had a dual personality.

Of course, now that I understand that most of the "gods" of ancient people were actually ETs who were using their vastly superior technological and metaphysical knowledge to hoodwink and control humans, it doesn't surprise me that most of the Old Testament is documentation that Jehovah was just such an ET. At least since Erich von Däniken began to publish his books in the 1960s and 1970s, UFO groups have been exuberant in their contention that Jehovah was a UFOnaut (who I am referring to as an ET) - with good reason! The Old Testament is full of evidence that the ancient Hebrews were dealing with one, or probably a group of ETs. The hypothesis that Jehovah was the Captain and the "angels" were crew members of what we today call a UFO, is quite common in "UFO circles". The most quoted evidence of this hypothesis are the "visions" of Ezekiel, a 6th

Millennium B.C. Jewish prophet. While in exile in Chaldea (southern Mesopotamia) near the river or canal of Kebar, "the hand of the Lord" came upon Ezekiel and he

> ...saw a storm wind coming from the north, a vast cloud with flashes of fire and brilliant light about it; and within was a radiance like brass, glowing in the heart of the flames. In the fire was the semblance of four living creatures in human form...[3]

I'd say, as have many before me, Ezekiel's description of what we would today call a UFO was pretty good for a man who was totally devoid of any of what we would today call mechanical knowledge. It has been suggested,

> ...that the Israelites had the inclination...to associate the space craft with clouds or to call them clouds directly, because of a deficiency of alternate and better terminology within the language itself.[4]

As we shall see, the cloud terminology is used frequently in the Old Testament when the Israelites are referring to Jehovah and his angels.

The ancient Hebrews were freed from slavery in Egypt, with the strong arm help of Jehovah. Right from the start of this mass exodus, the Jehovah entity demonstrated cruelty and lack of compassion for humans, as we have seen was characteristic of the ancient gods of the Sumerians centuries earlier. Also, as we shall see, "the Lord" in those days was not considered an ethereal being somewhere up in the heavens, but was often directly involved in the day to day lives of the ancient Hebrews.

The scholars are not certain of the exact time period of the exodus of the ancient Hebrews from Egypt. As we have seen, the dates of the ancient world have apparently been covered up and deliberately falsified, so much so that modern historians can hardly pin down an exact date for any of the events of the ancient world. Often dates of 1,200 BC to 1,300 BC for the exodus are given, but occasionally estimates of this date are closer to 1,500 BC. In any case, just to give the readers some time reference, remember that Sitchin gives a date of 2,024 BC for the nuclear destruction of Sodom and Gomorrah which Abraham witnessed.

As most readers know, Moses was the man who led the Hebrews out of Egypt under the supervision of the Jehovah entity. Many have claimed that Moses was a member of a secret Egyptian brotherhood, or an Egyptian priest or holy man of some sort. It is recorded in the book of Exodus that, "...Moses was a very great man in Egypt in the eyes of Pharaoh's courtiers and of the people." [5] Jehovah sent Moses to the Pharaoh several times to say:

> These are the words of the Lord: "Let my people go in order to worship me. If you do not let my people go, I will..."

He (Jehovah) will cause some sort of calamity to occur in Egypt, is the way these threats are completed. It is one thing to try to force the Pharaoh to let his people leave Egypt and the slave conditions in which the Hebrews apparently lived in that country, but it is another thing if a "supernatural" being somehow makes the Pharaoh stubborn so that he may rain more misery on the Egyptians:

> Then the Lord said to Moses, 'Go into Pharaoh's presence. I have made him and his courtiers obdurate, so that I may show these my signs among them, and so that you can tell your children and grandchildren the story: how I made sport of the Egyptians, and what signs I showed among them. Thus you will know that I am Lord.' [6]

Having made the Pharaoh and his courtiers obdurate, the Pharaoh again and again refused the request of Moses to let the ancient Hebrews leave Egypt. This gave Jehovah the opportunity he wanted to show his power and cause many calamities to occur in Egypt. These calamities included: causing the water of the Nile River, on which the Egyptians depended so much, to turn to blood, killing fish and making the water totally unfit to drink; a plague of frogs that covered Egypt; a plague of maggots that covered the land; a plague of flies that covered the land; a plague of locusts that covered the land; a violent hailstorm; a terrible pestilence that killed all of the domestic animals of the Egyptians (but not one domestic animal of the Israelites); Jehovah also caused a sickness of festering boils on the Egyptians. Jehovah's final calamity was that he killed the firstborn of every Egyptian, from the Pharaoh to the slaves (except the Israelites). [7]

Finally, the Pharaoh and his courtiers let the Israelites leave Egypt and the mass exodus began. Leading the horde of Israelites and their domestic animals was Jehovah; "...all the time the Lord went before them, by day a pillar of cloud to guide them on their journey, by night a pillar of fire to give them light, so that they could travel night and day." [8] This "pillar of cloud" never left the Israelites as they trekked from Egypt through the wilderness.

Then came the famous parting of the Red Sea so that the Israelites could cross. It might be expected that a powerful, omnipotent god would part the sea to help his people on their trek. But, Jehovah was not satisfied with merely helping his people; he said, "...I will make Pharaoh obstinate, and he will pursue them (the Israelites), so that I may win glory for myself at the expense of Pharaoh and all his army..." [9] So, the Red Sea parted, the Israelites passed through the parted waters, followed by the Egyptian army which pursued them, and the waters collapsed and the army was destroyed. "Not one man was left alive."

I remember watching this dramatic scene depicted in the movie, "The Ten Commandments" with some high school friends. One of them leaned over and asked, "You guys don't really believe this, do you?" Well, I did at the time. I was even a little offended by the question. Of course, the Lord would take care of those evil Egyptians. And, I still believe it, with some modifications. Any ET capable of travel in space would have enough material and metaphysical understanding to pull off such a stunt, it seems to me. Perhaps it was a somewhat narrow, shallow section of the Red Sea. Possibly the ancient and succeeding Hebrews have exaggerated the exploits of themselves and their god. On the other hand, perhaps there was, unfortunately, little exaggeration of the exploits of Jehovah. We have already reviewed some of the tremendous technological and metaphysical powers of the Sumerian gods, the Anunnaki. It is only logical that Jehovah, whatever stripe of rogue ET he was, was capable of performing what seems to us incredible feats. What is shocking is that an ET being would be so intent on the goal of self-aggrandizement at such a horrible cost to thousands of members of a younger, less evolved species - the human species.

The Israelites needed a food supply to survive the trek through the Sinai Peninsula wilderness that was to last 40 years. Undoubtedly they took some food with them when leaving Egypt, and they brought their domestic animals. But, there was little or no food available in the desert for the hundreds of thousands of people who made this astonishing trek. Jehovah provided food for them called manna. Manna "...was white, like coriander seed, and it tasted like a wafer made with honey..." The Israelites ate the manna for 40 years until they came to a land where they could settle; they ate it until they came to the border of Canaan.[10] The Jehovah entity apparently had a tremendous capacity of producing food, even if it was a bit monotonous.

Jehovah and the Ancient Hebrews

When the Israelites came to the wilderness of Sinai three months after leaving Egypt, Jehovah arranged with Moses to meet the multitude of "his people." The people were told to assemble around the mountain, Mount Sinai, but they were warned not to go too close to the mountain. The men were also warned not to go near a woman on the day of the Lord's visit. On the morning of the day Jehovah arranged to meet the masses,

> ...there were peals of thunder and flashes of lightning, dense cloud on the mountain and a loud trumpet blast; the people in the camp were all terrified...Mount Sinai was all smoking because the Lord had come down in fire; the smoke went up like the smoke of a kiln: all the people were terrified, and the sound of the trumpet grew even louder. Whenever Moses spoke, God answered in a peal of thunder.[11]

The version of the Bible, *The New English Bible*, from which this quote was taken, has a footnote indicating that the above quoted translation of "thunder" can also be translated as, "by voice." The manner in which Jehovah is depicted, or perceived as addressing the Israelites changes more than slightly with the substitution of "thunder" for "by voice" in the above quoted passage. Could this be another mistranslation of an ancient language, which is common in ancient texts of all peoples, as Sitchin often illustrates? This may be another example of a deliberate obfuscation of ancient texts done by a small group who wanted to influence and control a large part of humanity. "Thunder" may have been substituted for "by voice" to make the Lord Jehovah seem less human and more ethereal than the ET actually was.

It is understandable that the people were terrified. Chances are, however, that a first rate sounds and special effects group in Hollywood, perhaps from one of the major studios, given an unlimited budget, could put on a show that at least approximates the tremendous display created by the Jehovah entity on Mount Sinai.

After climbing up the mountain to confer with the Lord, Moses went down to the people and gave them the Ten Commandments. Let us examine them in order to gain more insights into the nature of Jehovah. Moses read,

> God spoke, and these were his words: "I am the Lord your God who brought you out of Egypt, out

of the land of slavery. You shall have no other God set against me."[12]

Again, *The New English Bible* has a footnote which may represent another mistranslation, as the last "God" in the above commandment can (or should?) be translated as "Gods". Given the nature of the ancient world, gods would seem to be the correct translation. Moses and Jehovah were some of the first entities to purport monotheism on this Earth. Some have suggested that Moses may have been taught monotheism under the Egyptian Pharaoh, Akhnaton, the first and only Pharaoh to advocate monotheism. Here again, there are conflicting dates and nobody knows exactly when the exodus took place. In any case, one can see how monotheism could confuse humankind's concept of the divine, especially that foisted on the ancient Hebrews by Jehovah.

Moses continued quoting Jehovah's commandments:

> You shall not make a carved image for yourself nor the likeness of anything in the heavens above, or on the Earth below, or in the waters under the Earth. You shall not bow down and worship them; for I am a jealous god. I punish the children for the sins of the fathers to the third and fourth generations of those who hate me. But, I keep faith with thousands, with those who love me and keep my commandments.[13]

Again, it is hard to imagine why a "lofty one" would behave so pettily and be so jealous that he punishes the descendants of "sinners," to the fourth generation after the sin had been committed! Why should a man or woman suffer punishment for the "sins" of their great-great-grandfather?

As we continue with the commandments, we see that Jehovah identifies himself with the Supreme Creator:

> You shall not make wrong use of the name of the Lord your God; the Lord will not leave unpunished the man who misuses his name. Remember to keep the Sabbath day holy. You have six days to labor and do all your work. But, the seventh day is a Sabbath of the Lord your God; that day you shall not do any work, you, your son or your daughter, your slave or your slave-girl, your cattle or the

> alien within your gates; for in six days the Lord made heaven and Earth, the sea, and all that is in them, and on the seventh day he rested. Therefore, the Lord blessed the Sabbath day and declared it holy.[14]

It is interesting that Jehovah fell into the third person when he said that "he rested" on the seventh day instead of "I rested." Nevertheless, one can see clearly that Jehovah wished to be identified as the Supreme Being, the Creator.

The commandments continue:

> Honor your father and your mother, that you may live long in the land which the Lord your God is giving you.
> You shall not commit murder.
> You shall not commit adultery.
> You shall not steal.
> You shall not give false evidence against your neighbor.
> You shall not covet your neighbor's house;
> you shall not covet your neighbor's wife, his slave, his slave-girl, his ox, his ass, or anything that belongs to him.[15]

If these commandments were obeyed at that time (or any time), it would represent a tremendous step up in the moral evolution of humankind. One could raise objections to the individual self-aggrandizement Jehovah demonstrates in giving the commandments, especially in claiming to be the Prime Creator, and claiming to be the Supreme and only deity. Also, the jealousy and cruelty Jehovah demonstrates should certainly not be glorified. And, of course, in today's modern world societies we don't need the frequent references to domestic animals and human slaves that a majority of us do not have, and sensitive people would strongly object to the extreme sexism of the ancient Hebrew culture that Jehovah supports. Aside from these objections, these commandments might be good moral guidelines to humankind today. All decent people would hope that all humans might someday come to an agreement that all of us would not covet, not bear false witness, not steal, not commit adultery, not murder anyone ever, and honor our fathers and mothers. If all the humans that live on Earth would follow these commandments, the world would be a much more pleasant place to live than it is now. But, the questionable behavior of Jehovah belies these commandments. Did not Jehovah cause the Egyptian Pharaoh and his attendants to be obstinate and hard-hearted in not letting the Hebrews leave Egypt, until he caused a host of calamities, including the killing of all the first-born children of families of Egypt? Did not Jehovah cause the Egyptian army to chase after the Hebrews to the Red Sea so that he could kill them, and thereby enhance his reputation (in his eyes)? Now, about three months later he is giving a commandment not to kill? Now that Jehovah had given the commandment to his people not to kill, one would logically expect that killing would be a small part of the culture of the ancient Hebrews in the future. Those who know the Old Testament know that the opposite is the case, as Jehovah himself instigated an astonishing amount of violence and killing, even on the Israelites, his chosen people.

Jehovah kept a very close, personal control over his people. Once, when Moses went up the mountain to receive more instructions from Jehovah, the remaining mass of Israelites became restless and broke one of the commandments by making a golden bull-calf, and began worshipping it and dancing around it. When Moses returned from the mountain he became furious, smashed the stone tablets on which the Lord had written, and destroyed the golden bull-calf. After this incident Moses began pitching a tent some distance from the main camp of the Israelites which he called the Tent of the Presence:

> Whenever Moses went out to the tent, all the people would rise and stand, each at the entrance to his tent, and follow Moses with their eyes until he entered the tent. When Moses entered it, the pillar of cloud came down, and stayed at the entrance to the tent while the Lord spoke to Moses. As soon as the people saw the pillar of cloud standing at the entrance to the tent, they would all prostrate themselves, every man at the entrance to his tent. The Lord would speak with Moses face to face, as one man speaks to another...[16]

Jehovah gave instructions for the Israelites directly to Moses who, in turn, passed them on to the rest of the people. Jehovah often demonstrates his own ruthlessness and jealousy in these instructions:

> The Lord said, Here and now I make a covenant. In full view of all your people I will do such miracles as have never been performed in all the world

> or in any nation. All the surrounding peoples shall see the work of the Lord, for fearful is that which I will do for you. Observe all I command you this day; and I for my part will drive out before you the Amorites and the Canaanites and the Hittites and the Perizzites and Hivites and the Jebusites. Be careful not to make a covenant with the natives of the land against which you are going, or they will prove a snare in your mist. No: you shall demolish their altars, smash their sacred pillars and cut down their sacred poles. You shall not prostrate yourselves to any other God, and a jealous god he is. Be careful not to make a covenant with the natives of the land, or, when they go wantonly after their gods and sacrifice, and to marry your sons to their daughters, and when their daughters go wantonly after their gods, they may lead your sons astray too.[17]

How does one go wantonly after their gods? The rivalry that the ETs had for each other and the close, often physical contact, the "gods" had with people are demonstrated in these sayings of Jehovah.

According to Sitchin, Jehovah's main rival at this time was the Canaanite god Baal. Baal, as Sitchin demonstrates, was almost certainly an ET whose domain Jehovah was to "give" to the ancient Hebrews.[18] Certainly there are many references to Baal in the Old Testament. For example, as the Israelites were still wandering through the wilderness on the plains of Moab before they entered the promised land,

> ...the people began to have intercourse with Moabite women, who invited them to the sacrifices offered to their gods; and they ate the sacrificial food and prostrated themselves before the gods of Moab. The Israelites joined in the worship of Baal of Peor, and the Lord was angry with them. He said to Moses, 'Take all of the people and hurl them down to their death before the Lord in the full light of the day, that the fury of his anger may turn away from Israel.' So Moses said to the judges of Israel, 'Put to death, each one of you, those of his tribe who have joined in the worship of the Baal of Peor.'[19]

More than 800 years after the exodus, Israelites were still killing the followers of Baal. The fanatical Hebrew prophet, Elijah, who prophesied in the Ninth Century BC., had a contest of faith with the prophets of Baal before many people (many men) of Israel. After Elijah and the Lord won the little contest, Elijah said to the people,

> 'Seize the prophets of Baal; let not one of them escape.' They seized them, and Elijah took them down to the Kishon and slaughtered them there in the valley.[20]

Because of the episode with the golden bull-calf and other incidents, Jehovah decided that the Israelites would have to wander through the wilderness for 40 years before they could enter the promised land. The Lord said:

> 'Because they (the Israelites) have not followed me with their whole heart, none of the men who came out of Egypt, from 20 years old and upwards, shall see the land which I promised...' The Lord became angry with Israel, and he made them wander in the wilderness for 40 years until that whole generation was dead...[21]

As the Israelites wandered through the wilderness those 40 years, Jehovah was very much a part of the day to day affairs of the Israelites. Moses made the Tabernacle in the Tent of the Presence, which, we recall, he set up some distance from the main camp:

> ...At every stage of their journey, when the cloud lifted from the tabernacle, the Israelites broke camp; but if the cloud did not lift from the Tabernacle, they did not break camp until the day it lifted. For the cloud of the Lord hovered over the Tabernacle by day, and there was fire in the cloud by night, and the Israelites could see it at every stage of their journey.[22]

Along the way, a man was caught in the act of gathering sticks on the Sabbath day, and the man was brought before Moses and all the community. But the Israelites wondered what was to be done in such a case, so they asked Jehovah, through Moses, what should be done:

> ...the Lord said to Moses, 'The man must be put to death; he must be stoned by all the community outside the camp.' So they took him outside the camp and all stoned him to death, as the Lord had commanded Moses.[23]

The Israelites had many violent encounters with other peoples who lived in the areas of the wilderness they wandered, under Jehovah's guidance. Once while in the plains of Moab, the Israelites were near a people, or a land (the text doesn't make clear) that was known as Midian. The Lord instructed the Israelites, through Moses, "...to exact vengeance for Israel on the Midianites and then you will be gathered to your father's kin."[24] One feels that much is left out of this story because we are not told why vengeance was "necessary." Nevertheless, Moses raised an army of 12,000 Israelites and "...made war on Midian as the Lord had commanded Moses, and slew all the men... They burnt all their cities, in which they had settled, and all their encampments. They took all the spoil and plunder..."[25]

Later, shortly before the Israelites crossed the Jordan, Jehovah helped provoke a conflict between them and the peoples of Heshborn and Bashan whose lands were located east of the River Jordan. Another bloody war was waged in which all of the people of these two lands were slaughtered with the help of Jehovah:

> ...the Lord our God...delivered Og, King of Bashan into our hands, with all his people. We slaughtered them and left no survivors, and at the same time we captured all his cities; there was not a single town that we did not take from them. In all we took sixty cities...Thus we put to death all the men, women, and dependents in every city, as we did to Sihon, King of Heshbon...[26]

The Israelites themselves were not exempt from suffering mass death. Once, while still in the wilderness, a small group of them under a man named Korah, challenged the authority of Moses, and indirectly, that of Jehovah. Basically, Korah and his followers objected to the authoritarian powers of Moses, and the fact that Moses had led them out of "...a land flowing with milk and honey to let us die in the wilderness..." Moses, furious at the questioning of his and the Lord's authority confronted the followers of Korah at their tents. The Lord caused the destruction of the followers of Korah, who "...holding themselves erect, had come out to the entrance of their tents with their wives, their sons, and their dependents." The Lord caused the ground beneath the followers of Korah to split. The "...Earth opened its mouth and swallowed them and their homes...they went down alive into Sheol with all they had; the Earth closed over them, and they vanished from the assembly..." [27] This act caused panic, turmoil and complaint among the rest of the assembly. The next day the Israelites complained to Moses and Aaron, the brother of Moses, about the violent deaths of some of the Lord's own people. This infuriated the Lord more and he caused a plague to fall on the people. Only the intervention of Moses and Aaron saved the people from being completely destroyed, according to the Bible. The plague was stopped, but not before 14,700 Israelites had died from it.

It is said in the Old Testament again and again that the ancient Hebrews must "fear their God." [28] Certainly Jehovah gave the Hebrews many reasons and examples of why they should be terrified of him.

Finally, the day came when the Israelites were to enter the promised land. Jehovah dammed up the River Jordan somehow and the whole nation entered Canaan from the east. A few days after they had entered Canaan, the Israelites ate the produce of the land. From that day on the "Israelites received no more manna; and that year they ate what had grown in the land of Canaan." [29] So the main staple of their diet for 40 years, manna from "heaven," was cut off forever.

Of course, Jehovah could not simply "give" the Israelites "the land of milk and honey." The promised land had to be violently taken from the Canaanites, the Amorites, the Hittites, the Perizzites, the Hivites, the Hebusites and whomever else lived in the promised land at that time. And the Israelites did this, with the heavy-handed help of Jehovah.

The first town or city the Israelites came to in Canaan was the ancient (even in those days) city of Jericho. After Jehovah, with the help of the Israelites(?), caused the walls that surrounded Jericho, or at least part of them, to collapse, the Israelites rushed in, and they "destroyed everything in the city; they put everyone to the sword, men and women, young and old, and also cattle, sheep, and asses." [30] One shudders to think of the enormous amount of blood, shrieking and screaming, shouting, dust, burning, crying, not to mention death, the "simple" extinguishing of life, this one sentence in the bible represents. The Lord had instructed the Israelites to be brutal and show no mercy to the Canaanites or any of the other occupants of the promised land. The Israelites destroyed everything in Jericho except, significantly, the most valuable material objects. They then set fire to the city and everything in it, except that they deposited the silver and gold and the vessels of copper and iron in the treasury of the Lord's house.³¹ Jehovah had demanded that the Israelites give the gold and silver and all cast metal to him. One named Achan secretly kept some gold and silver and a fine mantle from Sinar (Sumer). Jehovah apparently had what we call metaphysical ability - as do most more evolved entities that humans claim to have come into contact with today - of being able to be aware of what all humans are doing, including what is being thought. He knew that Achan was secretly holding back some valuables. Jehovah had Joshua, his chosen successor to Moses, seek Achan out. After Achan had confessed to hoarding the goods, "...all the Israelites stoned him to death: and they raised a great pile of stones over him, which remains to this day. So the Lord's anger was abated..." ³² The ancient Hebrews could not deviate, even a little bit, from doing exactly what Jehovah commanded. The penalty for deviation, as we see, was death.

The Israelites and Jehovah next turned their attention to the town (or city) of Ai. With a bit of deception masterminded by Jehovah, the Israelites managed to draw the men of Ai out of the city and ambushed them. After they killed these men, they turned to the city itself and killed everyone else. "...The number who were killed that day, men and women, was 12 thousand, the whole population of Ai..." ³³ Then they burnt the city.

And, so it goes through much of the Old Testament. There are many more examples of the violent, jealous and vindictive behavior of Jehovah. The Old Testament is one of the best documents we have, along with the ancient Mesopotamian texts, of ETs that are more evolved and much more powerful than humans, who interacted with humans with little respect or compassion for our species. Of course, people who lived after these ancient times and before the age of modern technology and the age of space flight would have trouble recognizing Jehovah for what he really was - an ET, who for reasons of his own, was intervening directly in the affairs of humans.

Who was Jehovah? Is there any way we can find out exactly what theme of ET he was? Boulay, in his reptoid interpretation of ancient history, believes Jehovah (Yahweh) was the Sumerian god of lightning and thunder, Ishkur, whose Semitic name was Adad. According to Boulay, Ishkur/Adad was trying to extend his sphere of influence among the gods from Anatolia (southern and central Turkey) southward throughout the area that became Israel, in his dealings with the Israelites.³⁴ Sitchin, in his writings, does not venture a guess as to who Jehovah might have been, although he assumes that Jehovah and all of the ancient ETs were of the Sumerian Anunnaki, which we have determined were possibly reptoids.

The Pleiadian Plus group supports Boulay's contention that Jeho-

Jehovah and the Ancient Hebrews

vah was a reptoid. In answer to a question about ETs other than the reptoids that may have been involved with genetic manipulations that created modern humans, the Pleiadian Plus group answered:

> There were many others...Many of them had human form. Presently, your greatest state of unrest, or discomfort, comes from those of a reptilian type of existence, because they seem the most foreign to you. And they will be the biggest shock to humans - to discover that is who the humans have been worshipping as Jehovah and Yahweh throughout the Bible...[35]

All of the actual sightings of Jehovah and/or his angels, such as the above-quoted observation of Ezekiel, mention only "living creatures in human form." We saw earlier that there are human eyewitness evidence that the reptoids are able to appear in human form, and certainly they must have had human-looking entities who worked for them.

The "Swiss Pleiadians" told Billy Meier that much of the Jehovah entity were Pleiadians. Throughout their communications with Meier in the 1970s, at least what has been provided to the public, the "Swiss Pleiadians" never mention the reptoids. They agreed with Meier that Jehav, as they refer to him, was very bloodthirsty and quite evil, at least at first. They say Jehav ruled with a small group of Pleiadians, and he shared his rule, to some extent, with a much more benign entity. The "Swiss Pleiadians" claim that the Pleiadian descendants of this little group of Jehav's, were much more spiritual and humane than was Jehav himself. Eventually, the spiritual evolution of the Pleiadian descendants of the Jehav entity,

> ...changed their minds and they decided to leave the development of the Earth beings to their natural course, and retired to their home-world, so they left the Earth and returned as peaceful creatures to the Pleiades, where their own mankind had reached advanced states...[36]

The more benign entity that the "Swiss Pleiadians" state ruled beside the cruel and bloodthirsty entity would seem to account for some of the more humane, less bloodthirsty aspects of the Old Testament. We've already discussed how a modified Ten Commandments would make a decent moral code for humankind. It is a curious fact that frequently in the Old Testament, right beside a law that is very severe and vindictive, more gentle and loving laws appear. For example, shortly after the 'eye for eye' law is given in Exodus,

> Whenever hurt is done, you shall give life for life, eye for eye, tooth for tooth, hand for hand, foot for foot, burn for burn, bruise for bruise, wound for wound.[37]

kinder and gentler laws appear, such as:

> You shall not spread a baseless rumour. You shall not make common cause with a wicked man by giving malicious evidence.[38]

And, of course, the Twenty-third Psalm and other beautiful writings of the Old Testament have been part of Western culture for centuries:

> The Lord is my shepherd; I shall want nothing. He makes me lie down in green pastures, and leads me beside the waters of peace; he renews life within me, and for his name's sake guides me in the right path. Even though I walk through a valley dark as death, I fear no evil, for thou art with me, thy staff and thy crook are my comfort. Thou spreadest a table for me in the sight of my enemies; thou hast richly bathed my head with oil, and my cup runs over. Goodness and love unfailing, these will follow me all the days of my life, and I shall dwell in the house of the Lord my whole life long.[39]

Yet, the chilling specter of a ruthless, bloodthirsty Jehovah pervades much of the Old Testament, especially beginning with Exodus and in the books immediately following Exodus. I am willing to concede the possibility that, given the level of morality in the ancient world at the time of the exodus of the Israelites from Egypt (1,500 ~ 1,200 BC.), the Old Testament might represent an overall step up in the moral behavior of humankind. However, I find it hard to believe that the Egyptians, whose first-born children were killed by Jehovah, or the Canaanites and all of the other thousands of people in the "promised land" that the Israelites, under the command of - and with the help of Jehovah, slaughtered in a most merciless and brutal manner, were so wicked as to deserve such a fate from the Lord God Almighty.

The Old Testament is believed to be much edited, with many deletions and additions made to the text by unknown authors for unknown reason for several hundred years after the time of the exodus. The version of the Old Testament contained in *The New English Bible* I have used is but one of many Jewish and Christian versions that have existed and been used over the past two to three thousand years. Beginning at around the middle of the last century, at about the time Darwinism was thrust upon human consciousness, some scholars and others assaulted the traditions and the very credibility of the Old Testament, especially books with historical content. This extreme view has been modified considerably by biblical scholars as the archeological finds made since this extreme view were introduced in the mid-Nineteenth Century have generally tended to support, rather than contradict, the historical narratives contained in the Old Testament.

Here we are following the point of view of Sitchin and others, that the Old Testament, no matter how much it is Hebrewized and otherwise edited, is one of the few historical documents we have of the origins and existence of humankind written in ancient times, and should be respected as such. There is more than enough veracity in the Old Testament to see Jehovah for what he is, or at least what he was during the exodus and the time period of the early occupancy of the promised land. He was a rapacious ET who apparently loved to intervene with the affairs of humans and demanded to be obeyed and worshipped. We have reviewed more than enough of the Old Testament to reaffirm one of the major points of this book:

> We can no longer accept, let alone worship, a superior entity simply because it has considerably more material and metaphysical powers than do we humans! A clearly more advanced being does not necessarily mean that such a being is more spiritually evolved and has the best interests of humankind in mind.

Any physical or non-physical entity, who attempts to divide humankind, and to represent humankind as anything other than variations of a common oneness - a unity of all living beings in the universe derived from a common creator, the Divine Creator, is not representing the best interests of humankind. Jehovah's entry into human affairs at the time of the exodus, plus the continued worship and reverential feelings towards this entity, represents just one more dividing spike that has been driven to separate humanity.

This brings up a curious question about Jehovah and his continued influence in human affairs. Given the clear-cut evidence of the rapacious, unscrupulous, violent nature of the Jehovah entity, at least the early entity of the exodus, why is Jehovah proclaimed by so many millions today, Christians and Jews alike, to be a holy moral leader - a god - of humankind? Much of the answer to the above curious phenomenon has to do with the primitive nature of the human species. We haven't evolved enough, most of us, to be able to determine if a superior being is a god or a God.

Maybe the fact that Jehovah was one of the last ET entities to have so openly and directly demonstrated his incredible powers on a portion of humankind, accounts for the fact that he is still respected and revered more than 3,000 years after the exodus of the ancient Hebrews from Egypt. We have seen the power biological imprinting can have on human consciousness and human belief systems. Part of the answer to Jehovah's continued influence into the Twentieth Century is the strong imprint of the myth of Jehovah as the one lord god established on the ancient Hebrews as a result of their direct interaction with this entity during the exodus and afterwards. This myth has been passed from generation to generation for hundreds and thousands of years by the Jewish cultures, as well as the Christian cultures, which grew out of the Jewish state of Israel beginning about 2,000 years ago. The very fact that the Old Testament makes up more than half of the Christian Bible, speaks of how important the Old Testament has been to Christians. Certainly all Christian sects pay more attention to the New Testament and the teachings of Jesus than to Jehovah and the Old Testament. However, much reverence is given to the god of the Old Testament by Christians, and some Christian sects pay considerable attention to the Old Testament. In fact, as was the case with me in my early years, no distinction is made between the god of the Old Testament and the god of the New Testament among most Christians. In fact, some Christian fundamentalist sects believe all of the Old Testament, every word, is the word of the lord god almighty and must be believed verbatim.

The very first Christians were Jews, of course. A few early Christians that had been Jews, such as Paul, and wrote about Jesus and their interpretation of his teachings, perhaps wanted to paint the Jewish religion and Jehovah in a good light so they could convert more Jews to the new religion. We will see evidence in later chapters that this, in fact, is almost certainly what happened. The strong Jewish imprint of Christianity might explain much of the fascination with Jehovah and the Old Testament that Christians have historically demonstrated.

Of course, it is probable that the biological process of imprinting has been much more solidly "wired" into the generations of Jewish minds, than the minds of Christians. After all, as we have seen, Jehovah impressed upon the ancient Hebrews of the exodus and violent settlement of the promised land to fear god. They were killed during this time if they did not do exactly what Jehovah commanded! This, I would suspect, caused the fear of god and the rituals necessary for the proper worship of him to very strongly imprint on the Hebrew mind.

This whole question of Jehovah's moral authority in the late 20th century would not be more than a matter of obscure academic and scientific interest if it were not for the fact that many of the crucial decisions of the world today are made by people who profess to be worshipers of Jehovah who, they claim, is the Lord God Almighty. Even as I write these words, Jews and Arabs are killing each other in southern Lebanon and northern Israel. As always in such conflicts, thousands of people are suffering extreme psychological and physical pain, not to mention those people who are actually killed. Roughly half of the people of this routine tragic human conflict swear by Jehovah.

Given the evidence we have uncovered of a conspiracy by regressive ETs to keep humankind ignorant and at each others throats, it is highly unlikely that an imprinting and conditioning theory can entirely explain why the entity Jehovah has an inordinately strong influence on the world of the 1990s. This whole subject deserves a much more thorough investigation than has been given here.

It is now time for me to try to assuage the feelings of the Jewish population in America and throughout the world, some of whose feelings I've brutalized. Let us explain once again that it is not one specific religion or belief system that has serious deficiencies. It is virtually all of the many formats or ritualized belief systems of the world I wish to expose as distorted. We, all of humankind, have been tricked into believing many types of reality systems that serve to keep us ignorant and divided. We've all been duped!

We will now move to point out some of the falsehoods and shortcomings of the religion of my childhood, Christianity. Am I anti-Christian? Do I look down upon those who profess and call themselves Christians? Certainly not! Here we are talking about my own family, my mother and my late father, my sisters, my aunts, uncles, and cousins. As with Jewish people and the Jewish religion, just because I am currently not much impressed with the Christian religion as it exists today, does not mean that I do not appreciate the basic loving goodness of many Christians. Let us examine the Christian religion to try to determine how much of it might represent spiritual truths, and how much of it might be distorted dogma.

[1] Sitchin, Z. 1985. *op. cit.*, pp. 281-309.
[2] *Ibid.*
[3] *The New English Bible* 1970. *op. cit.*, Ezekiel 1: 3-5
[4] Shellhorn, G. C. 1989. *Extraterrestrials in Biblical Prophecy*. Horus House Press, Inc., Madison, Wisconsin, p. 80.
[5] *The New English Bible*. 1970. Exodus 11: 3.
[6] *Ibid.*, Exodus 10: 1-3.
[7] *Ibid.*, Exodus 7,8,9,10.
[8] *Ibid.*, Exodus 13: 21.
[9] *Ibid.*, Exodus 14: 4.
[10] *Ibid.*, Exodus 16: 31 and 35.
[11] *Ibid.*, Exodus 19: 16-20.
[12] *Ibid.*, Exodus 20: 3.
[13] *Ibid.*, Exodus 20: 4-6.
[14] *Ibid.*, Exodus 20: 7-11.
[15] *Ibid.*, Exodus 20: 12-17.
[16] *Ibid.*, Exodus 33: 1-11.
[17] *Ibid.*, Exodus 34: 10-16.
[18] Sitchin Z. 1980. *op. cit.*, p. 293.
[19] *The New English Bible* 1970. Numbers 25: 1-5.
[20] *Ibid.*, 1 Kings 18: 39-40.
[21] *Ibid.*, Numbers 32: 10, 11 and 13.
[22] *Ibid.*, Exodus 40: 36-38.
[23] *Ibid.*, Numbers 15: 35-36.
[24] *Ibid.*, Numbers 31: 1-2.
[25] *Ibid.*, Numbers 31: 7-11.
[26] *Ibid.*, Deuteronomy 3: 3-7.
[27] *Ibid.*, Numbers 16: 27-34.
[28] *Ibid.*, Leviticus 25: 17, 36, 43; and Deuteronomy 6: 2.
[29] *Ibid.*, Joshua 5: 12.
[30] *Ibid.*, Joshua 6: 21.
[31] *Ibid.*, Joshua 6: 24.
[32] *Ibid.*, Joshua 7: 26.
[33] *Ibid.*, Joshua 8: 25.
[34] Boulay, R. A. 1990. *op. cit.*, pp. 70-73.
[35] Marciniak, B. 1990. *op. cit.*, p. 37.

[36] Stevens, W. 1988. *op. cit.*, pp. 129-130; and W. Stevens 1990. *op. cit.*, pp. 60-61.
[37] *The New English Bible* 1970. Exodus 21: 24-25.
[38] *Ibid.*, Exodus 23: 1.
[39] *Ibid.*, Psalm 23.
[40] Marciniak, B. 1992b. *op. cit.*, pp. 32-33.

"Seek and ye shall find, knock and it shall be opened unto you."
                                                               Holy Bible

Chapter 17

Christianity and the Teachings of Jesus

One of the hypotheses of this book is that all of the major and minor (minor in terms of the number of human followers) religions of the world have either been founded and/or distorted and/or otherwise influenced by different groups of ETs. In the last chapter we saw that there is considerable evidence from the Old Testament that an ET entity, Jehovah, interacted directly with the ancient Hebrews, and founded the Jewish religion. In this chapter we will deal with Christianity. We will examine some of the possible ET influences in Christianity in the next chapter, but now let us review some historical data that indicates that the teachings of Jesus, to an enormous extent, have been changed and distorted, and "teachings" added.

Almost all of the information we have about the life of Jesus and his teachings are recorded in the four Gospels of the New Testament, Matthew, Mark, Luke and John. The gospels are not believed by Biblical scholars to have been written until some time between 65 A.D. and 125 A.D., at the earliest, long after Jesus's disciples and all those who had known him, had died. In other words, there is no official account of Jesus written by a person who actually knew Jesus, and was a participant in the dramatic events surrounding his life. Right away, we can see the potential for distortion. Even if the writers of the official gospels took much of their material from a "proto-gospel" that had been written during the time of Jesus, as some scholars believe, there was much room for considerable alteration and rewriting of any "proto-gospel" that may have existed. Many scholars of the Bible have expressed opinions that this may have been the case. For example, the compiler of Matthew is believed by many scholars to have been a Jewish scribe who was converted to Christianity, and who, therefore, altered the sayings of Jesus in order to make the Jews look better than they had been depicted in a proto-gospel.[1] Some scholars contend that the Jewish writer of Matthew added the depiction of Jesus as the fulfillment of the prophecy of the Jewish religion, as the Messiah, or "Anointed One" of the lineage of David.[2]

The writings of the New Testament that were actually written down first were the Epistles, or letters, of Paul. Paul was the son of an upper class Hebrew Pharisee. The Pharisees were a conservative Hebrew religious party who wished to preserve the "Hebrewness" of the Hebrew culture. The early Christians were anathema to young, upper class, conservative Hebrews like Paul. Paul was apparently so zealous in his hatred of the small group of Christians that existed in the years shortly after Jesus' crucifixion, that he appealed to the high priest for special permission to persecute the followers of Jesus that lived beyond the Hebrew religious center of Jerusalem. On the road to Damascus, where he wished to carry out such a persecution of Christians, Paul experienced his famous, dramatic conversion. Not only did Paul become a Christian, but soon, he "became intoxicated by the image of himself in the role of the spiritual leader of a gigantic movement of the future."[3]

Paul traveled throughout much of the Roman Empire preaching, trying to convert people to Christianity. After visiting and preaching in a place, Paul often wrote letters to the converts of this place as he traveled throughout the land. These letters, plus the "Acts of the Apostles," believed to have been written by Paul, make up almost half of the writings of the New Testament. Paul, as much as anybody, was responsible for the founding of the "Church," that we know as the Roman Catholic Church. That is why the Catholics revere him and have canonized him as Saint Paul. By the Fourth Century A.D., the basic structure of the Catholic Church had been established, with a number of bishops, each bishop being the monarch of the church in his city, with the bishop of Rome claiming to be the head of all the churches. The organization of the very early church are obscure, but there is evidence that the basic organization of the church with bishops and a head bishop (Pope) in Rome, date from the Second Century A.D. Long before the church had become established in Rome, Paul and other early Christians were apparently engaged in a bitter debate concerning the interpretation of the teachings of Jesus. In Paul's second letter to the Corinthians he writes:

> ...I am jealous for you, with a divine jealousy, for I betrothed you to Christ, thinking to present you as a chaste virgin to her true and only husband. But, as the serpent in his cunning seduced Eve, I

> am afraid that your thoughts may be corrupted
> and you may lose single-hearted devotion to
> Christ. For if someone comes who proclaims an-
> other Jesus, not the Jesus whom we proclaimed,
> or if you then receive a spirit different from the
> Spirit already given to you, or a gospel different
> from the gospel you have already accepted, you
> manage to put up with that well enough...[4]

Paul goes on to urge the Corinthians to stay with the Jesus he gave them, not some new Jesus. The main "new Jesus" Paul opposed, was the Jesus of the Gnostics. The Gnostics were a group of Hebrew mystics that accepted and adopted the teachings of Jesus. Several Biblical scholars speculate that Jesus himself and his family may have been members of a Gnostic sect, known as the Essenes. The Essenes claimed that an individual by himself or herself could come to a Christ-like state of perfection and be in possession of "knowledge" and spirit.[5] The Essenes opposed any elaborate hierarchy, such as priests, bishops and a Pope. Not only was such a hierarchy unnecessary, the Essenes believed, but realizing one's own divinity might be more difficult with such a hierarchy coming between an individual and the Divine.

Clearly, Paul and his allies won the conflict between the Essenes and others who may have had a different interpretation of the teaching of Jesus, because much of the New Testament consists of his writings. The Essenes and others who opposed "Paulism" faded from the scene as the Church was established.

Paul in writing his many letters does not mention even one of the parables of Jesus. Paul does not use the actual teaching of Jesus at all, but rather spreads his own ideas and philosophy. Considering Paul's Hebrew background, it is not surprising that he characterized all people as subject to the wrath of God.[6] He stated that all people are quite lost and under the power of Satan, and he hung a sentence of damnation over everybody. Holger Kersten, a German Lutheran theologian, writes concerning the subject of death:

> Thus Paul as a human teacher made out of joyous
> tidings his threatening tidings and implied that
> only he could show the path to salvation. Of
> course, with such an attitude one can hardly ar-
> rive at a natural view of death, for it makes death
> a solution to sin.
>
> In no other religion we do (sic) find such cultiva-

tion of the fear of death as in Pauline Christianity. With Paul Christianity became a religion in which Christians, beset by fears, would bow docily under the yoke of threats. The religion was already *veering away from the concept of the kind and loving, all-forgiving God of Jesus, and reverting to the crudities of the wrathful Old Testament God...*[7] (his emphasis)

It is astonishing that one forceful, literate man could have had such an enormous influence on the belief-systems of so many millions of people who practiced some degree of Christianity in their lives over the centuries. There have been several critics who regard Christianity as the work of Paul with Jesus being only symbolic. Several other critics, such as Kersten quoted above, regard Paul as a perverter of Jesus' teachings. Without going into this matter further, I simply wish to emphasize that one man can, and apparently did, have an extraordinary influence on our perception of the spiritual teachings of Jesus.

The church was a sect of martyrs for the first 250 years of its existence. The Roman Empire persecuted all of those who did not follow the Roman pagan belief-system. Every Christian today knows that occasionally early Christians were fed to the lions in Rome.

Then the Christians found a friend in the Emperor Constantine (c.274 A.D. to 337 A.D.), often known as "Constantine the Great" by Christians. Constantine was tolerant towards Christianity, and in 313, through the Edict of Milan, forbade the persecution of all forms of monotheism in the Empire. However, contrary to tradition, Constantine did not become a Christian himself and did not make Christianity the official state religion of Rome. Constantine was, in fact, a member of the cult of *Sol Invictus*, which was a form of pagan sun worship. The cult of *Sol Invictus*, which was Syrian in origin, was a form of monotheism.[8]

The Church of Rome soon took advantage of their new found freedom from persecution and began to openly organize themselves. They organized the Council of Nicea in 325 A.D., where Rome was declared the official center of Christian orthodoxy.

The Church of Rome held a number of ecumenical councils in the first 1,000 years of its existence, in which the church leaders greatly modified and changed the teachings of Jesus. In this very first ecumenical council, the Council of Nicea, the divinity of Jesus and the precise nature of his divinity, was established by means of a vote among church leaders. Jesus himself never claimed to be an omnipo-

Christianity and the Teachings of Jesus

tent, all-powerful, single god, or the son of god, but he has become one in the Christian religion, beginning in Nicea. Many of the writings of early Christians which contradicted the radical changes that were made at the Nicean Council, were simply eliminated from Christianity forever. For example, any writings by the Essenes or any other group that denied the supremacy of the Church of Rome, were not only banned, but attempts were made to destroy them altogether. Fortunately, many of these early writings survived the purge of the early church fathers, and, as we shall see, give us some clues as to the real nature of the appearance on Earth of the man called Jesus. M. Baigent, R. Leigh and H. Lincoln, authors of *The Messianic Legacy* and other carefully researched books concerning the life of Jesus and the development of Christianity, state that "...Christianity as we know it today, derives ultimately not from Jesus' time, but from the Council of Nicea..." [9]

The Council of Constantinople of 553 AD. eliminated the concept of reincarnation from Christianity. Paul himself, when he had been known as Saul and was a Hebrew Pharisee, apparently believed in resurrection, not reincarnation. Resurrection is based upon the idea that the body is paramount, and the soul, a nebulous spiritual body, was attached to it. After death, the soul will exist in a heaven or hell or something in between, depending upon the merits of the life one has lived on Earth. With reincarnation, on the other hand, the soul of the physical body is the essence of the personality of an entity that attaches to one body after another in repeated lifetimes in order to learn spiritual truths. Closely associated with the concept of reincarnation, is the concept of karma. The concept of karma, put very simply, is that one reaps what one sows over one's lifetimes. For example, if one spends most of a lifetime bringing harm to others, this same "soul" will find itself in later lifetimes in a position where others will do harm to it. Of the modern religions, only Hinduism incorporates the concept of reincarnation, and Buddhism has a very similar view. Many ancient civilizations were familiar with the concept of reincarnation, including the Celts (the Druids), the Chinese, the Greeks, the Egyptians and in some respects, the Norse.[10] Through Paul and others, the concept of resurrection was already prevalent among Christians by the time the Council of Constantinople convened in 555A.D.

According to Kersten, the elimination of this important doctrine by the Council of Constantinople might have been mostly due to a whim of the wife, Theodora, of the Emperor Justinian. Kersten presents evidence that Justinian's ambitious wife began her career as a courtesan. In order to cover-up her "shameful" past, she as the Emperor's wife, ordered the death of five hundred of her earlier colleagues. According to the concept of karma, she knew that she would have to suffer the consequences of her deeds in future lifetimes, so she, with the help of her Emperor husband, set about to have the whole concept of rebirth abolished![11]

Could it be that important teachings of Jesus and other great spiritual teachers were so arbitrarily eliminated from a developing world religion? Of course, when the Council of Constantinople abolished this concept, they would have taken pains to eliminate all references to reincarnation from the Bible. The censors, however, missed eliminating every reference to reincarnation, as Jesus, in Matthew 11:14, states that John the Baptist, the prophet that was the herald of Jesus, had been the Jewish Ninth Century B.C. prophet Elijah. "...For all the prophets and the Law foretold things to come until John appeared, and John is the destined Elijah, if you will but accept it..."

An example of adding to the teachings of Jesus is seen in what was accomplished at the ecumenical councils of Lateran of the 12th century. There the Church fathers decided to add the tenet of the "Holy Trinity." With this addition, Jesus is further elevated to the status of "God" by becoming the "Son of God," a concept that was never taught by Jesus.[12]

In addition to the numerous ecumenical councils of the Church of Rome that "officially" determined Christian beliefs and modified the teachings of Jesus, there developed a number of splits among Christians that resulted in further modifications of beliefs over the centuries. The first major split and division of Christianity occurred when the eastern portion of the Church, which was centered in Constantinople, split from the Church of Rome. Ecclesiastical animosity developed between Eastern and Western Christendom until they split into two different churches by the 9th Century A.D. This split attained an official, permanent status in 1054 A.D., when Pope Leo IX excommunicated the patriarch of Constantinople, Michael Cerularius, for verbally attacking the pope. What we know today as the Eastern Orthodox Church of eastern Europe and western Asia has had a long history of physical and ideological separation from the Church of Rome.

Since this first major division, there have been many more divisions of Christianity. There developed a veritable explosion of divisions during the Protestant Reformation movements of the Sixteenth Century. The first split occurred in Germany, led by an Augustinian monk, Martin Luther (1483 - 1546 A.D.). Luther had become disillusioned with the church on account of the degree of spiritual laxity and corruption he perceived in the Church. He was especially upset with the manner the church was granting indulgences. Indulgences in the early 16th Century church were seen as a pardon of sins; a pardon of sins committed on Earth were forgiven "in the eyes of God", so said the church, in exchange for money paid to the church. Led by Luther in the first few decades of the 16th Century, Lutheranism was founded with the help of many German nobles, and has been a major "branch" of Protestantism to the present day.

Another split with Roman Catholicism formed in Switzerland about the same time Lutheranism was being born in Germany. Huldreich Zwingli (1484 - 1531 A.D.) and later John Calvin (A.D.1509 - 1564) led the Swiss Protestant reformation that has had a huge impact on the development of the Protestant movement. However, the Swiss and German Protestant movements could not come together because Luther disagreed with Zwingli, and then Calvin, over the ritualistic observance of the Last Supper of Jesus, the sacrament of the Eucharist. While the Swiss insisted upon regarding the sacrament as only being symbolic, the Lutherans retained the belief that celebrating the sacrament was a mystical undertaking in which the body and blood of Christ join with the bread and wine used in the service. This and other differences between Luther and Zwingli and Calvin, divided early Protestantism into the Lutheran Church and the "Reformed Churches."

Zwingli, in a general council of the church held in Zurich in 1523, argued for acceptance of the Bible as the sole basis of the truth, and the denial of practices and authority of the church not included in, or contrary to the scriptures. Zwingli was the first Protestant to refer to the Bible as the sole basis of the truth. It is common among Protestant fundamentalists in the United States today to claim the Bible to be inviolable as the word of God. One wonders whether Zwingli, or American Christian fundamentalists today knew and know of the heavy editing job the Hebrews did on the Old Testament and the Pauline Catholics did on the New Testament?

Many Protestant sects that have developed since the beginning of the 16th Century have not always based their splits and divisions on ecclesiastical concerns. A primary motivation that led to the founding of the Church of England (which eventually led to the development of the Episcopal Church of the United States), was the displeasure King Henry VIII felt toward the pope because the pope refused to grant him a divorce from his Spanish Catholic wife.

So we have seen in Christianity one division of humans after another since the time of Jesus. All of these divisions might not matter much if it were not for the pain, suffering and death caused by the wars so-called Christian men have fought against other so-called

Christian men, partially in the name of one religious cause or another. One such example was the Thirty Years War (1618 - 1648), fought in Europe, in which Catholics and Protestants slaughtered each other.

What about the life and teachings of Jesus? How much do Christians and non-Christians alike know today about the man who lived in Palestine, after 2,000 years of political and religious modification and editing of his views? Some scholars estimate that no more than five percent of the actual teachings of Jesus are found in the Bible and have survived to the present day.[13] Baigent et al. have written a radical interpretation of the life of Jesus and the beginnings of Christianity based on their interpretation of the historical evidence.[14] Baigent et al. use the standard Christian historical writings as well as apocryphal writings (hidden writings) of the time, the writings that were rejected and suppressed at the early ecumenical councils of the Church. Baigent et al. talk of the fairy-tale nature in which Jesus and his ministry are presented in the canonized Gospels of the New Testament:

> The Gospels are documents of a stark, mythic simplicity. They depict a world stripped to certain bare essentials, a world of a timeless, archetypal, almost fairy-tale character...[15]

The authors point out that most Christians would rather believe the fairy-land, Disneyland-like version of Jesus presented as in the Gospels of the Bible. They point out that Biblical scholars, often using newly-found documents, such as the Dead Sea Scrolls discovered in 1947, have sometimes come up with conclusions as radical about Jesus and the beginnings of Christianity as they have, but these radical ideas have not reached the public-at-large.

Baigent et al. use a tremendous amount of historical documents, plus a strong dose of their educated speculation, to present a Jesus who was a more politically oriented Hebrew Messiah, than the mystical, mythical Jesus presented in the Gospels. Jesus, according to Baigent et al., after preaching that he was the Hebrew Messiah of the House of David, set out to restore the Hebrew State, which had been seized and become a territory of the Roman Empire in 63 B.C. The Jesus of Baigent et al. was a married man with children, who did not die on the cross during the crucifixion, but survived and possibly escaped to Europe with his family after the crucifixion. Much of the god-like, mythical nature of Jesus as depicted in the Bible, according to these authors, is a result of the Jesus of the world religion Paul and others were trying to create, incorporating some of the mythical characteristics of the gods of other religions that already existed, with which Jesus and Christianity were to compete. For example, many of the gods of the religions Paul's version of world-wide Christianity were competing with were supposed to have been given birth by a virgin. One of the religions in particular, Baigent et al. believe, that had a particularly strong influence on the formation and coalescence of Christianity, was known as Mithraism. Mithraism was the religion of the cultic god of ancient Persia and India, Mithra, who is dated back to the 6th century B.C. Mithraism became a world religion and was one of the most popular religions of the Roman Empire until the 2nd century A.D. Baigent et al. write that Mithraism,

> ...postulated an apocalypse, a day of judgment, a resurrection of the flesh and a second coming of Mithras himself, who would finally defeat the principle of evil. Mithras was said to have been born in a cave or a grotto, where shepherds attended him and regaled him with gifts. Baptism played a prominent role in Mithraic rites. So, too, did the communal meal...[16]

Needless to say, the works of Baigent et al. have not been enthusiastically received by those who have strong beliefs based on the traditional view of Jesus and Christianity. I am not endorsing their views, as there are many areas of their quite complex, but fascinating views of Jesus and early Christianity that are simply not known. The views of Baigent et al., however, along with the other information provided in this chapter, strongly suggest that the actual appearance of Jesus and Christianity on this planet may have happened in a radically different way than is depicted in the New Testament of the Bible. Historical evidence indicates that the early New Testament is the product of the adding and editing of a small group of Paulites, who were later known as Roman Catholics, as they tried to mold and maintain a world religion that would follow their views and be followed by most of the people who lived in the eastern and central Mediterranean world shortly after the public ministry of Jesus. Historical evidence indicates that modern-day Christianity has little to do with an actual man (or men?) who may or may not have taught spiritual truths in Palestine almost 2,000 years ago.

What about one of the main hypotheses of this book - that all current world religions were either founded by, or distorted by various groups of ETs that were positively or less-than-positively inclined toward humankind? Let us turn to the next chapter to review some of the possible ET connections to Christianity.

[1] For example, Beare, F. W. 1981. *The Gospel According to Matthew*. Harper and Row, San Francisco.
[2] Kersten, H. 1986. *Jesus Lived in India*. Element Book Ltd., Shaftesbury, England, p. 25.
[3] *Ibid.*, p. 28.
[4] *The New English Bible* 1971. *op. cit.*, 2 Corinthians 11: 1-4.
[5] Borkamm, G. 1960. *Jesus of Nazareth*. Harper, New York, p. 71.
[6] *The New English Bible* 1971. *op. cit.*, Ephesians 2: 3.
[7] Kersten, H. 1986. *op. cit.*, p. 213.
[8] Baigent, M., R. Leigh and H. Lincoln 1986. *The Messianic Legacy*. Dell Publishing, New York, pp. 39-43.
[9] *Ibid.*, p. 40.
[10] Deardorff, J. W. 1990. *Celestial Teachings*. Wild Flower Press, Tigard, Oregon, p. 45.
[11] Kersten, H. 1986. *op. cit.*, p. 215.
[12] Bramley, W. 1990. *op. cit.*, p. 133.
[13] *Ibid.*
[14] Baigent, M., R. Leigh and H. Lincoln 1983. *Holy Blood, Holy Grail*. Dell Publishing; and Baigent, Leigh and Lincoln 1986. *op. cit.*
[15] Baigent, Leigh and Lincoln 1986. *op. cit.*, p. 17.
[16] *Ibid*, p. 79.

> "I am fully convinced that the soul is indestructible, and that its activity will continue through eternity. It is like the sun, which, to our eyes, seems to set in night; but it has in reality only gone to diffuse its light elsewhere."
> Goethe

Chapter 18

Possible ET Influences on Christianity

There have been several "UFO"-oriented books written, beginning in the 1950s, that have suggested that Jesus was either a messiah-ET himself or, at least, was working very closely with ETs of another planet. Most of the authors of these books have made such suggestions only in passing, and their books are now out of print and difficult to obtain.[1] No professional, academic Biblical scholar, to my knowledge, has even touched the interpretations of the Bible suggesting strong ET involvement in Biblical affairs, as this would lead to professional suicide. Nevertheless, there is much evidence that there has been a very strong ET influence in the formation of the New Testament, which, as we have seen, was the case in the formation of the Old Testament.

Much of the evidence that Jesus was involved with ETs is centered around his purported virgin birth. G.C. Schellhorn who has written an extensive account of what he sees as extraterrestrial influences in the Bible, in *Extraterrestrials in Biblical Prophecy*, sees two unusual births recorded in the Bible, besides the virgin birth of Jesus, as having been the result of genetic manipulation of some sort by ETs. The first, Schellhorn claims, was the birth of Isaac to Sarah and Abraham in Genesis. Sarah and Abraham were approaching 100 years of age, and Sarah had been barren all her life and had not been able to give Abraham an heir. As we have already seen, Abraham was visited by three men, or the Lord and two angels, or representatives of the Anunnaki of Sumer, or a spaceship captain and two crewmen, or some other brand of ETs. Abraham immediately recognized them as being special when he saw them:

> ...when he saw them, he ran from the opening of his tent to meet them and bowed low to the ground. "Sirs," he said, "if I have deserved our favor, do not pass by my humble self without a visit. Let me send for some water so that you may wash your feet and rest under a tree; and let me fetch a little food so that you refresh yourselves..."[2]

When the gods were sitting and enjoying Abraham's hospitality, they asked about his wife Sarah, and one of the strangers said:

> ...About this time next year I will be sure to come back to you, and Sarah your wife shall have a son...[3]

Sarah, who had been listening from a nearby tent, laughed at this because she was well past childbearing age. Then one of the visitors, the one Abraham recognized as "God," said, essentially, that an act causing an old woman to give birth to a son is easy for "God."

> The Lord said to Abraham, "Why did Sarah laugh and say, 'Shall I indeed bear a child when I am old?' Is anything impossible for the Lord? In due season I will come back to you, about this time next year, and Sarah shall have a son."[4]

A year later Sarah did have a son, Isaac, even though she was well past childbearing age.

Zechariah, the "father" of John the Baptist, while attending to one of his priestly duties, there

> ...appeared to him an angel of the Lord, standing on the right of the altar of incense. At this sight, Zechariah was startled, and fear overcame him. But the angel said to him, Do not be afraid, Zechariah; your prayer has been heard: your wife Elizabeth will bear you a son, and you shall name him John.[5]

The angel told Zechariah that his son would be filled with the spirit of the Lord. Zechariah, however, had doubts: "Zechariah said to the angel, 'How can I be sure of this? I am an old man and my wife is well on in years.'"[6] Zechariah was struck dumb for a period of time by the

Possible ET Influences on Christianity

angel for doubting that his wife would have a baby. We are told little more in the Bible except that his wife did conceive and give birth to John the Baptist.

The Bible provides rather more evidence of a possible genetically manipulated birth in the case of Jesus to a young girl and a virgin. An angel was sent "from God" with a message for a young girl, Mary, who was betrothed to a man named Joseph:

> "Greetings, most favoured one! The Lord is with you." But she was deeply troubled by what he said and wondered what this greeting might mean. Then the angel said to her, "Do not be afraid, Mary, for God has been gracious to you; you shall conceive and bear a son, and you shall give him the name Jesus. He will be great; son of the Most High; the Lord God will give him the throne of his ancestor David, and he will beking over Israel forever; his reign shall never end." "How can this be?" said Mary; "I am still a virgin." The angel answered, "The Holy Spirit will come upon you, and the power of the Most High will overshadow you; and for that reason the Holy child to be born will be called, Son of God. Moreover your kinswoman Elizabeth has herself conceived a son in her old age; and she who is reputed barren is now in her sixth month, for God's promises can never fail." [7]

The New English Bible has a footnote to explain that this last passage is also "read" or could be translated as, "for with God nothing will prove impossible".

Something as simple as artificial insemination, or in vitro insemination, where a female ovum is nurtured and fertilized in a test tube, then transplanted to the womb of Mary, certainly would not have proved impossible to ETs. As we have discussed before, such technology has been available to modern humans since the 1970s and would have surely been available to ETs who had mastered space travel.

From the apocryphal writings, the writings that were eliminated by the Pauline Church of Rome during their various councils, we get another version of Gabriel's visit that gives a better picture of a possible in vitro, or artificial insemination of Mary. For example, in the Gospel of Mary, after Mary was told by Gabriel that she would give birth:

> She said, How can that be? For seeing, according to my vow [of chastity], I have never had sexual contact with any man, how can I bear a child without the addition of a man's seed?[8]

To this the angel replied and said,

> Think not, Mary, that you will conceive in the ordinary way. For without sleeping with a man, while a Virgin, you will conceive and while a Virgin you will give milk from your breast. For the Holy Ghost will come upon you, and the power of the Most High will overshadow you, without any of the heats of lust. So that to which you will give birth will be only holy, because it only is conceived without sin, and being born, shall be called the Son of God...[9]

The esoteric source, Pleiadian Plus, states and implies that a radically different genetic pattern - a radically different DNA pattern than most humans have now, is characteristic of entities who are more "spiritually evolved." Perhaps the particular group of ETs involved with the birth of Jesus was making sure that the genetics of Jesus were right for a man they wished to become a teacher of the Light to humans. Certainly, in vitro insemination of a human woman by a species of humanoids for whom such inseminations are well known in their own culture would be a way to be certain of the genetics of the baby that was to be born of an Earth woman.

The reader undoubtedly noticed some of the differences of the two very similar passages quoted above, one from a canonized version of a gospel of Jesus, and one an excised version. While the story or event related is the same - the initial visit of the angel Gabriel to Mary - subtle differences in wording might yield profound differences for the religion that would become Christianity, as we noted in the last chapter. The major difference is the emphasis in Luke of Jesus being of the lineage of David, or of the Hebrew lineage of messiahs, which meant a lineage of holy secular kings of Israel to the Hebrews of the time of Jesus. The suppressed and excised Gospel of Mary did not have this emphasis.

The evidence that the birth of Jesus was either caused or greatly influenced by at least one group of ETs goes beyond the possible in vitro, or artificially inseminated birth to a young virgin. When Joseph went to meet his bride-to-be and learned that she was already pregnant, naturally he wanted to call off the marriage and have nothing to do with Mary. However, as Joseph was contemplating calling off the marriage,

> an angel of the Lord appeared to him in a dream. "Joseph son of David," said the angel, "do not be afraid to take Mary home with you as your wife. It is by the Holy Spirit that she has conceived this child. She will bear a son; and shall give him the name Jesus (Savior), for he will save his people from their sins. All this happened in order to fulfill what the Lord declared through the prophet: 'The virgin will conceive and bear a son, and he shall be called Emmanuel', a name which means 'God is with us.' " Rising from sleep Joseph did as the angel had directed him; he took Mary home to be his wife, but had no intercourse with her until her son was born. And he named the child Jesus.[10]

The official and unofficial accounts of Jesus tell us that he was born in Bethlehem under auspicious circumstances. Of the canonized gospels, Mark and John have no account of Jesus' birth in Bethlehem, Matthew makes but a very scant reference to the fact "...Jesus was born at Bethlehem in Judea...", while from Luke we learn that Mary in Bethlehem, "...gave birth to a son, her first-born. She wrapped him in swaddling clothes, and laid him a manger, because there was no room for them to lodge in the house."[11] So it has come down to us that Jesus was born in an animal stall where Jesus was placed in a manger. Some apocryphic accounts of Jesus' birth, however, state that Jesus was born in a cave. Bramley quotes from a work that was eliminated from Christianity by the early church fathers, a book called *Infancy*:

> And when they came by the cave, Mary confessed to Joseph that her time of giving birth had come, and she could not go on to the city, and said, Let us go into this cave. At that time the sun was nearly down. But, Joseph hurried away so that he might fetch her a midwife; and when he saw an old Hebrew woman who was from Jerusalem, he said so her, Please come here, good woman, and go into that cave, and you will see a woman just ready to give birth. It was after sunset, when the old woman and Joseph reached the cave, and they

> both went into it. And look, it was filled with lights, greater than the light of lamps and candle, and greater than the light of the sun itself. The infant was then wrapped up in swaddling clothes, and sucking the breast of his mother Saint Mary.[12]

Bramley believes this evidence of high-tech lighting is evidence for the presence of high-tech people at Jesus' birth.

The evidence for the association of high-tech, or highly-evolved people with Jesus' birth is also found in the star of Bethlehem. Again in Mark and John there are no accounts of the star. Luke recounts a visit to nearby shepherds by angels amongst a lot of light, - "...the splendor of the Lord shone round them..." - but no mention of a special star. In Matthew we learn that a special star guided astrologers (*The New English Bible* translation) from "the east" as far as Jerusalem. There they made inquiries as to where the child who would be "king of the Jews" was located saying, "...we observed the rising of his star, and we have come to pay him homage..."[13] This news alarmed Herod, the puppet king of Judea who was reigning at the time of the birth of Jesus. Herod consulted his own "...chief priests and lawyers of the Jewish people...," and summoned the astrologers from the east to ask them what they knew of the birth of a new king of the Jews and the star that had appeared. Told to check in on the child and report back to Herod, the astrologers

> ...set out at the king's bidding; and the star which they had seen at its rising went ahead of them until it stopped above the place where the child lay. At the sight of the star they were overjoyed...[14]

The astrologers entered the house over which the "star" was apparently hovering, found the baby Jesus, left gifts, and

> ..being warned in a dream not to go back to Herod, they returned home another way.[15]

Scientists, of course, dismiss the tale of the star of Bethlehem as a myth, suggesting that a very bright exploding star, or some other "natural" phenomenon can account for the star that supposedly led the wise men of the east. Most Christians over the ages have believed that the star was created by God and was a supernatural phenomenon. Since the dawning of the "UFO-age," several UFO writers have speculated that the star was a sort of UFO often seen today, - a "flying saucer" complete with hovering ability. Schellhorn points out that, new

> ...bright stars don't appear in the sky. But a spacecraft can. Stars don't lead people. They do seem to move as the Earth rotates on its axis. But they don't come to a stop and hover over a town and a manger. A spacecraft can...[16]

Bramley argues that apocryphal writings give the strongest support that the star of Bethlehem was some sort of ET craft. He quotes an apocryphal book known as Protovangelion, where one of the astrologers, or wise men, states:

> We saw an extraordinarily large star shining among the stars of heaven, and so outshined all the other stars, that they became not visible...So the wise men began their travel, and look, the star which they saw in the east went before them, until it came and stood over the cave where the young child was with Mary his mother.[17]

Certainly these stories of the conception and birth of Jesus, as well as the subsequent flight of Joseph, Mary and the baby Jesus to Egypt to escape the cruelty of Herod, during which they were guided by "angels," give more indication of ET involvement with the Jesus phenomenon. Those who do accept the possibility that the Jesus drama was closely linked to at least one group of ETs, fall into roughly two categories. First, there are those who believe that the whole Jesus phenomenon was "created" by regressive ETs in order to further divide humankind along religious lines and to distort spiritual truths. Then there are those who believe that Jesus was a product of spiritually advanced, progressive ETs, and that his life and teachings were subsequently distorted almost beyond recognition by regressive ETs with the help of humans they misled.

As we have seen, the fact that the teachings and life of Jesus were distorted and twisted is well-established. We have seen that modern scholars have interpreted the accounts of Jesus that are available to us today, as indicating anything from the concept that Jesus was a very human, a mostly political Jewish messiah[18], to the concept that Jesus was the Son of God Almighty, the supreme deity. Certainly it could be argued, and has been argued, that humans, filled with their own sometimes passionate agendas, do not need any help from regressive ETs to distort anything that might smack of spiritual truths. However, we have seen evidence that, in fact, ETs have been involved with many, if not most, historical events of human history.

We have seen both terrestrial and esoteric sources that claim that regressive ETs were involved with the Jesus drama. In Chapter Twelve we briefly reviewed evidence that the members of the U.S. secret government learned that the ETs, known as the greys, claim to have been secretly manipulating human affairs through the use of secret societies, occult practices, and organized religions! Cooper writes that when the secret government learned how the aliens were deceiving humanity, when they discovered the hologram nature of the events in Fatima in Portugal in the early part of this century, the "aliens showed a hologram, which they claimed was the actual crucifixion of Christ." [19] The secret government filmed this hologram, according to Cooper! The Pleiadian Plus group has also stated that one of the realities - one of the parallel realities - of the crucifixion of Jesus was a holographic insert! [20]

As for the distortion of any true spiritual teachings Jesus may have taught, I know of no direct evidence that certain humans may have had transmissions given to them from ETs that would result in such distortions. Since Paul was so instrumental in the formation of the Pauline Church of Rome that seems to have altered and even subjugated any spiritual teaching of Jesus, one might suspect that Paul or one of his close followers may have been in contact with regressive ETs. I know of no evidence that he was channeling information from progressive or regressive entities, but a passage from I Corinthians suggests he knew about such transmissions, and that he valued them. He advised his Corinthian converts about the "gifts of the spirit":

> In each of us the Spirit is manifested in one particular way, for some useful purpose. One man, through the Spirit, has the gift of wise speech, while another, by the power of the same Spirit, can put the deepest knowledge into words. Another, by the same Spirit, is granted faith; another, by the one Spirit, gifts of healing, and another miraculous powers; another has the gift of prophecy, and another ability to distinguish true spirits from false;...[21]

One can't help but wonder, given the highly hierarchical church he founded and the Gnostic interpretations of Jesus' teachings he helped suppress, if Paul himself knew how to distinguish true "spirits" from

Possible ET Influences on Christianity

false "spirits."

There is evidence that the cessation of Christian persecution by the Roman Empire and the subsequent consolidation of secular power by the Church of Rome may have been influenced by "supernatural" entities. We have already reviewed how Constantine I, in 313, forbade the persecution of all forms of monotheism. In 312 Constantine and his followers were fighting Maxentius and his followers for the imperial throne of Rome. Shortly before the crucial Battle of Milvian Bridge in which Maxentius was killed and his forces defeated by the forces of Constantine, Constantine is said to have had a vision of a luminous cross suspended in the sky. On this cross the sentence 'In Hoc Signo Vinces ('By this sign you will conquer') was inscribed. Constantine later had a dream of this vision. Schellhorn argues that prophetic dreams of humans in the Old and New Testaments are probably some sort of telepathic transmissions from ETs[22], and we have reviewed esoteric evidence of the same. After this vision and dream, Constantine is said to have ordered the sign of the cross to be put on the shields of his troops. The subsequent victory by Constantine's army was viewed by early Christians of the church as a miraculous triumph of Christianity over paganism inspired by God.[23] But remember, it was not necessarily a victory of all Christians, but rather a victory of the Church of Rome, as it was already by this time well organized. The church taking advantage of its newfound freedom, organized the first ecumenical council in Nicea in 325 and began the process of "officially" determining the nature of Jesus. This and subsequent councils "officially" eliminated Gnostic interpretations of the teachings of Jesus and canonized the "supremacy" of the Church of Rome. In the years following Constantine's vision and dream, Jesus "officially" became "God," who could only be reached through "faith" and the Church, not simply an advanced spiritual teacher, whose teachings could be known and understood through one's own personal effort.

We have already reviewed evidence that the Mormon Church was established by regressive ETs (Chapter 11) with the help of a hologram, purported to be similar to those of the hologram of the crucifixion of Jesus and the holographic visions of the children of Fatima reported by Cooper, as well as, possibly, the vision of Constantine I, in 312. Judging from the evidence we have managed to uncover, it is probable that all of the numerous divisions of Christianity were introduced by the greys and other regressive ETs through holograms, through dreams and other forms of "inspiration," and through other forms of their advanced technology of which we humans are not even aware.

Possible ET Influences on Christianity

The True Teachings of Jesus?

Every now and then a new version of the life and teachings of Jesus appears that is said to be more accurate and more complete in some way than the canonized gospels of the New Testament. For example, in the 19th century there appeared a book written (channeled) by someone named Levi called *The Aquarian Gospel of Jesus the Christ*. This *Aquarian Gospel* was said to be an accurate and complete rendition of the Jesus drama because it was taken from the Akashic Records, which are said to be a secret, mystical record of everything that has ever happened on Earth. There are no historical records of the life of Jesus aside from the canonized and apocryphal gospels and, therefore, no way to check the authenticity of the *Aquarian Gospel*, or any of the supposedly "more accurate" versions of the life of Jesus. However, many people have said or written that the *Aquarian Gospel* contains many "powerful" spiritual teachings that have inspired them very much. It is reasonable that the authenticity of the details of Jesus' life are not really so important so long as his written teachings inspire some people and lead them towards a more harmonious life. It is in this spirit that I review another, quite recent version of the "true" life and teachings of Jesus that has a decided ET slant.

This book is entitled, *The Talmud of Jmmanuel* (pronounced, Immanuel), which claims to be the true testament of Jesus (Jmmanuel).[25] This book claims that the man we call Jesus was conceived in vitro by benign Pleiadians from the planet Erra in this star cluster and implanted in Mary's womb. According to *The Talmud*, Jmmanuel (Jesus) then, as an adult, carried out his short spiritual ministry in the manner that it is generally depicted in the canonized gospels. *The Talmud* claims that the canonized gospels have been heavily edited and contain so much spiritual misinformation that the true spiritual teaching of Jmmanuel (Jesus) can hardly be recognized. *The Talmud* was written in Aramaic, the language of Palestine at the time of Jesus, supposedly by his disciple Judas Iscariot, who did not betray Jmmanuel (Jesus) as has been depicted for 2,000 years, as the betrayal actually was done by another Judas who was not a disciple of Jmmanuel (Jesus). The entity the Pleiadians of Erra call Jmmanuel supposedly had much to do with the ancient spiritual evolution of these Pleiadians, and is revered by them.

The document was found in 1963 by Eduard "Billy" Meier, the Swiss farmer who we have seen claims to have had many contacts with the "Swiss Pleiadians" in the 1960's and 1970's, and a Greek Catholic priest by the name of Isa Rashid. The document was found under a flat rock by Meier and Rashid in the very tomb to which Jesus

(Jmmanuel) was supposedly taken after his crucifixion. *The Talmud* was written on scrolls and were found encased in preservative resin. Rashid translated some of the scrolls from Aramaic to modern German and gave the translation to his friend, Billy Meier. Not all of the scrolls were translated as, in the first place, pieces of the scroll were decayed and completely illegible, and some of the scrolls were obviously missing.[26] Secondly, Rashid was unfortunately murdered in 1974, along with his family as they were fleeing from an Israeli raid on a Lebanese refugee camp where they had been living, before Rashid could finish translating the document. The document itself was either lost or destroyed during this raid. Rashid had only translated about one fourth of the document before he was killed.[27]

So, the authenticity of this possible true testament of Jesus (Jmmanuel) becomes questionable even to those of us who know of the ET presence throughout human history. But, the testament of Jmmanuel (Jesus) contained in the part of *The Talmud* that was translated and given to Meier is quite impressive and, in my opinion, deserves scrutiny and study by scholars and the general public alike.

James Deardorff, a former professor and researcher in the field of atmospheric sciences, has written a book, *Celestial Teachings*, in which he argues that the *Talmud of Jmmanuel* may have been the testament which the compiler of the Gospel of Matthew used as a sort of a template to create Matthew. The portion of *The Talmud* that remains, covers much of what is covered in Matthew, plus a bit more. The ordering of the contents of Matthew and *The Talmud of Jmmanuel* are closely matched as well. According to Deardorff, 21% of the verses of *The Talmud* are quite close in meaning to the verses of Matthew, and another 23% of the verses of the two testaments can be recognized as cognates, or verses related in origin and word content. A further 3.1% of *The Talmud*, according to Deardorff, are cognates of verses in Luke, John, or Acts.[28]

These cognates, especially those of Matthew, although similar in form and wording to *The Talmud*, are different in meaning, sometimes radically different. Deardorff believes the cognates with altered meanings represent the possibility that the compiler of Matthew took passages from *The Talmud* and changed it in order to suit his (their, her?) own purposes. One of the main purposes for these changes, Deardorff argues, was to transform Jmmanuel (Jesus) into a god, a figure of worship, instead of a teacher of wisdom, as he is depicted in *The Talmud*. For example, Deardorff quotes the following passage from Matthew:

> Mt 13:19-23 "When anyone hears the word of the

kingdom and does not understand it, the evil one comes and snatches away what is sown in his heart; this is what was sown along the path... As for what was sown on good soil, this is he who hears the word and understands it; he indeed bears fruit..."

Then Deardorff quotes a cognate passage from the *Talmud of Jmmanuel*:

> TJ 15:37-41 "If someone hears words of truth, of the spirit, or of the laws and does not understand them, the evil one comes and snatches away what was sown in his mind; that is what was sown on the pathway...The seeds that were sown on good ground are the ones who accept the word and seek and find the truth, so they can live according to the laws of truth. Thus they allow the fruit to grow and ripen, which brings forth a rich harvest; one person bears a hundredfold, another sixtyfold and another thirtyfold."

Deardorff then quotes a Biblical scholar who had written a criticism of the Gospel of Matthew, calling this whole quoted parable from Matthew as being unintelligible. Deardorff points out that the parable as it appears in *The Talmud of Jmmanuel* is more understandable from logical and spiritual points of view:

> ...(t)he compiler (of Matthew) here, as elsewhere, replaced "spirit" with "kingdom" and eliminated "truth." His purpose we again surmise was to encourage Christians to accept the teachings of the church as it reflected the kingdom of God, and not to rely upon one's own spirit and thinking for the truth. He could not, of course, state that the word of truth was the word of the Christian church and its emerging scriptures, since his gospel was supposedly written before the church was formed, so he had to delete "truth." [29]

In all cognates the Jmmanuel of *The Talmud* is more logical than the Jesus of Matthew. For example, two cognates are quoted:

> Mt. 5:3 "Blessed are the poor in spirit, for theirs is the kingdom of heaven."
>
> TJ 5:3 "Blessed are those who are rich in spirit and recognize the truth, for life is theirs."

and:

> Mt 5:5 "Blessed are the meek, for they shall inherit the Earth."
>
> TJ 5:5 "Blessed are the spiritually balanced, for they shall possess knowledge."

Then Deardorff, using the help of Biblical scholars who would probably not touch *The Talmud* because it is "tainted with UFOs", argues that *The Talmud* is more logical and just simply makes more sense than Matthew.[30] Deardorff concludes that Jmmanuel of *The Talmud*, compared to the Jesus of the canonized gospels, "...was much more a teacher, more logical, much more of a prophet, and more provocative. Above all, he (Jmmanuel) emphasized and demonstrated the existence and power of the individual spirit..."[31]

Deardorff believes *The Talmud* makes a good candidate for having been a "proto-Matthew" and a general template from which the synoptic gospels (Matthew, Mark and Luke) were compiled, after about 100 A.D.

The "UFO" connection with Jmmanuel is made manifest in *The Talmud*. For example, after John the Baptist baptized Jmmanuel,

> ...he soon came out of the water of the Jordan, and behold, a metallic light dropped from the sky and descended steeply over the Jordan.
> Consequently, they all fell on their faces and pressed them into the sand while a voice from the metallic light spoke, "This is my beloved son with whom I am well pleased. He will be the king of truth who will lift this human race to knowledge." Behold, after these words, Jmmanuel entered into the metallic light, which climbed into the sky, surrounded by fire and smoke, and passed over the lifeless sea, as the signing of metallic lights soon faded away. After that, Jmmanuel was no longer seen for forty days and nights.[32]

This version from *The Talmud* is radically different from that appearing in the gospels. In Matthew we read:

> ...After baptism Jesus came up out of the water at once, and at that moment heaven opened; he saw the Spirit of God descending like a dove to alight upon him; and a voice from heaven was heard saying, "This is my Son, my Beloved, on whom my favor rests."

Jesus was then led away by the Spirit into the wilderness to be tempted by the devil. For forty days and nights he fasted...[33]

While the Jesus of the gospels fasted and was tempted by the devil, Jmmanuel of *The Talmud* was taken by the guardian angels, the celestial sons, to an unknown location where, "...(t)hey taught him the wisdom of knowledge."[34]

There are many more differences between the story of Jesus of the four gospels of the Bible, and Jmmanuel of *The Talmud*. The most radical difference is that Jmmanuel of *The Talmud* survived the crucifixion, escaped Palestine, and eventually ended-up in India where he lived and taught several years before he died. I can imagine the outrage and disbelief most Christians must feel when reading the above statement. I simply wish to point out that there exists a few Biblical scholars and Christian theologians who base their research on strictly "Earth sources," who have contended that Jesus did not die on the cross and later lived in India, long before the *Talmud of Jmmanuel* was published in 1992. For example, researchers Baigent, Leigh and Lincoln point out that it was unusual for a person who was executed by the Roman custom of crucifixion to die in only a few hours. Usually a man crucified by Roman procedures would live for a couple of days, sometimes as long as a week, before he died.[35] In the Gospel of Mark, when Joseph of Arimathea asks Pilate for the body of Jesus after Jesus had been hanging on the cross for only a few hours,

> ...Pilate was surprised to hear that he was already dead; so he sent for the centurion and asked him whether it was long since he died. And when he heard the centurion's report, he gave Joseph leave to take the dead body...[36]

Baigent et al., also gives evidence that the crucifixion of Jesus itself was not a large public affair on a barren hill named Golgotha, but was a private affair that took place in a private garden. They point out, for example, that according to Matthew, Mark and Luke, most people, including the women, are depicted as witnessing the crucifixion from a considerable distance. Baigent et al., believe that the crucifixion of Jesus may have been a "fake crucifixion" where Jesus was not intended to actually die on the cross, but was made to appear to die on the cross.[37]

Holger Kersten, a German theologian, in his book *Jesus Lived in India*, also presents evidence and argues that Jesus survived the crucifixion, and he gives considerable physical evidence that Jesus eventually went to India.[38] Both Kersten and the team of Baigent et al., point out that the "Bible" of the Muslim religion, the Koran, claims that Jesus survived and lived several years after the crucifixion. I repeat that neither Kersten, nor Baigent et al., nor the other biblical scholars that have claimed that Jesus did not die on the cross knew of the *Talmud of Jmmanuel*, nor acknowledge in their books possible visitations and presence of ETs on Earth.

In any case, the teachings of Jmmanuel are quite radical relative to the Christianity practiced today by any denomination. It is interesting to note that the concept of god held by Jmmanuel is quite different than the concepts of god of the various Christian sects. In one instance Jmmanuel managed to calm a storm on a lake where he and his disciples were located. The disciples marveled at Jmmanuel's ability to calm a storm on command and were proclaiming him as the greatest prophet they ever knew.

> But Jmmanuel answered, "I tell you there are greater masters of spiritual powers than I,... And great are they also, those who came out of space, and the greatest among them is god, and he is the spiritual ruler of three human races.
> "But above him is Creation whose laws he faithfully follows and respects; therefore he is not omnipotent, as only Creation itself can be.
> "Thus there are limits for him who allows himself to be called god and who is above emperors and kings, as the word says.
> "People are ignorant and immature because they consider god as Creation and follow the false teachings that were adulterated by scribal distortions.
> "Thus, when people believe in god, they do not know about the truth of Creation, because god is human as we are.
> "There is a difference, however, that in his con-

sciousness and wisdom, logic and love he is a thousand times greater than we and greater than all people upon this Earth.

"But he is not Creation, which is infinite and without any form..."[39]

So there were and are many "celestial sons" from space who were greater masters of spiritual power than Jmmanuel, according to Jmmanuel himself. These spiritually advanced celestial sons, including the one known as God, were human like you and I.

I understand the "human" form of God in the above passage to mean that they were humanoid, not human as modern science has narrowly defined *Homo sapiens*. The celestial sons may have been quite different genetically and in other physical ways from Earth humans, and certainly they were much different from Earth humans in terms of higher consciousness and wisdom, and their abilities in logic and love. The celestial sons told Jmmanuel that he was on a mission that was instigated by them. And Creation, "...which is infinite and without any form..." is above all spiritually advanced celestial sons.

The God to whom Jmmanuel referred was apparently not the God Jehovah of the Israelites, for in another portion of *The Talmud*, Jehovah is indirectly criticized:

> Truly, I say unto you: The nation of Israel was never one distinct people and has at all times lived with murder, robbery and fire. They (the Israelites) have acquired this land through ruse and murder in abominable, predatory wars, slaughtering their best friends like wild animals.[40]

It is interesting to note what *The Talmud* has to say concerning the biological origin of humans. During the forty days after being baptized by John the Baptist, the celestial sons taught Jmmanuel:

> ..."People have come from the heavens to Earth, and other people have been lifted from Earth into the heavens, and the people coming from the heavens remained on Earth a long time and have created the intelligent human race..."[41]

Jmmanuel's teaching of the power of the spirit, he was told,

> "...will contribute to the well-being of the human races, though the road leading there to will be very difficult for them and you.
> "You will be misunderstood and renounced, because the human races are still ignorant and given to superstition...
> "Hence, following the fulfillment of your mission, centuries and two millennia will pass before the truth of your knowledge brought among the people will be recognized and disseminated by some humans.
> "Not until the time of space-traveling machines will the truth break through and gradually shake the false teaching that you are the Son of God or Creation.[42]

I could go on pulling quotes out of the *Talmud of Jmmanuel* for several more pages, because I am impressed with it, and, clearly, *The Talmud* fits in with the overall scenario presented in this book. But, I am not prepared to say that *The Talmud of Jmmanuel* is the true testament of Jesus (Jmmanuel) for certain, because I am not certain. Yes, I am impressed with the logic of Jmmanuel and am emotionally moved at times when I read *The Talmud...* . I fully agree with Deardorff in believing that the Jmmanuel of *The Talmud* is a more forceful, logical, more believable teacher - a more believable human being - than the Jesus depicted in the gospels of the Bible.[43]

Let me hasten to express some doubts and questions. In the first place there is so much "smoke" around the "Jesus phenomenon," and the real nature of the world in general, one would be foolish to accept the idea that any one view of "reality" is the correct view. Further, we have already reviewed the unusual circumstances surrounding the appearance and disappearance of *The Talmud...* in the latter half of the twentieth century. This alone casts some doubts on the veracity of *The Talmud of Jmmanuel*. I agree with James Deardorff that *The Talmud of Jmmanuel* makes a very good candidate for being the "proto-Matthew" from which Matthew was written (and distorted). Deardorff, in his careful comparisons of Matthew and *The Talmud...*, argues that if *The Talmud...* is a fake, it is a very clever fake. At this state of our (Earth human) evolution, I don't think we can rule out completely the possibility that the "Swiss Pleiadians" group, or some regressive ET group is using *The Talmud...* for their own unknown purposes. Furthermore, I am impressed with other arguments concerning the historical Jesus, such as that of Baigent et al., that the family of Jesus, and perhaps Jesus himself, may have gone to Europe after the "crucifixion".[44] Perhaps the entity Jesus (Jmmanuel) is actually a conflation of two or more entities; as we have seen, the church fathers at the Ecumenical Council of Nicea in 325A.D. "officially determined" the true nature of Jesus. Who knows what was really happening in Israel 2000 years ago?

Discussion

As with Judaism and other world religions, I do not wish to discredit the good that has come out of Christianity. Many Christians have felt the love of Jesus - or as I would prefer to call it, the love of the Divine - over the centuries. Countless Christian priests, ministers, nuns, and ordinary people have tried their best to love and help their fellow humans in the name of Jesus. One can feel nothing but love and honor for such Christians.

But, the history of Christianity is fraught with false teachings and abuse of power. Millions of people have suffered and died in the name of Jesus, and it is time - way past time - for this to cease altogether. Certainly, a person who has examined the history surrounding the rise of Christianity as a world religion must question the modern-day concepts of Jesus and God as presented by modern Christian denominations. As pointed out above, it has been argued that the "historical facts" of Jesus' life are not important, but that his spiritual teachings are. But, as has been demonstrated, the evidence is that only a very small part of the true spiritual teachings of Jesus have survived and appear in the teachings of any Christian denomination.

As with all the belief systems covered in this book, many of the negative aspects of Christianity have been emphasized at the expense of some of the good. It is extremely crucial that we humans of the late 20th century closely examine all current belief systems in view of the knowledge we are now gaining about our origins and spiritual existence. Hopefully all Christians everywhere will consider the evidence presented in these last two chapters, that most of the spiritual truths of our existence have been distorted and/or eliminated from Christianity, then embark on their own personal investigations of these claims.

Let us briefly turn our attention to other world religions and examine how spiritual truths may have been distorted in them.

Possible ET Influences on Christianity

[1] See the list of titles in, Schellhorn, G. C. 1989. *op. cit.*, Introduction, pp. vii-ix.
[2] *The New English Bible* 1970. *op. cit.*, Genesis 18: 2-5.
[3] *Ibid.*, Genesis 18: 10.
[4] *Ibid.*, Genesis 18: 13-15.
[5] *Ibid.*, Luke 1: 5-7.
[6] *Ibid.*, Luke 1: 11-14.
[7] *Ibid.*, Luke 1: 18.
[8] The quote of the Gospel of Mary VII: 16-21, from Bramley, W. 1990. *op. cit.* p. 135.
[9] *Ibid.*, Luke 1: 28-37.
[10] Bramley, W. 1990. *op. cit.*, p. 139.
[11] *The New English Bible* 1970, Matthew 1: 20-25.
[12] The quote of Infancy from Bramley, W. 1990, *op. cit.* pp. 137-138.
[13] *The New English Bible* 1970, Matthew 2: 2.
[14] *Ibid.*, Matthew 2: 9-10.
[15] *Ibid.*, Matthew 2: 12.
[16] Schellhorn, G. C. 1989. *op. cit.*, p. 253.
[17] Protovangelion quoted in Bramley, 1990, *op. cit.* pp. 138-139.
[18] Baigent, M., R. Leigh and H. Lincoln 1982. op. cit., and _____ 1886. *op. cit.*
[19] Cooper, M. W. 1991. *op. cit.*, p. 213.
[20] Marciniak, B. 1990. *op. cit.*, p. 13.
[21] *The New English Bible* 1970, 1 Corinthians 12: 7-10.
[22] Schellhorn, G. C. 1989. *op. cit.*
[23] Baigent, M., R. Leigh and H. Lincoln 1986. *op. cit.*, pp. 38-39.
[24] Levi 1968. *The Aquarian Gospel of Jesus the Christ*. DeVorss & Co., Los Angeles.
[25] *The Talmud of Jmmanuel* 1992. Wild Flower Press, Tigard, Oregon.
[26] *Ibid.*, foreword by Eduard Meier, p. XIX.
[27] Deardorff, J. W. 1990. *op. cit.*, Synopsis, X.
[28] *Ibid.*, p. 4.
[29] *Ibid.*, pp. 152-153.
[30] *Ibid.*, pp. 115-116.
[31] *Ibid.*, p. 233.
[32] *The Talmud of Jmmanuel* 1992, 3: 30-34.
[33] *The New English Bible* 1970, Matthew 3: 16-17; and 4:1.
[34] *The Talmud of Jmmanuel* 1992, 4: 6.
[35] Baigent, M., R. Leigh and H. Lincoln 1983. *op. cit.*, pp. 353-357.
[36] *The New English Bible* 1970, Mark 15: 44-45.
[37] Baigent et al. 1983, *op cit.*, pp. 352-357
[38] Kersten, H. 1986. *op. cit.*
[39] *The Talmud of Jmmanuel* 1992, 16: 50-57.

[40] *Ibid.*, 10: 26.
[41] *Ibid.*, 4: 7.
[42] *Ibid.*, 4: 36, 37, 48, 49.
[43] Deardorff, J. W. 1990. *op. cit.*
[44] Baigent et al. 1982. *op. cit.*

> *"There is only one religion, though
> there are a hundred versions of it."*
> George Bernard Shaw, *Pleasant and Unpleasant*

Chapter 19

Some Spiritual Distortions in Other World Religions

Now that we have reviewed some doubts and questions concerning Christianity, believed in one form or another by 1.5 billion humans, and some doubts and questions concerning Judaism, which has about seventeen million believers, let us look at some possible spiritual distortions with regard to the other major world religions. In this chapter we will briefly review what I consider the "spiritual misinformation" that has crept into Hinduism, Buddhism and Islam, world religions who have a combined total of almost two billion followers.[1] Not as much time will be spent discussing these three religions, as was spent discussing Judaism and especially Christianity, primarily because the author does not know as much about them. I was born in and spent most of my life living in a Judeo-Christian culture. Naturally, I have spent more time investigating the two religions that have most influenced the culture in which I was reared. I do, however, know enough about Hinduism, Buddhism, and Islam to know that the spiritual teachings of these religions have been distorted, as have those of all religions. I also know enough about all of the religions to know that they deserve our respect, like Judaism and Christianity. Millions of people today live by the precepts of one of these religions, and millions gain much spiritual comfort and growth from their religions. However, I again put forth the argument that we humans, in order to evolve spiritually must recognize the spiritual misinformation that is found in all world religions. As has been indicated, the evidence is that most of this spiritual misinformation has been deliberately perpetuated by regressive ETs. It is our privilege, as late 20th century humans, to closely investigate all of our current belief systems, religious or secular, then save from them what is good, and discard information that would divide humankind and otherwise inhibit the individual and collective spiritual evolution of humans towards the Light - towards the Truth.

Hinduism

"Hinduism" is the western term assigned to the religious beliefs and practices of the majority of the people of India. Hinduism today has about 700 million members worldwide, most of them living in India itself. While Hinduism has no fixed scriptural canon, several ancient writings have been the subject of elaborate theological commentary. The oldest of these are the Vedas, which are a group of ancient Sanskrit hymns, chants and ritual formulas used in ceremonies of piety and devotion to various sacred deities. However, the development of these ancient Vedic religious views into the historic Hindu views is quite obscure, yet is believed to have begun to have taken place around 1,000 BC.[2] It should be pointed out, however, that most Hindu spiritual teachers believe the Vedas are very ancient, if not the oldest records in existence, at least pre-deluge in age.

We recall that the first known civilization of India, the Harappan or Indus River Valley civilization flourished between 2,400 BC. and 1,800 BC. The ancient Mesopotamian texts claim the Anunnaki, the creator ET "gods" of the Sumerians, founded the Indus Valley civilization. Archeological records confirm that this first known Indian civilization engaged in extensive exports of grain and metals to ancient Mesopotamian civilizations. We also reviewed evidence that this Indus Valley civilization was brought to an abrupt halt through nuclear intervention by one of the ET groups. The largest city of this civilization, Mohenjo-Daro, shows evidence of having been destroyed by a nuclear blast.

Somewhere between 1,500 BC. and 1,200 BC., western historians believe India was invaded by tribes of people from the northwest known as "Aryans." Exactly who the Aryans were is still debated by historians today, but Aryans spoke an Indo-European language, Sanskrit. Modern Indo-European languages include Greek, Latin, German, English and Persian. Many are familiar with the racist "Aryan supremacy" idea supported by the Nazi regime of Germany. The Nazis promoted the idea that Aryans were non-Semitic, white- skinned people who were specially "chosen" or "created" to dominate the world.

The Aryans that invaded India supposedly several hundred years before Buddha and Jesus were born, made themselves the new ruling class of India and forced the Indian people that already lived there into lower servile classes. They imposed a caste system in India based upon color and the domination of the Aryans over the native

Indians. The Aryans themselves occupied the highest caste, the Brahman caste. Over the generations, of course, darker-skinned genes crept into the Brahman caste. Even today, however, in northern and parts of western India, lighter-skinned Indians, supposedly descendants of the original Aryan invaders, dominate the upper castes.[3]

Hinduism, or Brahmanism, had emerged as a religious-political system by the 6th century B.C. The rulers and priests were chosen from the Brahman castes. The largest spiritual distortion of Hinduism (Brahmanism) is expressed through the caste system and their belief in reincarnation.

We recall that the belief in successive lives, banished forever from Christianity and the teachings of Jesus during the Ecumenical Council of Constantinople in 553 A.D., is a central tenet of Hinduism. However, Hindus are taught that they are born into the caste of their father and they will remain in this caste throughout this particular lifetime. If a Hindu is good and obedient to the ruling caste, they are taught that one might be reincarnated into a higher caste in the next lifetime. The castes, therefore, became a measure of one's spiritual evolution in Hinduism. The belief that lower caste people were lower in their spiritual evolution than higher caste people gave the higher caste the justification they needed to mistreat the lower castes. Thus, members of the lowest caste, the "outcasts" and "untouchables", were forced to perform the most menial tasks of society and live their lives in abject poverty. The "untouchables" were treated with scorn and aversion throughout most of Indian history because of their presumed sinfulness in a prior lifetime. Such beliefs and mistreatment were attacked by Mohandas Gandhi in the first half of the 20th century, and are today against the law in India. The caste system itself is said to be less rigid than formerly in today's India, but it and the beliefs behind it are still very strong, as recent violence and other expressions of animosity between the castes in India indicate.[4]

Modern research into the reincarnation phenomenon indicates that the spirit or soul entity chooses its own life in which to incarnate. The entity chooses its own parents. The purpose of living these many lives in the material-physical plane is to learn lessons. The ultimate goal of learning these lessons through a succession of lives is to attain a state of love - of unconditional love of oneself, of all other people, and of all Creation.[5] Therefore, if this line of research is coming close to the truth of the matter, one cannot and should not try to measure one's level of spiritual evolution by the caste or position in society they occupy.

Yet, each Hindu caste was said to be a necessary step on a cosmic staircase. Bramley, in what he refers to as "Aryan beliefs," has summarized the effects of the belief of "staircase" reincarnation on Indian spiritual evolution:

> ...Hinduism stressed that obedience was the principal ingredient bringing about advancement to the next caste. At the same time Aryan beliefs discouraged people from making pragmatic attempts at spiritual recovery. The myth of spiritual evolution through a caste system hid the reality that spiritual recovery most probably comes about in the same way that nearly all personal improvement occurs: through personal conscious effort, not through the machinations of a fictitious cosmic ladder.[6]

Bramley argues that what he calls the Brotherhood, who have operated secretly since human civilization began with the "Custodians," or groups of ETs who wanted to keep humankind spiritually ignorant for control purposes, established a strong base of operations in India after the Aryan invasion.[7]

As we have seen, several writers have pointed out how many of the "gods" of ancient Hindu writings are depicted riding about the skies in remarkable and powerful "chariots" and firing tremendous weapons from these "chariots," which emit powerful beams of light. Several writers, from von Däniken in the 1960s and 1970s, to the present, have argued that the "gods" of the Vedas and other ancient Hindu writing represent ETs and many of their conflicts with other ET groups of the world.

Furthermore, the most important information we have gained from this very brief look at Hinduism is how, once again, humankind has been spiritually inhibited through false religious beliefs, almost certainly propagated by regressive ETs. This was the primary bit of information for which we were looking for evidence. So, again, we are going to leave a religion emphasizing the bad, at the expense of the good. Again, let us repeat that no religion is all bad, nor all good. It is understood very well that Hinduism has been and is a most wonderful religion for millions of people. But we can do better. Let us look at another religion that was "born" in India.

Buddhism

Buddhism of one form or another has about 320 million practitioners, mostly in central and southeastern Asia. Buddhism was founded by Gautama Siddhartha Buddha, who was born and lived in northern India between 563 and 483 B.C. Siddhartha was born into a high caste (Kshatryia caste), the son of a wealthy landowner and ruler. He lived a sheltered life of ease until, at the age of 29, he learned of human suffering. After he learned that the average, normal state of most humans often approximated misery, he renounced luxury and became an ascetic and wanderer. He sought spiritual Truth through spiritual enlightenment. He is said to have been rewarded for his ascetic living by being granted enlightenment under a Bo tree. "The great enlightenment" under the Bo tree gave Buddha knowledge of cosmic principles on which he founded Buddhism, according to Buddhist tradition. He then became a spiritual teacher, took in disciples, and tried to spread the content of his vision of enlightenment across India and abroad. Buddhism emphasized the "right-living," gentle, compassionate side of human nature. Buddhism opposed the caste system and did not support the Brahman doctrines in general.

Buddhism met with much early success in India, and for several hundred years it rivaled Brahmanism in popularity. Bramley calls Buddhism a "maverick religion," one which occasionally sprang up in human history that was very close to spiritual Truths, unlike "custodial religions." We've noted that Bramley believes that maverick religions eventually decay as they are infiltrated with custodial ideas that inhibit spiritual growth.[8] For example, he argues that Buddhism "...underwent a great deal of change, splintering and decay as the centuries progressed."[9] Bramley asserts that Siddhartha's teachings were distorted and others' ideas added as the years passed. For example, the concept of "Nirvana," as taught by most modern Buddhists, Bramley contends, is quite distorted. Buddhist teachings emphasize that all physical existence is bound up with pain. Therefore, they reason, the ultimate goal of a potential Buddha (an enlightened one) is to escape physical reality and move into a state of "blissful nonexistence" - Nirvana. Nirvana has also been translated as the "Void" or "Nothingness." In reality, Bramley argues, the Buddha taught that a true state of Nirvana, "...originally referred to that state of existence in which the spirit has achieved full awareness of itself as a spiritual being and no longer experiences suffering due to misidentification with the material universe."[10]

There is clearly a crucial difference between the modern Buddhist concept of Nirvana and the original concept that Siddhartha intended, according to Bramley. A spiritual entity would have a difficult time evolving past a certain stage if it did not remain conscious of its physical surroundings, and strove instead for a state of non-existence.

Some Spiritual Distortions in Other World Religions

Islam

Islam, which means submission to - or having peace with God, today has more than 950 million believers, who live mostly in north Africa and Asia. A follower of Islam is called a Moslem, or Muslim, which means, "one who submits." Muslims are today often divided on political and religious interpretation grounds, but they are united in the belief of one God and the prophet Mohammed.

Mohammed was born around 570 A.D. and died in 622 A.D., after which Islam was formalized in Arabia, from where it spread throughout much of the Old World. Not a lot is known about the childhood and early adulthood of Mohammed, but at age forty he emerged as a religious prophet and eventually founded a powerful new world religion. According to Muslim teaching, God gives humans revelations of divine nature and the proper course of living through prophets. Humans constantly fall away from the teachings of prophets, but God mercifully sends new prophets. Three major prophets of God that came before the time of Mohammed, according to Islam, were Abraham, Moses and Jesus. Mohammed is considered the last and greatest prophet, and Islam is considered the only "true religion" by Muslims, and all non-Muslims are considered "infidels." After the people fall away from Islam, according to Muslim teachings, the world will come to an end and a judgment by God will follow. Heaven awaits the faithful followers of Islam, and hell awaits the "infidels."

Islam is the third monotheistic religion of the world that contains prophecies of "end times," the "day of judgment," "Armageddon," or whatever one chooses to call the "end of-the-world" prophecies. Judaism and Christianity also contain end-time prophecies where the "chosen" or "faithful" are rewarded with eternal bliss in heaven, and the "un-chosen" or "un-faithful" suffer a horrible fate.

A feature of all types of practitioners of Islam is their devotion to a book, the Koran, which is believed to be a revelation of God channeled through Mohammed. Mohammed describes the messenger that brought him teachings from God, as an angel, which called itself Gabriel. In the Koran Mohammed describes this messenger:

> That this is the word of an illustrious Messenger imbued with power, having influence with the Lord of the Throne, obeyed there by Angels, faithful to his trust, and your compatriot is not one possessed by jinn (spirits); for he saw him in the clear horizon.[11]

Some Spiritual Distortions in Other World Religions

Islam spread rapidly throughout western Asia and north Africa after the death of Mohammed, primarily because the prophet himself raised a "holy" army and set off to "convert" infidels, a tradition continued by militant Muslims. Eventually the militant Muslim empire stretched from Spain in the west, across north Africa, through most of the Middle East, and as far east as India.

The early Muslims found the Koran inadequate as an authority of the new life of Islam, hence the Sunna arose and became fundamental to Islam. The Sunna is made up of moral sayings and anecdotes of Mohammed collected by Muslims of the 9th and 10th Centuries A.D. The Sunna is almost as important as the Koran in Islam. In addition to the Koran and Sunna, Islam also rests on a principle which is called Ijma. Basically, Ijma is a principle believed by Muslims in which every Muslim knows that any belief entertained by most Muslims throughout history is true beyond question. Thus, this belief in the infallibility of Muslim traditions, plus the Koran and Sunna, make up the foundations of Islam and help to hold the faith together despite frequent religious and political divisions over the years.

The primary division of Islam is between what are known as Sunnites and Shiites. Sunnites are much more numerous and practice a form of Islam thought to be closer to what the prophet taught, than the Shiites. Therefore, the Sunnites are usually treated as the norm of the Islamic faith. The Shiites themselves have split into numerous sects over the centuries, including the infamous Assassins, who fought the Christian Crusaders and developed the deadly tool of the "lone assassin." The Assassins and other Shiite sects are partially responsible for the bad name Islam has had in Europe over the years. To quote an encyclopedia account of Islam, "...Shiism has always lent itself in an extraordinary degree to bigotry and persecution of non-Shiites..." [12]

What about spiritual distortion in this religion followed by almost one billion souls? With respect for all Muslims everywhere, the militancy of the followers of Islam from its inception in the days of Mohammed, argues against Islam being the "true religion," a religion conceived to help bring the Light and Truth to humankind. Beginning with the army raised by Mohammed himself, untold lives have been lost, mutilated, and severely psychologically damaged by militant Muslims in order that they might spread what they consider the "true religion." As with many other religions, Muslims often did not allow the people they conquered, especially in the first centuries of its existence, the luxury of choosing whether they wanted to become members of the Islamic faith or not. It was either that the people of a conquered region became a Muslim or they would be exterminated. One can well understand how one's spirit could be attracted to some of the ritual, ceremony, and teachings of Islam, but one who wishes to spiritually evolve following the principles of love can not accept the premise that a person had to convert to Islam or be killed! Nor could one accept this tactic of "converting," practiced by several belief-systems, including Christianity, Judaism and some secular belief systems throughout the bloodstained history of humankind.

Bramley argues that Islam is but another Custodial religion.[13] As we have seen, by Custodial religion, he means a religion inspired and influenced by regressive ETs for the purpose of dividing and controlling humankind. Following this scenario, the "angel" who communicated with Mohammed was either put up by regressive ETs to deceive Mohammed into believing he was receiving the "true religion," or spiritual truths of the messages of the angel were subsequently distorted.

In any event, a thinking person who was not raised in a Muslim culture and imprinted by Islam, might have grave doubts concerning claims that Islam is the "true religion." In addition to the doubts expressed concerning the very foundation of Islam (the "angel" that communicated with Mohammed), the militant nature of historical Islam and the fact that some of the Sunnas are known to have been distorted[14], argue against the claims of Muslims that all of humankind should embrace Islam. And only a Muslim could believe in the principle of Ijma, the notion that the historical beliefs and practice of Islam are infallible.

Distortion and World Religions

This is a troubled world in which we live today. Political tyranny, starvation, ecological destruction and human pain and sufferings of all types are common on this Earth. Many of the billions of inhabitants of this troubled Earth have found some degree of comfort and relief from suffering through their belief and ritualistic practices of their own particular world religion. In this and the previous three chapters I have undoubtedly offended millions of believers of world religions - Judaism, Christianity, Hinduism, Buddhism, and Islam - and have deliberately not reviewed the good and beneficial aspects of all these religions. My major point in this very brief and slanted review of world religions is that all religions contain spiritual distortions and corruptions. This supports the overall claim of this book - that all our major belief systems here on Earth, be they religious or secular, are distortions of the Truth of our existence and were and are secretly promoted by regressive ETs in order to keep humankind divided and at odds with each other.

As mentioned earlier, it is my opinion that it is our privilege, honor, and duty as late 20th century A.D. humans to preserve the good of all these religions, and discard features that tend to divide and spiritually inhibit us. There is no "true religion," but there is much good in all world religions, which I have not touched upon here. I am particularly impressed with some of the more mystical traditions of all of these religions - the traditions that emphasize love of all humans and of all creation. Certainly no human should be despised and un-loved simply because they are or are not members of any particular religion. Even Darwinists and other agnostics should not be despised and un-loved by the faithful of any "true religion" or belief system that is close to the Truth.

Now, before we attempt to outline a belief system that is close to the Truth, let us look at a few examples of humans "dwelling in darkness," simply to demonstrate that we humans cannot hold the reptoids and greys and other regressive ETs fully responsible for the beastly behavior humankind has exhibited towards other humans, and all life here on Earth. The negative influence of regressive ETs cannot fully explain the vast amount of negativity that has been expressed on this Earth throughout human history. Humans must share much of this responsibility.

---

[1] The numbers of followers of the world religions quoted in this chapter are from, "Major World Religions", San Francisco Sunday Examiner and Chronicle, August 29, 1993, p. A-6.

[2] Starr, C. G. 1983. *op. cit.*, p. 166.

[3] Bramley, W. 1990. *op. cit.*, p. 108.

[4] Moore, M. 1994. "Caste animosity in India breaks into ugly violence." *The San Francisco Chronicle*, February 19, pp. A8 and A12.

[5] Albertson, M.L. and K.P. Freeman 1988. "Research related to reincarnation." Proceedings of the International Conference on Paranormal Research, Colorado State University, Fort Collins, July 7-10.

[6] Bramley, W. 1990. *op. cit.*, p. 109.

[7] *Ibid.*, pp. 103-111.

[8] *Ibid.*, pp. 116-118.

[9] *Ibid.*, pp. 116-117.

[10] *Ibid.*, p. 117.

[11] *The Koran* quoted in Bramley, 1990, *op. cit.* p.166.

[12] *The Columbia Encyclopedia* 1950. *op. cit.*, p. 979.

[13] Bramley, W. 1990. *op. cit.*, pp. 165-167.

[14] *The Columbia Encyclopedia* 1950. *op. cit.*, p. 979.

> *"His family raised a new lodge,*
> *and the shaman brought in his sacred pipe.*
> *The wise man blew smoke to the four*
> *winds, asking them to carry*
> *Joseph's spirit back to the spirit land."*
> Chief Joseph Thunder Rolling Down from the Mountains
> by Diana Yates

Chapter 20

Man's Inhumanity to Man

We humans cannot blame the negative ETs for all of the behavior of humankind that is directed away from the Light, toward some sort of darkness. Human history is replete with tales of abominable, despicable, loathsome behavior of humans interacting with humans of other groups. We need to be reminded of just how despicable humans have historically treated humans they consider different, and therefore, "inferior." Perhaps, the regressive ETs are somewhat, even mostly, responsible for the perpetuation of human conflicts by dividing humans along racial, religious, ethnic, economic and political lines of thought, but we humans do not have to agree to behave in a stupid uncompassionate manner toward our fellow human beings! "Man's inhumanity to man," as it is often referred to, is a characteristic of us modern humans throughout our history. In any time of human history, since the Sumerian civilization began almost 6,000 years ago, we can find examples of atrocities humans have committed on other humans. We have already reviewed some of this abominable behavior of humans, such as the heartless slaughter by the ancient Hebrews of the Canaanites and the other groups of people that lived in the "promised land." Here, let us briefly review a more recent tragic, sad episode of inhumanity in order to begin to bring the focus to the human animal, and away from the dastardly, uncompassionate regressive ETs. I do this because, ultimately, we humans are responsible for the relations we humans have with other humans, and we cannot spiritually evolve unless we begin to treat all of life, especially

Man's Inhumanity to Man

other humans, with compassion.

A brief review of how the Indians of the high plains and their cultures of the western portion of the United States were crushed and reduced to memories by the American government and the first American white settlers will suffice as an example of humans' inhumanity to humans. I chose the American example because I am an American.

The destruction of the Indian cultures in the United States is now well documented in terms of human numbers. There were an estimated 900,000 Indians north of Mexico when Columbus "discovered" America in 1492, and only an estimated 300,000 in this same area by 1870. Most of the decrease in the number of Indians in the area was due to "white man's diseases" that the Europeans brought with them, and not the wars the North American Indians fought with the colonial governments of France and England, and the young, developing countries of Canada and the United States. The Indians had not been exposed to these white-man diseases, especially smallpox, measles and cholera, and had not built up immunities in their bodies through natural selection, as had the Europeans. Consequently, these European diseases killed hundreds of thousands of Indians.[1]

Generally, the Indians were friendly and welcomed the Europeans upon initial contact. It is probable that almost all of the English Puritans (Pilgrims) who established the first successful English colony in North America in Massachusetts in 1620, would have died of starvation during their first winter there if they had not received help from friendly Indians.[2] However, as the European settlers became more and more numerous, as more and more roads were built through their lands, as more and more trees were cut down for more settlements, the Indians realized that their land and lifestyles, if not their lives were at stake. Then the interaction between the white settlers and the Indians became hostile. Understandably, the Indians did not want to give up the lands that they and their ancestors had grown up on and lived upon for thousands of years. The European settlers, on the other hand, believed the land to be theirs, because a European king granted them a charter to the land. The main reason, however, that the European settlers felt justified in depriving the Indians of the land and lifestyles was because the whites felt that they and their lifestyles were vastly superior to the Indians and their lifestyles. A few whites had compassion for the Indians and tried to accommodate them, but most whites considered Indians heathen savages. The Indians did not even know who Jesus was!

The young United States Government continued the treaty system established by Great Britain and other colonial governments, until

1871. The most important purpose of a treaty between the United States Government and one of the numerous Indian tribes, as far as the Government was concerned, was almost always to deprive the Indians of their lands.³ The Indians were generally removed from their lands to reservations which the whites considered worthless. By 1840, most of the Indians of the eastern United States were either dead, or had been displaced to reservations west of the Mississippi River. The forced migration of eastern Indians to the west resulted in thousands of deaths and untold hardships for the Indians. The United States Government hoped that the "Indian problem" would be solved by granting all land west of the Mississippi (except Missouri, Arkansas and Louisiana) to the Indians. However, this dream was shattered as whites settled in Texas, Oregon, California, Nebraska and Kansas in the 1840s, and 1850s. For example, when gold was discovered in 1848 in California, white men from all over the world poured into California. The Indians in California were as gentle as the climate in which they lived. The numerous California tribes were very small and there were no great war leaders among these non-warlike tribes. So, the whites that poured into California after 1848 had little trouble, with few exceptions, subduing and eventually exterminating the population of Indians. Most of the bones of now forgotten bands of Indians are now "...sealed under a million miles of cement freeways, parking lots and slabs of tract housing." ⁴

Attitudes of the whites toward Indians remained negative, to say the least. For a typical impression the whites had toward Indians, I turn to the impressions of a middle-aged English woman, Isabella Bird, who saw some "Digger Indians" on a train platform in California in 1873. "Digger Indian" was a term of contempt whites applied to Indians of California, and surrounding states (Oregon, Nevada, Utah, Arizona and Idaho), who derived part of their living digging and eating wild root plants. Miss Bird was a remarkable woman for her time, in that she traveled to remote areas of the world, including the American West, where she suffered hardships that hardly any world traveler of her era and ilk would deign to suffer. She wrote interesting and articulate accounts from those remote places. Miss Bird observed that the train in which she was traveling had Indians riding on the outside platform:

> The platforms of the four front cars were clustered over with Digger Indians, with their squaws, children and gear. They are perfect savages, without any aptitude for even aboriginal civilization, and are altogether the most degraded of the ill-

> fated tribes which are dying out before the white races... They were all hideous and filthy, and swarming with vermin... They were a most impressive incongruity in the midst of the tokens of an omnipotent civilization.[5]

Miss Bird had an attitude toward Indians that was shared by all nationalities of white persons that traveled through America in the latter half of the 19th Century, be the person English, American, French, Russian or whatever nationality. Whites of the American West, with very few exceptions, had very little understanding of, or compassion for, the Indians that were losing their cultures - their way of life - and their very lives, in droves, before the onslaught of the whites.

The Indians of the High Plains (see Figure 20-1) were different from the California Indians, in that they did have more war like traditions. The ancestors of the tribes that lived on the High Plains in the 1800s had been nomads that hunted the bison there on foot and eked out a precarious existence barely a few hundred years earlier. They used virtually every ounce of the bison they killed (called buffalo in the American West), including their meat, hides and hooves. They often traded buffalo meat and hides with the sedentary farming Indians that lived to the east of the plains, for corn and other crops the farming Indians grew. The Spanish explorers introduced horses to America in the early 1700s, and with the horse to help their nomadic movements and buffalo hunting, the Plains Indians became relatively prosperous.

The Plains Indians of the 1800s were divided up into several tribes who divided up the riches - to them the buffalo and other animals they depended upon for survival. Hostile tribes were usually able to avoid each other, but bloody encounters at times did occur when small bands of the hostile tribes did encounter each other on the Plains. Even though bloody encounters between factions of the Plains Indians were apparently not common, the Plains Indians were, in a sense, in a state of perpetual warfare with hostile tribes.

The Plains Indians had developed a system of "coups" they used to measure a warrior's prestige among their tribes. "Coups" were valued according to the degree of recklessness the warrior exhibited toward warriors of an enemy tribe. For example, the type of coup the Plains Indians valued the most was when a warrior struck an armed enemy warrior with his bare hands, not killing or harming the enemy in any way. This was considered the highest act of bravery a warrior could accomplish. Coups were also awarded to a warrior for killing an enemy warrior, wounding him, scalping him and stealing his horse.

So, a warrior tradition was strong among the Plains Indian tribes when the American Government tried to deprive them of their lands in the period of time between 1850 and 1876. The Kansas-Nebraska Act of 1854 opened up the territories of Kansas and Nebraska for white settlers. These territories contained vast tracks of the High Plains. The discovery of gold in the mountains of Colorado and Montana in 1858 and 1862, respectively, put more pressure on the Plains Indians, as white gold miners sought passage across the plains to get to the gold camps in the mountains. Soon the United States Government decided to build a transcontinental railroad which cut right through the plains. The pressure on the Plains Indians to give up their lands and retire to a reservation became stronger and stronger as whites flocked into and through their lands.

Atrocities were committed by both the Indians and the whites, but the bottom line of this tragic conflict of cultures was that the whites wanted to settle on the land that had been the Indians' land for several generations. The Americans negotiated treaties with the Indians guaranteeing them their traditional lands if they would not make trouble. Unfortunately, these treaties were broken time after time by the Americans as more and more whites wanted to move west. In order to justify breaches of treaties and a greedy land-grabbing policy, the Government invented and followed what was called the Manifest Destiny. According to the Manifest Destiny, the

> ...Europeans and their descendants were ordained by destiny to rule all of America. They were the dominant race and therefore responsible for the Indians - along with their lands, their forests, and their mineral wealth...[6]

The Plains Indian tribe known as the southern Cheyennes, whose major leader was known as Black Kettle, had endeavored to keep the peace and get along with the whites. However, a few of Black Kettle's young warriors and their warrior friends of the Sioux tribes further north, had in 1864 attacked white settlers, wagon trains, and stagecoach stations along the South Platte River route in eastern Colorado, the main route used by the whites to enter Colorado territory at that time. This provoked fury and fear by the white settlers, and a militia force was formed by the Governor of Colorado Territory and placed under the command of one Colonel Chivington. The primary purpose of this militia force of Colorado volunteers was to drive the Plains Indians from Colorado. The Indians, including Black Kettle's Cheyenne group, were ordered to remove themselves to the vicinity of

Fort Lyon in eastern Colorado in November of 1864, because of the attacks on whites along the South Platte River route. Colonel Chivington and his 600 volunteers, anxious for "retaliation" against Indians for their attacks on whites, arrived at Fort Lyon intent on attacking nearby Indians on November 27, 1864.[7] Not all of the regular United States Army troops at Fort Lyon wanted to accompany Colonel Chivington in his planned attack on the camp of Black Kettle's Cheyennes, camped about 40 miles from the Fort at Sand Creek. Some of the regular Army officers protested that Black Kettle's camp was peaceful, and that they had been promised safety by some of the officers. They proclaimed that any officer participating in an attack on Black Kettle's camp would be committing murder and dishonor the uniform of the U.S. Army. At this, Chivington became furious and shouted, "Damn any man who sympathizes with Indians!... I have come to kill Indians, and believe it is right and honorable to use any means under God's heaven to kill Indians."[8]

Figure 20-1   *The approximate area occupied by Plains Indians in 1850*

The Cheyenne camp consisted of about six hundred Indians, more than two-thirds of them older men, women and children. The Cheyennes were so sure of their safety, most of the warriors were several miles east hunting buffalo. Chivington's 600 Colorado volunteers and Army regulars from Fort Lyon attacked Black Kettle's Cheyenne group at dawn on November 28th. Black Kettle at first could not believe that the American soldiers would attack his camp, because his group of Cheyennes had remained peaceful and on good terms with the Army. He encouraged his followers to stay by him under an American flag he had been given, with the promise that he and his people would not be harmed as long as he flew the flag. When he and his followers realized that the soldiers were going to try to kill them under any circumstances, they fled, fighting rearguard actions. Chivington's men butchered many defenseless women and children that terrible morning. When the smoke had cleared, one hundred and five Indian women and children were killed by the soldiers and 28 Indian men were dead. Chivington's official report claimed that his attack had resulted in the deaths of between 400 or 500 Indian warriors. Nine soldiers were killed and 38 were wounded, many of these casualties being the result of careless firing by soldiers eager to kill Indians.[9]

The atrocities committed by the Colorado volunteers on the Indians were documented by a few regular Army officers. Dee Brown quotes in his book, *Bury My Heart At Wounded Knee*, the report of Lieutenant Connor:

> In going over the battleground the next day, I did not see a body of man, woman, or child but was scalped, and in many instances, their bodies were mutilated in the most horrible manner - men, women and children's privates cut out...(sic) I heard one man say that he had cut out a woman's private parts and had them for exhibition on a stick; I heard another man say that he had cut the fingers off an Indian's to get the rings on the hand; according to the best of my knowledge and believe these atrocities that were committed were with the knowledge of J. M.Chivington, and I do not know of his taking any measures to prevent them; I heard of one instance of a child a few months old being thrown in the feeding box of a wagon, and after being carried some distance left on the ground to perish...[10]

It is hardly surprising that the men under Colonel Chivington should commit such atrocities. Not long before the Sand Creek massacre, in a public speech given in Denver, Chivington advocated the killing and scalping of all Indians, even the infants. "Nits make lice!" he is quoted as having said.[11]

The Sand Creek massacre led to the Cheyenne and the Arapahoe tribes of Indians abandoning all claims to the Territory of Colorado. They had already lost most of the claims to hunt in their traditional territories in Kansas.

Chivington's senseless massacre at Sand Creek destroyed any power retained by any Cheyenne or Arapahoe chief who advocated peace with the white man. Most of the surviving Cheyenne and Arapahoe warriors turned to their war leaders in an effort to save themselves from extermination.

After a few more bloody clashes with soldiers and other whites in Colorado and Kansas, many Southern Cheyennes moved north where they were welcomed by their kin, the Northern Cheyennes and their allies, the Sioux. The Sioux and Cheyennes still controlled much buffalo hunting grounds in larger parts of what are now five states, Wyoming, Montana, North Dakota, South Dakota, and Nebraska. In the 1860s, the Sioux and their Cheyenne allies had actually thwarted an attempt by white Americans to build a road across their Powder River hunting grounds in Wyoming. In 1865, some seventy-three gold-seekers tried to move through the Indian territory along the Bozeman Trail in Wyoming to the gold camps of Montana. At the same time a large contingent of the U.S. Army invaded the Powder River country with the intent of breaking the hold the Sioux and Cheyennes had on these buffalo hunting lands. The leader of this invading force, General Patrick Connor, told his officers to accept no peace overtures from Indians, but to "(a)ttack and kill every male Indian over twelve years of age."[12] The Sioux and Cheyenne warriors, who were furious at the intrusion of the whites into the country they had been promised would remain theirs in previously negotiated and signed treaties, successfully defended their lands. Under the leadership of the Sioux leader, Red Cloud, they forced a withdrawal of "blue coat soldiers," who attempted to build a couple of forts along the Bozeman Trail. But, the pressure by the United States Army continued until finally, in 1868, Red Cloud and other Sioux and Cheyenne chiefs signed a treaty with the United States Government, which required them to live in reservations in western Nebraska and South Dakota, but gave them some access to their traditional hunting grounds.

The Black Hills of South Dakota and northeastern Wyoming were granted to the Indians in the treaty of 1868. The Black Hills is a large area of heavily forested hills and low mountains which the Indians considered sacred. This area of mountains that rose from the surrounding plains, was considered much more than hunting grounds. For a few years mostly peaceful relations existed in the center of their world, considered a holy place of the Great Spirits by the Sioux,

Cheyenne and Arapahoe. The Plains Indians, whose form of spiritual beliefs was expressed as a type of shamanism, often went to the holy mountains to speak with the Great Spirit and await visions. The treaty of 1868 stated specifically that the Americans were not allowed in territory that had been granted to the Indians, without the consent of the Indians. But, white gold miners began to invade the Black Hills in the early 1870s to look for the yellow metal in the clear streams of the hills. Unfortunately for the Indians, the whites found gold in the Black Hills, and by 1874, gold-hungry Americans put so much pressure on the government that the Army was ordered to make a reconnaissance expedition into the Black Hills. Neither the Army officers involved in the expedition nor any U.S. Government officials asked permission from the Cheyenne or Sioux to explore the Black Hills, as had been required by the 1868 treaty. The leader of the Army expedition, Lt. Colonel George Custer, reported that the Black Hills were full of gold. After this, hundreds of gold miners flocked to the Black Hills, even though they knew it was illegal to do so. The Army made no serious effort to remove these miners, who were in the process of building wooden houses and gold camps that became small towns in the area the northern Plains Indians considered the most sacred of all of their lands. The Indians were furious at the white man's intrusion into their sacred land.

In 1875, the United States Government sent a commission made up of politicians, military officers, missionaries and traders, "to treat with the Sioux Indians for the relinquishment of the Black Hills." [13] This commission was to negotiate a price the Sioux and Cheyenne would be paid to give up their claims to the Black Hills. There was no negotiating with the Indians as to whether they wanted to give up the Black Hills to the American Government; this was already a fait accompli, as far as the government was concerned.

The Indians had had enough. They could see that the "words" of the treaties meant nothing to the government and that the white man could come and take any of their land they wanted, regardless of what had been said and agreed to in earlier treaties. Many of the younger warriors decided to leave the reservations in defiance of this same 1868 treaty and live with the Indians who had never accepted the reservation system and never accepted the hand outs of the whites, such as the small bands led by Sitting Bull and Crazy Horse. Thus, the stage was set for the largest and best known of the hundreds of bloody conflicts between the native population of America and the United States Government.

By early 1876 the relationship between the Sioux and Cheyenne and the United States Government had deteriorated to the point of war. The government declared all of the 10,000 or so Sioux and Cheyenne Indians that were living outside of their assigned reservations as "hostile" to the United States. The Army was sent after these "hostile Indians" in the spring and early summer of 1876. In late June of that year, a three-pronged search for the "hostile Indians" was launched by the Army into Indian territory.

The Indians at that time had all come together in a huge camp next to the Little Bighorn River in what is now southern Montana. All the bands of Sioux, plus several bands of Cheyenne and a few Arapahoe were together in this large camp. These Indian groups did not normally all camp together, but the hostilities with the Americans had thrown them all together.

Leading the Seventh Cavalry Regiment that was probing for Indians from Fort Abraham Lincoln in North Dakota towards the west was Lt. Colonel George Custer. Custer was a veteran of the Indian wars and had led the army contingent into the Black Hills in 1875. During the American Civil War (1861-1865), Custer had served with much distinction and bravery. During the war he had been awarded the brevet (nominal) rank of Major General, and he is remembered today as General Custer. Custer had political ambitions, one of which may have been the Presidency of the United States.

Custer found the huge Indian encampment on the Little Bighorn on June 25th and decided to attack at once. To attack this huge body of Indians that contained between three and four thousand warriors with only a regiment of cavalry, can only be considered foolhardy. It is believed by many, that Custer's political ambitions interfered with his ability to make logical military decisions on that June day. In any case, Custer divided his regiment into three groups, two of which were to attack the Indian village from the south, while he took the remaining group of over two hundred cavalrymen to attack the village from further north.

The Indians were taken by surprise at first, but through sheer force of numbers, which greatly complemented their bravery, the Indians killed every one of the two hundred plus men under Custer's personal command, including Custer himself. A few miles up the river, the other two groups of soldiers of the regiment were holding out on the top of a hill which the Indians had completely surrounded. On the next day, the Indians broke camp and headed to the southwest because scouts had seen a large Army column moving down from the north.

When the Army troops under the command of General Terry arrived on the scene, they found all of Custer's men stripped and many of them mutilated. This, of course, added to the reputation of Indians among the whites as "pure savages," but the whites did not remember the Sand Creek massacre of Black Kettle's band of Cheyennes in 1864 by the Colorado volunteers and the atrocities committed there.

When the white men of the east heard of Custer's defeat at the Little Bighorn, they called it a massacre and were very angry. The idea that a group of "savages" had "massacred" troops of the United States Army infuriated many whites, including white politicians. They sent more Army troops out to hunt down the Indians that were off the reservations. By 1877, all of the Plains Indians were either dead, in a reservation, or had, like Sitting Bull, fled to Canada. They had lost their lands and their way of life, their culture, to the whites, who felt they deserved the Indians' land because they belonged to a culture that was materialistically and religiously superior to those of the heathen Indians. It was destiny - Manifest Destiny. Maybe it was destiny. Maybe there was some greater plan behind this human tragedy. One certainly doesn't want to judge either side of this tragic conflict, as they both exhibited barbaric behavior. Still, a strong case has been made that the white culture could have accommodated the Indians, even if the whites felt compelled to take much of the land on which the Indians lived, for themselves. What the white culture did in fact was strip the native Americans of all but a fraction of their land, while at the same time strip the Indians of any of their traditional means of preserving their culture - their very way of life - and their pride in themselves.

Putting aside all temptations of judging this historical event, we can at least say that this is one of the more abominable, horrible examples of two human groups - two human cultures - interacting with each other. And, we all know that hundreds of other examples of cultures clashing could be given, that are at least as lacking in love as we saw in this brief review of the clash of the Plains Indian culture with that of 19th century United States culture. The systematic attempt of German Nazis to eliminate the European Jews before and during World War II, the attempt of genocide by the Khmer Rouge on groups they considered bourgeois Cambodians in the 1970s and, the current example of attempted genocide in Bosnia, are but a few of the examples of interactions of human groups that are without love, and where one group was/is forcibly trying to control another.

Why should we humans be upset at the attempts of the regressive ETs to dominate and control our cultures? Is not what the United States Government did to the Plains and other Indian cultures, similar to the control the regressive ETs are trying to exert over humans through deceitfulness? Is not the behavior of the Americans versus the Indians, and the hundreds of other examples of "man's inhumanity to man" one could name, similar in many ways to the relationship the greys and reptoids have with the whole human race? The regressive ETs are a reflection of ourselves. We do not have to accept the negative "rules" or the negative imprints of the dominator-victim game of the regressive ETs. Extremely negative interactions humans commonly have with others would not happen if humans did not act out these scenarios themselves. We did not write this book to incite readers to act physically against the less-than-positive ETs. There is nothing we could do physically, because the greys and reptoids are much more technologically advanced than are we humans. There is no hope of physically overpowering the regressive ETs. But, if we are aware of the dark forces on this planet and their machinations to control us, and we take responsibility for our own spiritual evolution and logically make a conscious decision to turn away from darkness and turn toward the light, we do not have to play the victim role to the regressive ETs anymore. The truth of our existence is that we are potentially tremendously powerful beings of Light. We can escape the machinations of those ETs who do not have our best interests in mind by simply developing our potential to hold on to the Light, the Truth of our existence. We are capable of spiritually evolving a tremendous amount in a short period of time. The major key to this rapid evolution is to develop our emotional and spiritual selves in the direction of love and respect for all life. Certainly we cannot evolve in this direction if we think of any group of humans with whom we share this planet, with anything less than consideration and compassion. This is a great leap up from any of the standard religious or secular belief systems we humans have inherited from our ancestors. Let us take a closer look at this belief system that is based upon knowledge, compassion and unconditional love.

---

[1] Brues, A. M. 1977. *People and Races*. Macmillian Publishing Co., Inc., New York, p. 186.

[2] Brown, D. 1970. *Bury My Heart at Wounded Knee*. Holt, Rinehart and Winston, Inc., New York, p. 3.

[3] Kvanika, R. M. 1988. "United States Indian treaties and agreements." In: *Handbook of North American Indians, Vol. 4*. Smithsonian Institution, Washington, p. 195.

[4] Brown, D. 1970. *op. cit.*, p. 220.

[5] Bird, I. L. 1960. *A Lady's Life in the Rocky Mountains*. University of Oklahoma Press, Norman, p. 6.

[6] Brown, D. 1970. *op. cit.*, p. 8.

[7] The account of the Sand Creek massacre is taken from Brown, D. 1970. *op.*

cit., pp. 67-102.
[8] *Ibid.*, pp. 86-87.
[9] *Ibid.*, p. 91.
[10] *Ibid.*, p. 90.
[11] *Ibid.*
[12] *Ibid.*, p. 105.
[13] *Ibid.*

"Love ye one another."
Jesus

Chapter 21

Humanity's Golden Moment

We have now, in this introductory book, reviewed the origins and existence of our species. We saw how this Earth and life's evolution on it was apparently originally conceived and supervised by *loving* entities of the Light, but that forces of darkness raided and have controlled most of the reality-existence of humans for the past at least 300,000 years. We have seen how the ET representatives of the forces of darkness have sought to control humans by keeping us humans divided along political, racial, monetary, and especially religious lines. I have argued and tried to demonstrate that all human belief systems that purport to explain human origins and existence, from the anthropological-Darwinian point of view of the scientists, to that of any of our world religions, are part of the worldwide cover-up of the fact that we humans have, for the most part, been manipulated and controlled by entities, extraterrestrial and terrestrial, who do not have the best interests of the human species in mind. We humans have been duped, especially with regard to our spiritual potential and purpose. We have been programmed and taught that the potential of humans on Earth is but to survive, perhaps raise a family, and maybe live out a somewhat happy retirement.

Despite the grim news of the presence of the regressive ETs and other dark forces on the Earth for thousands of years, this last decade of the 20th century is a marvelous time for us humans to be alive. Finally the truth of human origins and existence is becoming known. We humans no longer have to be confused by the seemingly innumerable religious beliefs of the different sects of the major and minor religions, as well as the beliefs of the powerful materialistic scientific community, which seems to contradict and negate all belief in a spiritual existence for humankind.

During the medieval times of Europe, many lost faith in the seemingly naive explanations of the world and of human origins and existence offered by religious dogma. So materialistic science was born, and for centuries much of humankind relied on science to explain the world and human existence. Now, in the late twentieth century after decades of intensive investigations of "natural systems" of the world by science, the natural systems of the planet are on the verge of collapse. Ecological, economic, nuclear, social dysfunction, and a host of other calamities threaten the very existence of humankind. Furthermore, humankind is as divided as it ever was. In the decade of the 1990s, however, more and more people, through significant coincidences and insights, are recognizing that there is something else besides mere physical existence.[1] A spiritual existence beyond material, physical existence is now quite obvious to millions of people, but an understanding and integration of such an existence still eludes most. Knowing of the regressive ET influences on the development of humankind over the past several thousand years will help us understand and integrate our spiritual nature into our physical existence. Finally we humans will be able to understand the seemingly bewildering world in which we live, and to discard the misinformation that has been heaped upon us and integrate the knowledge necessary to make the decisions to grow and evolve spiritually.

So what are we supposed to do to escape the control and machinations of the ET representatives of the forces of darkness and their human allies? First, we must become aware of our true origins and the human predicament throughout history. Simply knowing about the genetic manipulations the regressive ETs carried out to create and control a slave species full of fear and ignorance, is the first step in expanding our consciousness. Then we must learn and know how the regressive ETs and their human allies have negatively manipulated and influenced our civilizations throughout history. After becoming aware of these historical facts, and then by turning to the spiritual path characterized by a tradition of unconditional love, we are capable of expanding our consciousness to the point where we do not have to play victim in the dominator-victim game that the forces of ignorance and darkness have been playing here on Earth for the past thousands of years.

Since I have presented evidence here that the regressive ETs have worked especially hard to keep the human species ignorant of its spiritual existence and spiritual potential, I now feel obligated to give a brief, introductory statement of my understanding of our true spiritual existence, and the probable course of our spiritual evolution over the next few years. I am happy to finally assume this task because it is the happiest, most joyful part of the book. The really astounding information revealed during the process of researching the material for this book, is that there is going to be, within the next couple of decades, a very intensive expansion of human consciousness toward a paradigm shift. There is really no event or series of events in our human history to prepare us for the tremendous amount of spiritual knowledge that many, perhaps most, if not all we humans will acquire in the next few years. This is astounding, astonishing, happy news, indeed!

Human Spiritual Existence

In the modern, materialistic, western Judeo-Christian cultures not much is said about human spiritual existence, except that our souls will go to heaven or hell upon death, depending upon whether we have lived a good or bad life here on Earth, according to the "laws" of God. However, there is much evidence that the souls of entities incarnate into several physical bodies during the course of their spiritual evolution.

Most readers have experienced something in their own lives (or know of people who have had such experiences) that indicated that humans have an existence beyond the physical one. An overwhelming majority of Americans believe in some sort of spiritual existence of humankind, although understanding and interpretations of the nature of this spiritual existence vary widely. Spiritual information from esoteric sources indicate that the understanding humans have of spiritual matters is, in general, quite limited. What could one expect from a young primitive species that has been exposed to so much darkness for the past thousands of years?

A large part of the problem of spiritual understanding by Americans is that the official government position, seen through the official government support given to the materialistic scientific view, advocates that there are no spiritual aspects to humans. Generally speaking, there is very little investigation into possible paranormal phenomena by government supported scientists, or other scholars. There are, however, a very few inquiries into paranormal phenomena by academicians. One such investigation is by former colleagues of mine at Colorado State University (whom I did not know of while I was a professor at this institution), and their research is a good place to begin this kind of investigation of human spiritual aspects, and in particular, reincarnation.

Engineering professor, Dr. M. L. Albertson, and psychology professor, Dr. K. P. Freeman, have reviewed former research into the reincarnation phenomenon in a paper entitled "Research Related to

Reincarnation."[2] Albertson and Freeman point out that, even though reincarnation is generally not considered an appropriate subject of scientific inquiry, evidence of reincarnation will not go away, and such evidence has "...come to the fore in an enormous surge that has defied suppression..." during the past few decades.[3] Some of the evidence the authors review, that supports the notion that most humans have incarnated into several lives on Earth and elsewhere, are the hundreds of documented cases of young children actually remembering a previous life, usually the immediate past life. Several cases have been documented where a child, usually under the age of six, remembers with great clarity details of a former physical existence. Often, in India or other countries where belief in reincarnation is prevalent, the memories of a child are actually checked out by visiting the claimed physical location of a former life, and details remembered by the child are found to be astonishingly accurate.[4]

Albertson and Freeman also quote evidence of people who have been regressed, usually through hypnosis, under which they remember one or more former lives. The authors state that more than 1,000 such cases have been investigated scientifically. It is interesting to note that most of the regression hypnosis cases involve people from western societies, where the prevalent view is that there is no such thing as reincarnation.

The primary evidence Albertson and Freeman offer to support the claim that we are spiritual beings, that we do not perish upon the death of our physical bodies, are near-death-experiences (NDEs). NDEs often occur when a person is being operated upon during which time he or she "dies" - the person is seen as clinically dead for a period of time before they revive. Research has produced more than 400 studies of NDEs which report remarkably similar experiences during the minutes the patient was considered clinically dead. Most people who have reported NDEs, state that a much lighter version of themselves lifted out of their "dead" bodies. Their lighter bodies were able to float through walls and the bodies of living people without being noticed. The people who had near-death-experiences usually felt very peaceful during their NDEs and experienced "...a very pleasant sensation of relief and relaxation."[5] Some of those that experienced NDEs saw relatives and friends that had already died, while others saw religious figures. Most of the people who experienced NDEs made the conscious decision to return to their physical body, at which point they were observed to remarkably revive by attending medical personnel. The authors report a Gallup poll that estimates that as many as 8,000,000 Americans have experienced a NDE.[6]

Albertson and Freeman also review reincarnation beliefs of the world religions, including ancient religions, that have or had belief in some form of reincarnation. As we've seen, Buddhism and Hinduism are the current major world religions where reincarnation is accepted as fact. Furthermore, the authors point out, even world religions that deny reincarnation have doctrines of reincarnation in more mystical traditions of their religion, or may have had beliefs in reincarnation in the past. In this regard, we have briefly reviewed in Chapter 18 the history of how belief in reincarnation was eliminated from early Christianity, even though there is evidence, even in the "official" gospels, that Jesus believed in and taught reincarnation.

Orthodox Judaism, like Christianity, rejects any doctrine of reincarnation, and the official position of the Jewish religion is that reincarnation had no part in the early Jewish religion. Yet, Albertson and Freeman point out that there are a few passages in the Old Testament that refer to some form of rebirth (Proverbs 8:22-51; Jeremiah 1:45; and Malachi 4:2-6).[7] Furthermore, the Kabala (Kabbalah), the mystical tradition of Judaism that has been traced to the 3rd Century B.C. and purports to be the hidden wisdom behind the Old Testament, contains many references to reincarnation.

Islam is another world religion whose vast orthodox population does not accept any doctrine of reincarnation. However, some of the less well-known, more mystical sects of Islam, such as the Sufis and the Druses, have advocated doctrines of reincarnation. Also, many Islamic mystics claim that the Koran, when translated correctly, contains references to reincarnation.[8]

Thus, despite the distortions that have been perpetrated on our world religions that have no belief in several physical lives of a soul, there is considerable information about reincarnation. Even if the concepts of reincarnation held by the Hindus and Buddhists are corrupted, at least the millions of humans of these two religions have some sort of concept of this spiritual awareness and understanding. And, much can be discovered about reincarnation in the three world religions that publicly deny having any form of the concept, Judaism, Christianity, and Islam, if one looks for it.

Albertson and Freeman point out that some research into reincarnation indicates that one's own spirit chooses its parents for each incarnation.[9] The soul, in fact, chooses the circumstances of incarnations in order that the soul might learn and grow. This concept of a soul being able to choose the circumstances of its physical incarnations is found in many of the more mystical traditions of spirituality of the world.[10] The basic lessons the higher self or spirit is striving to learn during its numerous incarnations on Earth and elsewhere in the universe are lessons of unconditional love. Lessons of expanded knowledge, compassion and unconditional love are apparently difficult lessons for a soul to learn, especially on a planet where the frequency has been controlled by ET entities who do not believe in the power of love. Yet, there are some who still hold the traditions sacred of a compassionate, loving state of being.

According to the scenario supported, in part, by world religions and modern research, the human species has been physically trapped in a cycle of deaths and rebirths in order that our greater entities, our spirits, can learn the lessons of unconditional, non-judgmental love through experiences in physicality, or in what we call the third dimension. There are supposedly many characteristics of being physical that are very valuable in the course of one's spiritual evolution.[11] The Earth has been likened to a "spiritual boot-camp," where our spiritual beings can expand and grow. Eventually, as both esoteric information and ancient Earth traditions claim, we grow out of the need to learn lessons through physicality and move on to other dimensions - other frequencies - as we will move beyond the need to experience physicality and begin to experience higher dimensions and frequencies. A whole new consciousness is awakened and manifested!

Academicians Albertson and Freeman state that reincarnation is closely related to many so-called paranormal phenomena, including "...possession...poltergeists; mediumships and channeling; UFOs, walk-ins, extraterrestrials, aliens and guides; and life before life, life between lives, and the experiences of the unborn child..."[12]

The paranormal phenomena, including ETs, are somehow expressions of other dimensions, about which we understand very little. However, even our brief review of the existence of humankind has uncovered much evidence that not only do higher dimensions and frequencies exist, virtually all ETs who have in one way or another interacted with humans on this Earth, have much experience with other dimensions. As we have noted, there is some evidence that we sentient beings may be capable of evolving capabilities of maneuvering in two frequency dimensions, or density levels, above the third density without evolving the capacities of Unconditional Love. However, if we humans choose the way of Love, the way of devoting one's life completely to the path of Unconditional Love and respect for all life, humans can evolve to fourth and fifth density levels of frequencies and beyond, into a mental state resonating bliss.

An accurate understanding of dimensions and density levels beyond our current three dimensions is beyond the comprehension of most humans at present. This should be a stark reminder to us of just how new and primitive a species modern *Homo sapiens* are. And we have remained quite primitive, duped and manipulated by our creator gods and their allies for thousands of years. Again, there are many indications that things are about to change for much of the human species. Many Earthly and esoteric sources are predicting dramatic changes that are to occur as we approach the end of the 20th century. Everything from earthquakes, to severe hurricanes, devastating floods, very destructive fires and other natural disasters, as well as nuclear wars, the end of time, an Earth axis tilt, the second coming of Jesus - almost everything and anything has been predicted to occur towards the end of this century. There might be something to the plethora of predictions of dramatic changes in the near future, but the evidence indicates that the most important change that will occur, will be the dramatic change in the consciousness of us humans. Within a very short time, according to predictions, hundreds, thousands, and millions of humans will begin to remember who they are, and begin to heal their relationships with what we call God, themselves, and each other. Knowing of our true origins and existence will add immensely to our spiritual understanding and evolution. Why this particular time is more propitious for this astounding jump of consciousness than other times, I don't know - but we might be in the midst of a natural cycle.

We have apparently been genetically programmed to now evolve very quickly to the point where we remember our origins and the Prime Creator, towards whom we are evolving. As Jesus and other great spiritual teachers have said, the human potential is to be like gods.[13]

According to esoteric sources, the efforts of all of us who tried to understand the truth of human origins and existence in the early part of the 1990s, will seem primitive when we reach that advanced stage of our evolution, where we will know that we are multidimensional beings that not only have lived in many past lifetimes, but have simultaneous existences at the present time. Apparently, we will be able to remember past lives we have lived, and current lives we are simultaneously living. Among the most important things we will learn will be that Unconditional Love involves compassionate non-judgment of all that is around us, including the regressive ETs, past and present. We will learn that parts of ourselves have been, or are, as manipulative and regressive as the most manipulative and regressive ETs. When we realize that we have assumed all types of physical existences in other incarnations, it will make it easier for us to be non-judgmental of the actions of other humans and regressive ETs in the coming changing times. A mass healing of consciousness must take place, beginning with a healing of all relationships. Some of us will be afraid of changing and will resist change at every turn. Others will be deceived by the holograms and other trickery put forth by the regressive ETs. Some people still need the regressive ETs! Other humans will want to take advantage of the guidance coming in from the positive forces of the universe, and evolve to a higher, more blissful state of consciousness and existence. Since this is a freewill universe, the *choice* for healing and integration is up to each individual.

Despite any contravention the members of the secret government or regressive ETs have perpetrated and committed and are committing on humankind, we must develop the capacity of compassionate non-judgment and forgiveness if we are to survive, let alone evolve. This is an especially tough lesson to learn, as a normal biological human reaction would be to wish no good on those who are treating us in a considerably less than honorable fashion. This is not to say that we humans do not have the right to stand up to these dark universal forces and insist on our rights, and not allow them to distract us from our goal of evolving toward the Light. This is humanity's awakening moment in time. It is now time for all of us to wake up! First, we must invite these regressive ETs and their human allies to join us in following the paths of Light and Love. Esoteric sources hint that a few reptoids and greys have actually turned to more compassionate emotions, so we must pray that this is possible for them. As for the human allies of the dark forces, our fallen brothers and sisters, as long as they still have an abundance of primate-mammalian genes in their bodies, they would seem to have a greater capacity than their regressive ET masters to turn toward the Divine Light and Love. If the dark forces now on Earth do not want to turn toward the Light, those of us who are turning toward this radiance must insist they go exist on another planet, or in another dimension, in a more conducive environment for developing and maintaining a slave race. Failing this, if the negative forces on this planet refuse to leave and cannot be forced to leave, those of us evolving towards this radiance will hopefully evolve to other dimensions and leave the dark forces in this third density to carry out their machinations. This is quite definitely humankind's moment of choice to stand up and be counted.

We pray that the regressive ETs and all negativity on Earth and elsewhere, is part of a Divine plan. We must learn to accept this negativity, and our own personal negativity, expressed in this or other lifetimes, as part of the whole of creation - of universal existence. In past physical incarnations our souls may have themselves assumed a lizzie or grey species existence. And, while we are to learn to accept the dark side of the universe without judgment, we must also learn to emphasize the positive aspects of creation centered around the ability to love.

The personal choice to lead a life of Divine Love, which is the path of Light, is ours to make. It is a logical choice. It is much more logical to choose the paths of Love and Light, than to agree to continue to play by the dominator-victim rules of the dark world, which have been set up by the regressive ETs and their human allies. The realization of just how dark this world is, has helped me in my own personal decision, and hopefully it will help you readers make your decisions as well.

Both esoteric and exoteric sources talk about the importance of a much lesser emphasis on the logical mind, at least as we know it, in the evolution of a higher, more expanded spiritual consciousness. Many of the so-called "spiritual truths" defy the "logic" of the current consciousness of most humans of this planet. Humans who have come to rely the most on logic, such as scientists, might have more difficulty than people who try to rely on their intuition and awaken feelings of their hearts, to some extent, when attempting to understand and accept spiritual truths.

Many spiritual teachings emphasize that as one's spiritual self awakens, one develops a "knowing" - sort of a very acute form of intuitive guidance, where a person knows which actions and events in one's life are in the name of ignorance and darkness, and which events are for the purpose of Truth and Light. Oftentimes, events appear in our everyday lives that require choices to be made, and by making these small (or large) choices of everyday life, we have the opportunity to develop our discernment and knowing. As one commits to the paths of Divine Love and Light - commits every moment of every day to following the paths of Love and Light - one's discernment develops more and more. So, the logical mind seems to take an honorable back-seat to intuition, when expanding into some of the higher spiritual realms.

From one point of view, the regressive ETs and the secret governments would seem to have an overwhelming advantage over the primitive human species. Regardless of the exact relationship between the reptoids and the grey species, they both have an enormous technological advantage over human cultures at this time. They also have much knowledge and capability in other dimensions, in time travel, mind control and other aspects of ourselves we crudely label "metaphysical phenomena." They know much about our multidimensional selves, and have programmed us to even deny the available knowledge of the existence of multidimensionality. The regressive ETs are much more intelligent, at least in a logical sense, than all humans, even our resident geniuses. As we've seen, they can assume different physical forms that conceal their true identities and purposes. They have the technology that causes most humans to forget what they wish them to forget, such as the fact that they abduct humans for their own experimental purposes. They have the ability to create holograms and dramas that appear totally real and "true" to most humans.

While we hope that some of the more negatively inclined ETs might be moved to join the legions of Love and Light in this time of rapid evolution, the historical records we have, distorted though they may be, indicate that many of the regressive ETs will continue to try to control humans and feed from humans' fearful emotions, as the ETs themselves spread fear and chaos. In that case, those of us who wish to follow the path of Love and Light, will have to part ways with our reptoid creator gods and their grey-skinned allies.

Apparently we humans have complete control over all of the circumstances of our lives, if we would but recover and awaken to the facts of our spiritual potential and believe it, know it and live it! Until we do awaken to the fact that we control all the events of our lives, we will be subjected to whatever anyone wishes to do to us. *The choice is ours.*

As individuals discover and live by integrity, their rapidly developing consciousness will teach them that they are in charge of their lives. They will know, moreover, that each individual chooses the reality system in which they prefer to participate, and that this always relates to the level of consciousness that the individual is expressing at the time. So, if we want to avoid the greys, for example, one might expand one's consciousness until one clearly recognizes the greys for what they are, discern that humanity does not need nor want these distracting aliens, and thereby, exclude their presence from their lives. I know it is not as simple as that, but there are cases of individuals being able to avoid the greys by becoming more spiritually oriented. And, as we have seen, there is some evidence that some greys prefer not to abduct individuals who are seriously involved in a spiritual practice.

In addition to our enormous untapped powers we have been and are discovering, we are receiving help from the more spiritually oriented ETs. We've seen that the forces of The Great Spirit are pouring Love, or related information, onto our planet at the present time. For those who accept and cultivate this Divine Love, the Love acts to heighten the devotional frequency, the loving energy, more compassionate attitudes, and a lessening of judgments toward others to create the healing and purification necessary. We are in the process of healing the relationship with our Mother Earth, as well as ourselves. There even seems to be a possibility that there might be direct intervention from the "white shirted" ETs in our struggle to bring light to this planet. Lynette has understood and practiced that there is a Divine Plan to transform the Earth into a world expressing Love and Compassion. Quite possibly many of us have taken the responsibility *to choose* to evolve consciously and spiritually with a deeper knowledge. Humans are part of the plan. There is no precedent in human history, as we know it, for the rapid spiritual evolution we are about to experience and witness on Earth, although there may have been such periods occurring more than 10,000 years ago, that have been almost completely covered up. However, there is no guarantee that this Divine Plan will be accepted; that is, there is no guarantee that humanity will, in fact, become more loving and much less controlled at this time. Esoteric sources tend to indicate that this whole healing-consciousness transformation that is to occur on Earth, ultimately, depends on us humans. Only if enough of us commit to a life of complete integrity and spiritual practice will the Earth become a planet of Light. So, with our own tremendous, and largely untapped inner powers, like courage, strength, and most of all forgiveness, plus a lot of help from the forces of universal light, we can tip the state of this planet's consciousness toward Love and Light and heroically take the path away from the powers of darkness.

The Pleiadian Plus group claims to be from our future, and according to them, the future is not set, but contains a number of possibilities and probabilities. These beliefs appear to express the multidimensionality of ourselves and our existence. As we understand the circumstances, there may be an actual split of worlds between those that emphasize regressive control and those that emphasize love and compassion. This is quite more than most of us can logically comprehend at the present time. It is up to us to decide whether we want to live in a world dominated by darkness and misinformation, or one dominated by light and truth. It is up to all of us humans to make the choice toward making a total commitment to a life of love and light, or pass on the chances to make such a commitment.

Some who, perhaps, have not chosen as many learning experiences as others over the centuries, and are, consequently, not as ready as others for this extremely rapid evolution of consciousness, have incarnated into the world programmed for rapid change as observers. These entities are hoping for a rapid development of their own consciousness just by being present at this time. Remember the evidence indicates that the higher the aspects of one's soul, one's higher self, that we choose to realize, the more profound and even, perhaps more difficult the incarnations the soul is to assume. And the soul will choose its lives in such a way as to allow it to learn the lessons of life, including the lesson of love and compassion.

As mentioned, quite possibly many of us incarnated into this time of rapid change of consciousness that the Earth is programmed for, though they are not encoded for it. To make such a decision perhaps denotes rather a brave soul. Maybe the higher consciousness of human souls who incarnate under such circumstances understands that people can be catapulted into higher consciousness, simply by being present during the rapid changes of consciousness that are to come.

We are all in this together. We are all here that we may hopefully and consciously evolve very quickly. Certainly learning and understanding our origins and situation vis-à-vis the rest of our universe will help us evolve. I am not suggesting that we become obsessed with the regressive ETs, especially with the implications of their controlling and distorting knowledge throughout our history. I do think, however, that we should prostrate ourselves at their feet, at least symbolically, because they can illuminate so clearly and so forcefully the regressive aspects of ourselves. Because of them, many of us will rush towards the light. We will know that we have had enough of misinformation, non-information, and other forms of darkness, and turn towards Light and the Truth.

As mentioned earlier, the emotions of humans are apparently the key to our spiritual evolution. No wonder that our emotions have been stuffed and controlled in the various belief systems of the world. Learning to be more emotional at a feeling level in our everyday lives will help our spiritual evolution, because our emotions connect us with our spiritual bodies.[14] Specifically, our feelings in life, our sensitivity to ourselves and to others, the sincere kindnesses that one gives freely and receives from others, connect us to the Spirit. As we honor the Spirit, we choose all of the above! And, of course, love is one of the emotions we must be most concerned with experiencing. We can all begin to cultivate love by learning to unconditionally love our "loved ones" - our parents, our sisters and brothers, our spouses and children, and our friends. Then we can learn to love others with whom we associate in our lives. As we cultivate the emotion of love, eventually we can even learn to love the regressive forces on Earth, which, as has been argued, are part of our total being.

Those who honor a commitment to a life of interconnected love by learning to love will experience bliss far beyond the love and bliss most of us have experienced in our present-day lives, and will prevail. Most of us will evolve very rapidly, some to the point, in this lifetime, where rebirth into a physical, three-dimensional world, such as the one we currently live in, will no longer be necessary. The "heaven" of Judeo-Christian and Islam cultures is apparently an existence, probably non-physical in our terms, where one's soul experiences what we term love and bliss much more frequently and more intensely than we do in our present lifetime, wherein we continue to evolve. As we are reminded so often, a brief examination of our history as we know it will reveal to us very few examples of individual humans who have had truly uplifting lives on this Earth. Most of us have suffered damage of some sort in our personal lives, as a result of abuse during one's immature years, or as a result of having lived through wars, or a starvation diet, or any number of dreadful calamities that are perpetuated on Earth. Even if the circumstances of our lives are generally favorable, almost all of us have grown-up in a belief system, whatever it is, that is ultimately false and misleading, especially with regard to Divine Love and our Spiritual Existence. Even those of us who had loving parents have probably not experienced much Love, as this seems to have been a rare commodity on this Earth. Even if we humans have not experienced unconditional love, we can learn it with our own concentrated efforts, and with the help of God - the Divine forces in this universe.

How much we will have to deal with the regressive ETs in the coming years is unclear, but probably all of us will at least have to know enough about them in order to know not to choose to be with them. As mentioned, we will ultimately thank them for their existence, so that we might more clearly and quickly choose the paths of love and light. We can thank the regressive ETs and their human allies for reinforcing spiritual teachings that de-emphasize the physical, the material plane here on Earth. As we've seen, some religions, such as the Buddhist and other religions, teach that all on this material plane are part of an illusion. The basic information of this teaching is that the material world, including all of one's own personal dramas of this lifetime, are infinitesimally small when viewed from the knowledge of one's total multidimensional existence.

It behooves us all to begin immediately to investigate the Truth of our existence, spiritual and otherwise. Several people are already well along their spiritual journeys, and I hope this very brief introduction has convinced those that have not formally begun their own healing journey, to begin investigating some of the immense information that is being revealed at this time about other dimensions and our spiritual existence. At the very least, I would hope to convince readers that there is much more substance to this universe we inhabit than most of us ever dreamed possible.

The good news about the spiritual journeys each of us are to take, is that each individual is in charge of one's own spiritual evolution. Furthermore, we humans have the power to accelerate our own evolution by simply mentally desiring to evolve in the connected direction of one's pure will and loving thoughts. Certainly as members of a young and still quite primitive species, we need much help in understanding the multidimensional aspects of ourselves. In this regard, for some, following, at least temporarily, a master teacher or guru might be a wise thing to seek. Each individual should know that if their teacher or guru, or friend, does not have the highest intentions of your own soul and everybody's soul in mind, then he or she is not worth following. If your master teacher or guru insists that you give up some of your own power to them or some god or some "higher" purpose, beware. Do not give up your power to anyone or anything. Maintain your own power so that you may evolve in the direction you choose, and so that you are in charge of what type of world in which you want to live.

It is not clear that one can totally eliminate the influence of the regressive ETs simply by becoming spiritual. These regressive ETs and their human allies are very much part of the world we have chosen to live in at this time. One can only throw guesses at the details of any regressive ET agendas they may have designed for humankind at this time. Whatever agenda they may have planned for humankind, it almost certainly involves their own power and control over the humans on Earth.

Most likely, however, we humans are soon going to have to deal with the agendas of the secret governments (see Chapter 12) before we might have to deal directly with the regressive ETs. And, the agenda of the secret world governments would seem to be more power and control for the few "elite" under the auspices of a New World Order,[15] which they and the regressive ETs would control. This would be the same type of one world government that ETs instructed French contactee, Claude Vorilhon, in 1973, and others before and after Mr. Vorilhon, to advocate to the world (see Chapter 12). Apparently power and control are more addictive than some of the more addictive drugs, such as nicotine. The more power one has, the more one wants. On the other hand, we can hope that the humans who have fallen for this diabolical plot, will turn to the light and try to save their souls. Certainly we humans who were intended to be victims of this diabolical plot of power and control by the "elite," could forgive our misguided bothers and sisters, especially if they finally began to reveal the "above top secret" information of what has been going on between the regressive ETs and themselves.

In the meantime, those on a true spiritual path will develop their senses of intuition and knowing, and they will know what parts of the agenda of the new world government they wish to resist and avoid.

They will know what to do in order to evolve spiritually to higher dimensions and frequencies.

Conclusion

Many who are strongly attached to their imprints and conditioning and are not ready to change could see this upcoming time of change as a time of chaos and fear. Those who are learning who they are, learning much more of the true history of the human species on Earth, and are spiritually evolving, will welcome this coming time of change and purification. In any case, I just want to wish all my fellow humans, including my brothers and sisters in the secret governments of the world, and even the reptoid and grey species - all entities, both physical and non-physical - the best of luck and best wishes during this upcoming time of major Earth-cleansing. May your spirits evolve rapidly toward a greater and more loving understanding of the Creation by the Prime Creator of this universe.

Before we end this little book, we would like to briefly summarize the main points of my tentative, evolving perceptions of human existence.

1. The current scientific (Darwinian-anthropological) story of the chance, exclusively materialistic origins and development of humans is full of holes.

2. Human historical records, especially Mesopotamian historical records, and the evidence of the current "UFO-ET" phenomena, indicate that the planet Earth has been visited by what we refer to as ETs for thousands and millions of years, and has influenced every stage of human biological, cultural, spiritual development.

3. ETs who do not have the best interests of humans in mind have been overtly and covertly, with a small number of human allies, controlling information and frequency on this planet for at least the past 300,000 years.

4. The main information the regressive ETs and their human allies have endeavored to keep from humans, is that humans are spiritual beings with a tremendous potential of evolving to more loving, blissful existences.

5. The world and its inhabitants have possibly been somehow programmed by positive entities for a rapid evolutionary jump in consciousness and spiritual evolution during the next twenty or so

years.

6. Despite evidence of considerable ET influence on this planet, during this time of rapid purification and change, each individual human is ultimately in charge of their own lives, their own reality, especially their own spiritual evolution.

This is one person's version of the Truth of our origin and existence, at least as I understand it now. Skeptics and debunkers, "official" and unofficial, will try to discredit this book by claiming that one or more sources used are unreliable, or that one or more details are incorrect, and, therefore, insist that the entire book and all of its premises be dismissed and ignored. This is especially the method the secret government has used to try to debunk hundreds of books concerned with UFOs and any purported governmental conspiracy. In this regard, throughout this book I have frequently alluded to the fact that, due to the nature of the "beast," some of the sources and details used cannot be absolutely correct. Nevertheless, I maintain that the basic premises of the book, summarized above, are close to the truth.

On the other hand, I don't blame any of the readers who are immersed in any of the prevalent religious or secular belief systems of today for being incredulous of this bizarre scenario in the mid 1990s. My only advice to those whom the information contained in this book strikes as incredible, is to hold onto your hats and be as observant as you can as we approach the beginning of the twenty-first century. Despite tremendous efforts by those who want to control this planet by limiting the information available to humankind, much information will continue to surface to support the basic hypotheses presented here. There will be paleontological, anthropological and other scientific discoveries that indicate the poverty of the materialistic scientific view of human origins, and the historic ET influence in the affairs on Earth. There may be discoveries in the realm of religious history that throw light on the origins of Earth religions. The truth about the spiritual nature of humans will be revealed in ways we cannot imagine. In this regard, we should look for evidence that some humans are beginning to hold onto very high frequencies and evolve consciously and spiritually to rich understandings. In short, I expect a flood of information, as well as the usual flood of misinformation, to descend on humankind in the next five to ten years. This is "the proof of the pudding," as my wife Lynette is fond of saying. As it becomes more apparent that the hypotheses presented in this book are close to the Truth, I hope all of my brothers and sisters will be vigilant, by looking carefully and intently after their own souls.

We congratulate all of you for incarnating at this time. This is a great time to be alive - a great time to commit to learn and evolve. We wish all of you love and bliss as we pass through the coming drama/ Dharma.

[1] See Redfield, J. 1993. *The Celestine Prophecy*. Satori Publishing, Hoover, Alabama
[2] Albertson, M. L. and K. P. Freeman 1988. *op. cit.*
[3] *Ibid.*, p. 355.
[4] *Ibid.*, p. 360.
[5] *Ibid.*, p. 358.
[6] *Ibid.*, p. 360.
[7] *Ibid.*, p. 366.
[8] *Ibid.*, p. 368.
[9] *Ibid.*, p. 361.
[10] e.g., Blavatsky, H. P. 1967. *The Secret Doctrine* (an abridgment, E. Preston and C. Humphreys, editors). The Theosophical Publishing House, Wheaton, Illinois, p. 146.
[11] Marciniak, B. 1992b. *op. cit.*.
[12] Albertson, M. L. and K. P. Freeman 1988. *op. cit.*, p. 355.
[13] The Holy Bible, The New Testament.
[14] Marciniak, B. 1992b. *op. cit.*, p. 32
[15] Jasper, W. F. 1993. *Global Tyranny...Step by Step: The United Nations and the Emerging New World Order*. Western Island Publishers, Appleton Wisconsin.

Bibliography

Albertson, M.L. and K.P. Freeman 1988. "Research related to reincarnation." Proceedings of the International Conference on Paranormal Research, Colorado State University, Fort Collins, July 7-10.

Anonymous 1992. "Tragedy found in Cambrian carnival." Science News 142, p. 109.

Augros, R. and G. Stanciu 1988. The New Biology: Discovering the Wisdom in Nature. Shambhala, Boston.

Baigent, M., R. Leigh and H. Lincoln 1983. Holy Blood, Holy Grail. Dell Publishing, New York.

_____ 1986. The Messianic Legacy. Dell Publishing, New York.

Bakker, R.T. 1986. The Dinosaur Heresies. William Morrow and Co. Inc., New York.

Bauchard Jr., T.J., D.T. Lykken, M. McGue, N.L. Segal and A.Tellegen 1990. "Sources of human psychological differences: the Minnesota study of twins reared apart." Science 250, pp. 223-228.

Beare, F.W. 1981. The Gospel According to Matthew. Harper and Row, San Francisco.

Binford, L.R. 1982. Comment on: "Rethinking the Middle/Upper Paleolithic transition", by R. White; Current Anthropology 23, pp. 177-181.

Bird, I.L. 1960. A Lady's Life in the Rocky Mountains. University of Oklahoma Press, Norman.

Bishop III, J. 1991. "Reptilian-Grey Data". In, Valerian, V. Matrix II: The Abduction and Manipulation of Humans Using Advanced Technology. The Leading Edge Research Group; Yelm, Washington, pp 96-101.

Blakemore, C. and G.F. Cooper 1970. "Development of the brain depends on the visual environment." Nature 228: pp. 447-448.

Blavatsky, H.P. 1967. The Secret Doctrine (an abridgement, E. Preston and C. Humphreys, editors). The Theosophical Publishing House, Wheaton, Illinois.

Bonner, J.T. 1980. The Evolution of Culture in Animals. Princeton University Press, Princeton.

Borkamm, G. 1960. Jesus of Nazareth. Harper, New York

Boulay, R.A. 1990. Flying Serpents and Dragons: The Story of Mankind's Reptilian Past. Galaxy Books, P.O.Box 8542, Clearwater,

Bibliography

Florida 34618.

Bower, B. 1994. "Asian hominids make a much earlier entrance". Science News 145: p. 150.

Bowlby, J.A. 1969. Attachment and Loss, Vol. 1. Basic Books, New York.

Boylan, R.J. with L.K. Boylan 1994. Close Extraterrestrial Encounters: Positive Experiences with Mysterious Visitors. Wild Flower Press, Tigard, Oregon, p. 155.

Bramley, W. 1989. The Gods of Eden: A New Look at Human History. Dahlin Family Press, San Jose, California.

Brough, J. 1958. "Time and evolution." In Studies on Fossil Vertebrates, T.S. Westoll (ed.). University of London Press, London.

Brown, D. 1970. Bury My Heart at Wounded Knee. Holt, Rinehart and Winston, Inc., New York.

Brues, A.M. 1977. People and Races. Macmillian Publishing Co., Inc., New York.

Bryant, A., and L. Seebach 1991. Healing Shattered Reality: Understanding Contactee Trauma. Wild Flower Press, Tigard, Oregon.

Calhoun, J.B. 1962. The Ecology and Sociology of the Norway Rat. U.S. Department of Health, Education, and Welfare, Bethesda, Maryland.

Cann, R.L., M. Stoneking and A.C. Wilson 1987. "Mitochondrial DNA and human evolution." Nature 325: pp. 31-36.

Chalmers, N.R. 1968. "Group composition, ecology and daily activities of free living mangabeys in Uganda." Folia Primatologica 8: pp. 247-262.

Colinvaux, P. 1978. Why Big Fierce Animals are Rare: An Ecologist's Perspective. Princeton University Press, Princeton.

Cooper, M.W. 1991. Behold a Pale Horse. Light Technology Publishing. Sedona, Arizona.

Coven, R. 1992. "Hunting planets with a gravitational lens." Science News 141: p. 327

Cravens, H. 1978. The Triumph of Evolution. University of Pennsylvania Press, Philadelphia.

Crick, F. 1981. Life Itself. Simon and Schuster, New York.

Crowell, C. 1990. "The government factor." In: Valerian, V. 1991. Matrix II: The Abduction and Manipulation of Humans Using Advanced Technology. Leading Edge Research Group, Yelm, Washington, pp. 129-132.

Darwin, C. 1872. The Origin of Species, 6th ed. Reprinted, (1958) Mentor, New York.

Darwin, F. (ed.), 1888. The Life and Letters of Charles Darwin, 3

Bibliography

Volumes. John Murray, London.

Darwin, F. and A.C. Seward (eds.) 1903. More Letters of Charles Darwin, vol. II. Murray, London.

Dawkins, R. 1976. The Selfish Gene. Oxford University Press, New York.

_____ 1985. "What's all the fuss about?" Nature 316, p. 683.

Deardorff, J.W. 1990. Celestial Teachings. Wild Flower Press, Tigard, Oregon.

DeLoach, J.S. 1987. "Rapid change in the symbolic functioning of very young children." Science 238: pp. 1556-1557.

Dene, H.T., M. Goodman and W. Prychodko 1976. "Immunodiffusion evidence on the phylogeny of the primates." In, Molecular Anthropology, M. Goodman, R.E. Tashian and J.H. Tashian (eds.), Plenum Press, New York.

Denton, M. 1986. Evolution: A Theory in Crisis. Adler and Adler, Bethesda, Maryland

Douglas-Hamilton, I., and O. Douglas-Hamilton 1975. Among the Elephants. Viking Press, New York.

Eldredge, N., and S.J. Gould 1972. "Punctuated equilibria: an alternative to pyletic gradualism." In, Models in Paleobiology, T.J.M. Schopf (ed.), Freeman, Cooper and Co., San Francisco, pp. 82-115.

Fillion, T.J. and E.M. Blass. 1986. "Infantile experience with suckling odors determine adult sexual behavior in male rats." Science 231: pp. 729-731.

Fleagle, J.G. 1988. Primate Adaptation and Evolution. Academic Press, San Diego, California.

Fowler, C. 1981. "Comparative population dynamics in large animals". In Dynamics of Large Mammal Populations. C. Fowler and T. Smith (eds.). Wiley, New York, pp. 444-445.

Fowler, R.E. 1980. The Andreasson Affair. Bantam Books, New York.

____ 1991. The Watchers: The Secret Design behind UFO Abduction. Bantam Books, New York.

France, D.L. and A.D. Horn 1992. Lab Manual and Workbook for Physical Anthropology, 2nd ed. West Publishing Co., St. Paul, Minnesota.

Freud, S. 1953. "Three essays on the theory of sexuality." In: The Standard Edition of the Complete Psychological Work of Sigmund Freud, Vol. 7. Hogarth, London, pp. 125-248.

Fuller, J.G. 1966. The Interrupted Journey. Dell Publishing Co., Inc., New York.

Gallup Organization. 1993. "Half of U.S. believe creationism." San Francisco Chronicle, Sep 13, p. A5.

Geist, V. 1971. Mountain Sheep. University of Chicago Press, Chicago.

Gibbons, A. 1994. "Rewriting - and redating - prehistory." Science 263: pp. 1087-1088.

Gingerich, P.D. 1990. "African dawn for primates." Nature 346: p. 411.

Ginsburg, H. and S. Opper 1979. Piaget's Theory of Intellectual Development (2nd ed.). Prentice-Hall, Englewood Cliffs, New Jersey.

Goetz, D. and S.G. Morley 1950. Popol Vuh, The Sacred Book of the Ancient Quiche Maya. University of Oklahoma Press, Norman.

Good, T. 1988. Above Top Secret: The Worldwide UFO Cover-Up. William Morrow and Co., Inc., New York.

Goto, H. 1971. "Auditory perception by normal Japanese adults of the sounds of 'L' and 'R'." Neuropsychologica 9: pp. 317-323.

Gould, S.J. 1983. "Nature's great era of experiments". Natural History 92: pp. 12-21.

_____ 1989. Wonderful Life: The Burges Shale and the Nature of History. Norton Press, New York.

Gould, S.J., and N. Eldredge 1977. "Punctuated equilibria: the tempo and mode of evolution reconsidered." Paleobiology 3: pp. 115-151.

Goy, R.W.K., K. Wallen and D.A. Goldfoot 1974. "Social factors affecting the development of mounting behavior in male rhesus monkeys." In: Reproductive Behavior, W. Montagna and W. Sadler (eds.). Plenum Press, New York.

Greenough, W.T. 1975. "Experimental modification of the developing brain." American Scientist 63: pp 37-46.

Greenough, W.T., J.E. Black and C.S. Wallace 1987. "Experience and brain development." Child Development 58: pp.539-559.

Hastings, A. 1991. With the Tongues of Men and Angels: A Study of Channeling. Holt, Rinehart and Winston, Inc., Fort Worth, Texas.

Healey, R.F., F. Cooke and P.W. Colgan 1980. "Demographic consequences of snow goose brood rearing traditions." Journal of Wildlife Management, 44: pp. 900-905.

Herman, E.S. and N. Chomsky 1988. Manufacturing Consent: The Political Economy of the Mass Media. Pantheon Books, New York.

Hess, E.H. 1973. Imprinting: Early Experience and the Developmental Psychobiology of Attachment. Van Nostrand Reinhold Co., New York.

Hilts, P.J. 1993. "Battle heating up over veterans' mysterious Gulf War ailments." San Francisco Chronicle, November 25, p. A 17.

Hopkins, B. 1988. Intruders: The Incredible Visitations at Copley Woods. Ballantine Books, New York.

Horn, A.D. 1987. "The socioecology of the black mangabey, Cercocebus aterrimus, near Lake Tumba, Zaire." American Journal of Primatology, 12: pp. 165-180.

Bibliography

Horn, G. 1985. Memory, Imprinting and the Brain. Clarendon Press, Oxford.

Housfater, G., J. Altman and S. Altman 1982. "Long-term consistency of dominance relations among female baboons (Papio cynocephalus)." Science 217: pp. 752-755

Howe, L.M. 1991. "The 'alien harvest' and beyond". In, UFOs and the Alien Presence: Six Viewpoints, M. Lindemann (ed.). The 2020 Group, Santa Barbara, California, pp. 61-84.

Hoyle, F., and C. Wickramasinghe 1978. Life Cloud. J.M. Dent and Sons, London.

―― 1981. Evolution from Space. J.M. Dent and Sons, London.

Hubel, D.H. and T.N. Wiesel 1963a. "Receptive field of cells in striate cortex of very young, visually inexperienced kittens." Journal of Neurophysiology 26: pp. 994-1102.

―― 1963b. "Single-cell responses in striate cortex of kittens deprived of vision in one eye." Journal of Neurophysiology 26: pp. 1003-1017.

Hulen, D. 1992. "Radioactive soil had secret burial in Alaska in 1962." San Francisco Chronicle, October 12, p. A5.

Hynek, J.A. 1977. The Hynek UFO Report. Dell Publishing Co., New York.

Immelmann, K. 1972. "Sexual and other long-term aspects of imprinting in birds and other species." Advance Study of Behavior, 4: pp. 147-174.

―― 1975. "Ecological significance of imprinting and early learning." In: Annual Review of Ecology and Systematics, 6, R.F. Johnston, P.W. Frank and C.D. Michener (eds.), Annual Reviews Inc., Palo Alto, pp. 15-37.

Immelmann, K. and S.J. Suomi 1981. "Sensitive phases in development." In: Behavioral Development, K. Immelmann, G.W. Barlow, L. Perinovich and M. Main (eds.). Cambridge University Press, Cambridge, pp. 395-431.

Jackson, R.L. and R.J. Ostrow 1994. "Iran-Contra report blames Reagan, Bush". San Francisco Chronicle, January 19, p. A 1 and p. A 13.

Jasper, W.F. 1993. Global Tyranny... Step by Step: The United Nations and the Emerging New World Order. Western Island Publishers, Appleton, Wisconsin.

Jerison, J.H. 1973. Evolution of the Brain and Intelligence. Academic Press, New York.

Johnson, P.E. 1991. Darwin on Trial. Regnery Gateway, Washington D.C.

Jung, C.G. 1971. The Portable Jung. The Viking Press, New York.

Kettlewell, H.B.D. 1959. "Darwin's missing evidence." Scientific American 200, pp. 48-53.
Kemp, T.S. 1982. Mammal-like Reptiles and the Origin of Mammals. Academic Press, London.
Kersten, H. 1986. Jesus Lived in India. Element Book Ltd., Shaftesbury, England.
Kinder, G. 1987. Light Years: An Investigation into the Extraterrestrial Experiences of Eduard Meier. Atlantic Monthly Press, New York.
Knudsen, E.I. 1987. "Early experience shapes auditory localization behavior and the spatial running of auditory units in the barn owl." In: Imprinting and Cortical Plasticity, J.P. Rauschecker and P. Marler (eds.), John Wiley and Sons, New York.
Kottack, C.P. 1979. Cultural Anthropology, 2nd Ed. Random House, New York.
Krill, O.H. (No Date) "Orion Based Technology, Mind Control and other Secret Projects". Arcturus Book Service, Stone Mountain, Georgia.
Kroeber, A.L. 1917. "The Superorganic". American Anthropologist, 19: pp. 163-213.
Kuhn, T.S. 1970. The Structure of Scientific Revolutions, 2nd ed. University of Chicago Press, Chicago.
Kvanika, R.M. 1988. "United States Indian treaties and agreements." In: Handbook of North American Indians, Vol. 4. Smithsonian Institution, Washington.
Lack, D. 1954. The Natural Regulation of Animal Numbers. Oxford University Press, Oxford.
Lamberg-Karlovsky, C.C. and J.A. Sabloff 1979. Ancient Civilizations: The Near East and Mesoamerica. The Benjamin/Cummings Publishing Co., Inc., Menlo Park, California.
Lambert, W.G. and A.R. Millard 1969. Atra-Hasis, the Babylonian Story of the Flood with the Sumerian Flood Story by M. Civil. Clarendon Press, Oxford.
LaSalle, M. 1993. "Film on rabble-rouser Chomsky and media." San Francisco Chronicle, April 9, p. D5.
Lazar, B. 1991. "Alien Technology in Government Hands." In UFOs and the Alien Presence: Six viewpoints. M. Lindemann (ed.), The 2020 Group, Santa Barbara, California, pp. 87-127.
Lear, J. 1988. "The UFO Cover-Up?' In, Valerian, V. 1991. Matrix II: The Abduction and Manipulation of Humans Using Advanced Technology. Leading Edge Research Group, Yelm, Washington, pp. 239-252.
Lenneberg, E. 1987. Biological Foundation of Language. John Wily and Sons, New York.

Levi 1968. The Aquarian Gospel of Jesus the Christ. DeVorss & Co., Los Angeles.

Levinton, J.S. 1992. "The big bang of animal evolution." Scientific American 267, pp. 84-91.

Lieberman, P. and E.S. Crelin 1971. "On the speech of Neanderthal." Linguistic Inquiry 2: pp. 203-222.

Lindemann, M. 1990. "UFOs and the Alien Presence: Time for the Truth." Manuscript from The 2020 Group, Santa Barbara, California.

____(ed.) 1991. UFOs and the Alien Presence: Six Viewpoints. The 2020 Group, Santa Barbara, California.

Lippman, T.W. 1993. "Utah survivors remain bitter about Nevada bomb tests." The Oregonian (Portland), May 19, p. A13.

Lorenz, K. 193b5. "Der Kampan in der Umwelt des Vogels." Journal of Ornithology, 83: pp.137-213, 289-413.

Mainardi, D. 1980. "Tradition and social transmission of behavior in animals." In: Sociobiology: Beyond Nature/Nurture? G. Barlow and J. Silverberg (eds.), Westview Press, Boulder, Colorado, pp. 227-255.

Marciniak, B. 1990 "The Pleiadians". Manuscript of a tape of material channeled on Nov 15 and 16 at Stanford University, J. Horneker, (ed.). Bold Connections, Raleigh, North Carolina.

____ 1992a. The Harmonics of Frequency Modulation and the Human DNA, Part I: The Pleiadians through Barbara J. Marciniak at Stanford University, Palo Alto, California, Nov 15, 1990. Sedona 2, No. 2, pp. 17-33.

____ 1992b. Bringers of the Dawn: Teachings from the Pleiadians. Bear and Company Publishing, Santa Fe, New Mexico.

Marler, P. 1984. "Song learning: innate species differences in the learning process." In: The Biology of Learning, P. Marler and H.S. Terrace (eds.). Springer-Verlag, Berlin, pp. 289-309.

Matousek, M. and I. Peterson 1973. "Frequency analysis of the EEG in normal children and adolescents." In: Automation of Clinical Electroencephalography, P. Kellaway and I. Peterson (eds.). Raven, New York, pp. 75-102.

McGrew, W.C., C.E.G. Tutin, and P.J. Baldwin 1979. "Chimpanzees, tools, and termites: cross-cultural comparisons of Senegal, Tanzania, and Rio Muni." Man, 14: pp. 185-214.

Meier, E. 1992. Foreword in The Talmud of Jmmanuel. Wild Flower Press, Tigard, Oregon, pp. xiii-xxi.

Monastersky, R. 1992. "Giant Crater linked to mass extinction." Science News 142, p. 100.

____ 1993. "Fire beneath the ice", Science News, 143, p. 107.

Moore, M. 1994. "Caste animosity in India breaks into ugly violence." The San Francisco Chronicle, February 19, pp. A8 and A12.

Montagu, A. 1962. Man: His First Million Years. The New American Library of World Literature, Inc., New York.

Morgan, M. 1991. Mutant Message Downunder. MM Co., Lees Summit, Missouri.

Nelson, H. and R. Jurmain 1988. Introduction to Physical Anthropology, 4th ed. West Publishing Co., St. Paul, Minnesota.

The New English Bible 1971. Oxford University Press, New York.

Perkins Jr., D. and P. Daly 1974. "The beginning of food production in the Near East." In: The Old World: Early Man to the Development of Agriculture. R. Stigler (ed.), 1974. St. Martin's Press, New York, p. 74.

Petit, C. 1992. "Controversy over age of Egypt's Sphinx". San Francisco Chronicle, Feb. 8, p. A4.

Pilbeam, D., M.D. Rose, J.C. Barry and S.M. Ibrahim Shah 1990. "New Sivapithecus humeri. Pakistan and the relationship of Sivapithecus and Pongo. "Nature, 348: pp. 237-239.

Redfield, J. 1993. The Celestine Prophecy. Satori Publishing, Hoover, Alabama.

Renfrew, C. 1979. Before Civilization: The Radiocarbon Revolution and Prehistoric Europe. Cambridge University Press, Cambridge.

Rensch, B. 1959. Evolution Above the Species Level. Columbia University Press, New York.

Revelations of Awareness 1991. Issues 90-9, 90-14, 91-16, 92-4, 92-11, 93-5, 93-7, and 93-8. Cosmic Awareness Communications, P.O.Box 115, Olympia, Washington 98501.

Richard A. 1977. "The feeding behavior of Propithecus verreauxi." In Primate Ecology, T.H. Clutton-Brock (ed.), Academic Press, London, pp. 72-96.

Rindos, D. 1984. The Origins of Agriculture: An Evolutionary Perspective. Academic Press, Orlando, Florida.

Romer, A.S. 1933. Man and the Vertebrates. University of Chicago Press, Chicago.

Royal, L. and K. Priest 1989. The Prism of Lyra: An Exploration of Human Galactic Heritage. Royal Priest Research, Sedona, Arizona.

Ruppenthal, G.C., G.L. Arling, H.F. Harlow, G.P. Sacket, and S.J. Suomi 1976."A ten-year perspective of motherless-mother monkey behavior." Journal of Abnormal Psychology 85: pp. 341-348.

Sachs, M. 1980. The UFO Encyclopedia, Perigee Books of G.P. Putnam's Sons, New York.

Schellhorn, C.C. 1990. Extraterrestrials in Biblical Prophecy. Horus House Press, Inc., Madison, Wisconsin.

Bibliography

Schloeth, R. and D. Burchardt 1961. "Die Wanderung des Rotwildes (*Cervus elaphus L.*)" Rev. Suisse Zool. 68: pp. 146-155.
Shapiro, R. 1986. Origins: A Skeptic's Guide to the Creation of Life. Summit Books, New York.
Sheldrake, R. 1982. A New Science of Life. Thacher, Tiburon, California.
Sitchin, Z. 1976. The 12th Planet. Avon Books, New York.
_____ 1980. The Stairway to Heaven. Avon Books, New York.
_____ 1985. The Wars of Gods and Men. Avon Books, New York.
_____ 1987. Introduction. In Breaking the Godspell, by N. Freer, Falcon Press, Phoenix, Arizona.
_____ 1990. The Lost Realms. Avon Books, New York.
_____ 1991. Genesis Revisited: Is Modern Science Catching Up with Ancient Knowledge? Bear and Company Publishing, Santa Fe, New Mexico.
_____ 1992. When Time Began. Avon Books, New York.
Simpson, G.G. 1949. The Meaning of Evolution. Yale University Press, New Haven.
Speiser, E. A, 1969. "The Sumerian Problem Reviewed," in The Sumerian Problem, T. Jones (ed.). John Wiley and Sons, Inc., New York. pp. 93-102.
Stahl, B.J. 1974. Vertebrate History: Problems in Evolution. McGraw-Hill Book Co., New York.
Stanley, S. 1979. Macroevolution. W.H. Freeman and Co., San Francisco.
Starr C.G. 1983. The History of the Ancient World (3rd. ed.). Oxford University Press, Oxford.
Stebbins, G.L. 1951. "Cataclysmic evolution." Scientific American, 184: pp.54-59.
Sternberg, R.J. 1985. "Human intelligence: the model is the message." Science 230: pp. 1111-1117.
Stevens, W.C. 1982. UFO Contact from the Pleiades: A Preliminary Investigation Report. UFO Photo Archives, Tucson, Arizona.
_____ 1988. Message from the Pleiades: The Contact Notes of Eduard Billy Meier. UFO Photo Archives, Tucson, Arizona.
_____ 1989. UFO Contact from the Pleiades: A Supplementary Investigation Report. UFO Archives, Tucson, Arizona.
_____ 1990. Message from the Pleiades: The Contact Notes of Eduard Billy Meier 2. UFO Archives, Tucson, Arizona.
Stigler, R. 1974. "The later neolithic in the Near East and the rise of civilization." In: The Old World: Early Man to the Development of Agriculture, R. Stigler (ed.), St. Martin's Press, New York.
Stoneking, M. and R.L. Cann 1989. "African origin of human mito-

chondrial DNA." In: The Human Revolution: Behavioural and Biological Perspectives on the Origins of Modern Humans, vol. 1. P. Mellers and C.B. Stringer (eds.), Edinburgh University Press, Edinburgh.

Stringer, C. and P. Andrews 1988. "Genetics and fossil evidence for the origin of modern humans." Science 239: pp. 1263-1268.

Stryker, M.P., J. Allman, C. Blakemore, J.M. Greuel, J.H. Kaas, M.M. Merzenich, P. Radic, W. Singer, G.S. Stent, H. Vanderloes, and T.N. Wiesel 1988. "Group report: principles of cortical self-organization." In: Neurobiology of Neocortex, P. Radic and W. Singer (eds.), John Wiley and Sons, Limited, New York, pp. 115-136.

Summers, A. 1980. Conspiracy. McGraw-Hill Book Co., New York.

Swisher III, C.C., G.H. Curtis, T. Jacob, A.G. Getty, A. Suprijo, Widiasmoro 1994. "Age of the earliest known hominids in Java, Indonesia." Science 263: pp. 1118-1121.

The Talmud of Jmmanuel 1992. Translated from Aramaic to German By I. Rashid and E.A. Meier, and into English by J.H. Ziegler and B.L. Greene. Wild Flower Press, Tigard, Oregon.

Teleki, G. 1974. "Chimpanzee subsistence technology: materials and skills." Journal of Human Evolution, 3: pp. 575-594.

Temple, R.K.G. 1987. The Sirius Mystery. Destiny Books, Rochester, Vermont.

Templeton, A.R., S.B. Hedges, S. Kumar, K. Tamura, and M. Stoneking 1992. "Human origins and analysis of mitochondrial DNA sequences." Technical comment. Science 255: pp. 737-739.

Thatcher, R.W., R.A. Walker and S. Guidice 1987. "Human cerebral hemispheres develop at different rates and ages." Science 236: pp. 1110-1113.

Trinkaus, E. 1984. "Western Asia." In: The Origins of Modern Humans: A World Survey of the Fossil Evidence, F.H. Smith and F. Spencer (eds.). Alan R. Liss, New York, pp. 251-291.

Twigs, D.R., and B. Twigs. 1992. Secret Vows: Our Lives with Extraterrestrials. Wild Flower Press, Tigard, Oregon.

Valerian, V. 1991. Matrix II: The Abduction and Manipulation of Humans Using Advanced Technology, 3rd ed. Leading Edge Research Group, Yelm, Washington.

Vallée, J. 1965. Anatomy of a Phenomenon: UFOs in Space, a Scientific Appraisal. Ballantine Books, New York.

_____ 1980. Messengers of Deception: UFO Contacts and Cults. Bantam Books, Inc., New York.

Von Däniken, E. 1989. In Search of the Gods: Chariots of the Gods?; Gods from Outer Space; Pathways to the Gods. (Three volumes in one). Avenel Books, New York.

Bibliography

Ware, D. 1991. "The 'Larger Reality' Behind UFOs". In UFOs and the Alien Presence, Lindemann, M.(ed.), The 2020 Group, Santa Barbara, California, pp. 195-218.

Waser, P. 1977. "Feeding, ranging and group size in the mangabey, Cercocebus albigena," In, Primate Ecology: Studies of Feeding and Ranging Behavior in Lemurs, Monkeys and Apes. T. Clutton-Brock (ed.), Academic Press, London, pp. 183-222.

White, L.A. 1973. The Concept of Culture. Burges, Minneapolis, Minnesota.

Wilson, A.C., L.R. Maxon, and V.M. Sarich 1974. "Two types of molecular evolution: evidence from studies of interspecific hybridization." Proceedings of the National Academy of Sciences 71.

Wilson, E.O. 1975. Sociobiology, The New Synthesis. Harvard University Press, Cambridge, Massachusetts.

_____ 1977. Forward of Sociobiology and Behavior by D.P. Barash; Elsevier North-Holland, Inc., New York, pp. xiii-xv.

_____ 1978. On Human Nature. Harvard University Press, Cambridge, Massachusetts.

Winson, J. 1985. Brain and Psych. Anchor Press/Doubleday, New York.

Wolpoff, M.H. 1989. "Multiregional evolution: the fossil alternative to Eden." In: The Human Revolution: Behavioral and Biological Perspectives on the Origins of Modern Humans, vol. 1. P. Mellers and C.B. Stringer (eds.). Edinburgh University Press, Edinburgh.

Woolley, C.L. 1929. The Sumerians. W.W. Norton & Company, Inc., New York.

Zailan, M. 1993. "Dissident gets a voice." San Francisco Examiner, "Datebook", April 4, p. 33.

Index

Abraham 147
Abzu 56, 59
Adamu 57
African genesis 61, xx
Agriculture 113
AIDS 200
Akkadian empire 51-52
Albertson, Dr. M.L. 340
Amphibians 4, 18, 23, 27
Angiosperms 74
Animal mutilations 190
Antarctic ice cap 103
Anunnaki 55, 63, 68, 78, 81, 95, 160, xxx
Anu 56, 96, 99
Apes 36
Arapahoe 332
Archaeopteryx 19
Arnold, Kenneth 184
Aryans 142, 146, 317
Ashurbanipal 54
Assyrian empire 51-52, 58
Atra-Hasis 55
Audry, Robert xx
Augros, Robert 16
Australopithecus 40-41, 76, xx
Baal 274
Babylonian empire 50, 52
Binford, Lewis 221
Birds 3, 17-20, 26
Bishop, Jason 249
Bonner, J.T. 224
Book of Mormon 155, 174-175
Boulay, R.A. 88-89, 100, 104, 149, 246, 252-253, 278
Boylan, Dr. Richard 247
Braidwood, Robert 112
Brain Development 220
Bramley, William 155, 167-171, 175, 178, 300, 318, 320

Index

Brown, Dee 331
Bush, George 205
Calvin, John 292
Cambrian period 16, 73
Cenozoic period 35
Central Intelligence Agency (CIA) 194
Chariots of the Gods xxiii
Cheyenne 329, 332
Chimpanzee xxi
Chomsky, Noam 201
Comet 14, 25, 55, 99, 105, 172
Condon Report 187
Constantine 289
Cooper, William 194-198, 200, 303
Council of Constantinople 290
Council of Nicea 289-290, 313
Council on Foreign Relations 198
Crick, Francis 7, 14, 66
Cuneiform writing 53
Cylinder seal 52, 57
Daly, Patricia 109, 120
Dark Ages xxiii
Darwin, Charles 3, 28
Davenport, David 146
Dawkins, Richard 10, 24, 28
Deardorff, James 306, 312
Deluge (see Great Flood)
Denton, Michael 22, 25, 28
Descartes, René 11
Dinosaurs 19, 25, 35
DNA 7-9, 31-32, 76, 90, 160-161, 299
Dogon tribe 61
Draco 250
Egypt 36, 50, 54, 62, 105, 126, xxii
Eisenhower, President 186, 196, 198
Eldridge, Niles 23
Enki 56-57, 59, 96-99, 102, 115, 137-138, 148-149, 159-162, 248
Enlil 56, 59, 97-99, 102, 115-116, 137, 149, 152, 161-162
Eocene period 36
Eridu 96, 125
Eve 85, 88, 91
Evolution 1, 10, xv
Ezekiel 266

Index

Fatima 179
Flannery, Kent 109, 112
Fleagle, Dr. John 37
Fowler, Raymond 210
Freedom of Information Act 192
Freeman, Dr. K.P. 340
Freud, Sigmund 228
Genesis 2, 29, 58-59, 90-91, 95, 100, xvi, xxi
Genetic engineering xxix
Gilgamesh 96
Glacial period 104
Gnostic 89
Good, Timothy 192, 208
Goodall, Jane 222
Gorillas 38
Gould, Stephen J. 15, 23
Great Flood 95, 117-118, 120, 136-137, 245
Greys 190, 256
Haggadah 88-89
Haldane, J.B.S. 12
Harappan civilization 122, 130
Hastings, Dr. Arthur xxvi
Hebrew 81, 88
Higher primates 36
Hinduism 143, 317
Hollin, Dr. John T. 103
Hologram 174, 178
Hominids 44, 46, 221
Homo erectus 42-43, 56, 78, xxii
Homo habilis 41, 76
Homo sapiens 41, 45, 61, 78
Hopkins, Budd 208
Howe, Linda Moulton 207, 246
Hoyle, Frederick 14
Human abductions 207
Hunter-gatherers 108
Hynek, J. Allen 185, 187-188, 193
Illuminati 198
Immelmann, K. 226, 228
Imprinting 220, 225
Imprint xv, xxvii
Inanna 153
Indus Valley 130, 137, 142-143, 154, 317

369

Index

Islam 321-323, 342
Jehovah 265, 268-282, 311
Jesus 170, 180, 286-289, 291-294, 296, 299-307, 309-310, 312, 321, xix, xxiii
Jmmanuel 180, 305-308, 310-312
Johnson, Phillip 29
Judaism 176
Jung, Carl 237
Junk DNA 32
Justinian's Plague 171
Kemp, Dr. T.S. 21
Kennedy, John F. 140, 194, 202
Kersten, Holger 288-290, 310
Kinder, Gary 83
Kish 125
Kottak, Conrad 220
Kramer, Samuel N. 59, 115
Kuhn, Thomas 29
Lagash 129
Lamarck, Jean Baptiste 5
Lamarckism 5-6, 31
Lambert, W.G. 96
Lazar, Robert 67, 209
Lear, John 207
Lenneberg, Eric 229
Levinton, J.S. 16
Lindemann, Michael 199
Lizzies (see Reptoids)
Lorenz, Konrad 225, 228
Lyra, Lyran 68-74, 84, 143, 160
Machiavelli 168
MacNeish, Richard 112
Macroevolution 30-32, 139
Mahabharata 143
Majestic Twelve (MJ-12) 198
Mammals 4, 17, 19-22, 26
Marciniak, Barbara 87, 90
Marduk 148
Martin Luther 291
Mass Extinction 25
Maya, Maya civilization 91-92, 110, 167
Meier, Eduard 83-84, 86, 305-306
Mendel, Gregor 6
Mesoamerica 110

Index

Mesopotamia 48, 52, xxix, xxv
Millard, A.R. 96
Miller, Stanley 12
Miocene period 38
Mithraism 294
Mitochondria DNA 47-48, 61
Modern Synthesis 6
Mohammed 321
Mohenjo-Daro 130
Molecular Biology 7
Moses 267, 270, 273, 276-277
Nabu 148
National Security Agency (NSA) 193
Natural selection 4, 6-7, 17, 19-20, 24, 26, 28, 30
Nature/nurture debate 218-220, 235
Neanderthal 46-48, 85
Near-death-experiences 341
Nefilim 54, 82
Neo-Darwinism 1-2, 6, 10-11, 16-18, 24, 27-28, 30-31, 74
New Testament 286, xx
Nibiru 55, 104
Ninhursag 137
Ninti 57
Ninurta 115
Nippur 125
Nirvana 320
Noah 96
Nohle, Johannes 172
Nuclear 145-146, 151-152, 154-155
Nur, Amos 157
Old Testament 50, 54, 59, 80-81, 89, 155, 265-266, 274, 277, 279-282, xviii
Olmec civilization 110, 132-136, 139
Oparin, Alexander 12
Orangutan 39, 75
Orion 70
Paul, Apostle 282, 287-289, 294, 303
Perkins, Dexter 109, 120
Phyla 16
Piaget, Jean 229-230
Plague 171
Plains Indians 326-331, 333-335
Pleiadians 71, 75, 82-87, 104-105, 163, 177-178, 180, 211, 248, 279, 305
Popul Vuh 91-92

371

Index

Priest, Keith 68
Prime Creator 87, 178, 239, 244, 254, 344, 352
Project Blue Book 187
Project Grudge 187
Project Sign 187
Prosimians 35-37
Proteins 8-9, 12-13
Radiocarbon dating 118-119
Reagan, Ronald 205
Reincarnation 290-291, 318-319, 340-341
Renfrew, Colin 117
Rensch, Bernhard 31
Reptiles 17, 19-21, 30, 73, 224, 249-251
Reptoids 244, 246, 248, 250, 252-254
Rindos, David 113
RNA 8, 13
Roman Empire xxiii
Romer, Alfred 19
Royal, Lyssa 68-69, 71
Sagan, Dr. Carl 188
San Lorenzo 133
Sand Creek massacre 331
Schellhorn, G.C. 296, 302, 304
Scherer, Reed 103
Schoch, Robert 156
Secret societies 168-169, 176, 197-199, 203, 211, 257, 263, 303, xxx
Shapiro, Robert 13-15
Sheldrake, Rupert 238
Simpson, G.G. 12
Sioux 329, 332-334
Sirius, Sirian 61
Sitchin, Zecharia 54-55, xxv
Sivapithecus 39, 75
Smith, Joseph 155, 174-175
Sodom and Gomorrah 138, 147-149, 152, 157, 163, 191, 267
Solomon, Esther Abraham 146
Speiser, E.A. 125
Sphinx 156
Stahl, Barbara 20
Stanciu, George 16
Stanley, Steven 17
Starr, C.G. 129
Stevens, Wendelle 84

Index

Stigler, Robert 124-125, 128
Sumerian King List 89-90, 95
Sumer 51-55, 57, 59-60, 62-63, 123, 125, xxii, xxix
Takauti Documents 164
Talmud of Jmmanuel 305
Tarsiers 37, 75
Temple, R.K.G. 62
The Awareness 161-162
Titans 154
Trilateral Commission 198
Upper Paleolithic xxii
Urey, Harold 12
Uruk 125, 153
Ur 125
Utnapishtim 96
Valerian, Valdamar 146, 164, 208, 210, 212, 257
Vallée, Jacques 190
Vedas 143
Vega 84
Virgin Mary 179
Von Däniken 108, 117, 143-144, 146, 185, 266, xxiii
Vorilhon, Claude 191
Wallace, Alfred Russel 2-3, 6
Watson, James 7
Weismann, August 6
Wickramasinghe, N.C. 14
Wilson, Edward 11
Wood, Bernard 43
Writing 52
Zeus 154
Zoroastrianism 176
Zoukoutien 44, 47
Zwingli, Huldreich 292

Notes

Notes